The Comprehensive HANDBOOK of HAZARDOUS MATERIALS

Regulations, Handling, Monitoring, and Safety

Hildegarde L.A. Sacarello

LEWIS PUBLISHERS
Boca Raton Ann Arbor London Tokyo

Library of Congress Cataloging-in-Publication Data

The comprehensive handbook of hazardous materials: regulations, handling, monitoring and safety / by Hildegarde L. A. Sacarello

p. cm.

Includes bibliographical references and index.

ISBN 0-87371-247-1

1. Hazardous waste – clean-up. 2. Regulation. 3. Waste management – training. I. Sacarello, Hildegarde L.A. II. Title.

TD1401.L65 1993

636.8'27 – dc20 93-28372

CIP

Lewis Publishers is an imprint of CRC Press

International Standard Book Number 0-87371-247-1

Library of Congress Card Number 93-28372

Printed in the United States of America 1 2 3 4 5 6 7 8 9 0

Printed on acid-free paper

INTRODUCTION

The inspection, correct handling and cleanup of hazardous materials or waste is a relatively new profession. Experience gained, especially in the area of personnel safety, came originally from the chemical industry and the U.S. Army's Chemical Warfare Program. As the magnitude of the hazardous chemical and waste problem became evident, the numbers of persons employed in hazardous waste investigation, remedial response activities, cleanup, and emergency response increased.

The main objective of the new government environment regulations and programs is to assure the right of every worker to have a safe and healthy environment in which to work, and to assert the right of each individual to a clean and healthly environment in which to live. Figure 1.1 in Chapter 1 illustrates the volume of federal environmental regulations. Safe working conditions translate into increased worker productivity, savings in insurance benefits, and less off-the-job time for illness and injury. Many of the environmental regulations overlap and intertwine (see Figure 1.2, Chapter 1).

The major problem with any safety program is the reluctance of the employee to accept the perceived inconvenience/inefficiency associated with the proper safety prac tices and equipment. Proper safety techniques and the use of safety equipment should be considered as part of the job responsibilities. Failure to follow the protocols or failure to utilize the safety equipment should be regarded as failure to adequately complete the job assignment and would be grounds for an official reprimand. Poor safety habits, especially in the event of a hazardous chemical spill, not only endangers the life and health of the offender, but also jeopardizes the well-being of other employees and could possibly cause serious damage to the work site and to the surrounding public and environment.

In order to address the subject of hazardous materials training and related information, it is imperative that the trainee be able to distinguish between the terms hazardous materials, hazardous substances, and hazardous waste.

Hazardous material(s) is a broad term encompassing any material, including substances and wastes, that may pose an unreasonable risk to health, safety, property, or the environment, when they exist in specific quantities and forms. Hazardous substances are a subset of hazardous materials which are identified and regulated by EPA under various environmental regulations. Hazardous wastes are discarded materials that pose a risk to human health, safety, property, or the environment.

The various topics covered in this book are essential to the proper management of hazardous materials and their use. Compliance guidelines for comprehensive occupational safety and health programs, site safety, accident investigations, training, decontamination, and emergency response on and off-site will be addressed, too.

The various chapters covered in this text are essential for the proper management and reduction of exposures to hazardous materials. One must, first and foremost, know the "rules of the game".

DEDICATION

This book is dedicated to my father, Rudolf J. Staninger, who encouraged my thirst for the sciences as a child; when my sister Vina Lisa, our friend Merry Gale Bozeman and I would go “scrounging” in the rural sandy-pine areas of Mims, Florida in search of blackberries, bugs and all mother nature’s living wonders. Also, this book is dedicated to the memory of my mother Florence, who taught me perseverance, so I may do all of the tasks God has chosen for me to complete in this lifetime.

And last, but not least to my daughter Aurora and my Godfather from Tartus, Syria – Shar F. Aboudraah, both of whom were the answers to my prayers. May God bless you both.

Hildegarde L. A. Sacarello, Ph.D.

THE COMPREHENSIVE HANDBOOK OF HAZARDOUS MATERIALS: REGULATIONS, HANDLING, MONITORING, AND SAFETY

Hildegarde L. A. Sacarello, Ph.D.

TABLE OF CONTENTS

LIST OF FIGURES

LIST OF EXHIBITS

LIST OF TABLES

CHAPTER 1

Major Governmental Environmental and Training Regulations

THE FEDERAL INSECTICIDE, FUNGICIDE AND RODENTICIDE ACT (FIFRA)

Overview

The benefits of pesticides, herbicides, rodenticides, and other economic poisons are well known. They have done much to spare us from the ravages of disease, crop infestations, noxious animals, and choking weeds. Over the past two decades, however, beginning with Rachel Carson's *Silent Spring*, there has been a growing awareness of the hazards as well as the benefits of these chemicals, which may be harmful to man and the balance of nature. The first regulation for the regulations of pesticides was the Federal Insecticide, Fungicide and Rodenticide Act (FIFRA).

FIFRA is intended to act as a guideline in providing safety for persons, animals, organisms, and environments that come into contact with pesticides. The Act regulates the production, packaging, marketing, transportation, application, and effects of pesticides in relation to possible adverse conditions such activities may create. For the purpose of this regulation, a pesticide is broadly defined as almost anything intended to be used to kill, repel, or control any nonhuman form of life. Only microorganisms or drugs that affect man or animals are excluded from regulation by FIFRA.

FIFRA identifies Environmental Protection Agency (EPA) and state enforcement activities such as pesticide regulation through registration, analysis, inspection, record keeping, and administrative review. In addition, FIFRA identifies corrective and/or punitive action which may be taken in the event of regulation violations.

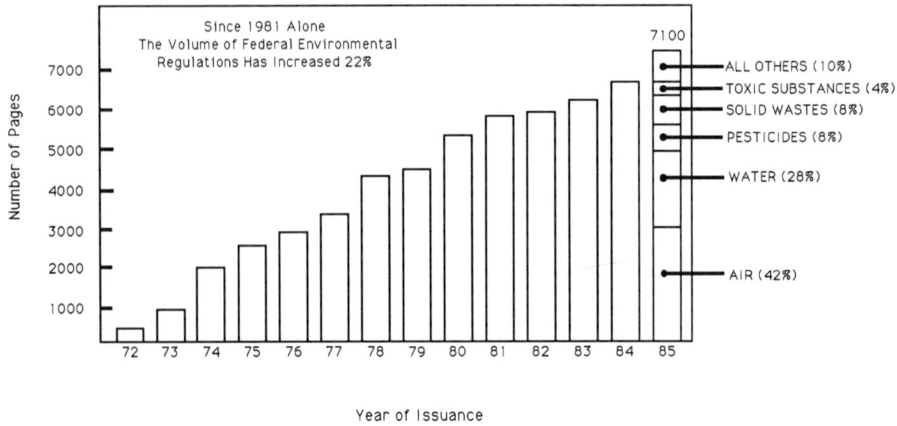

Figure 1.1. Volume of federal environmental regulations up to 1985.

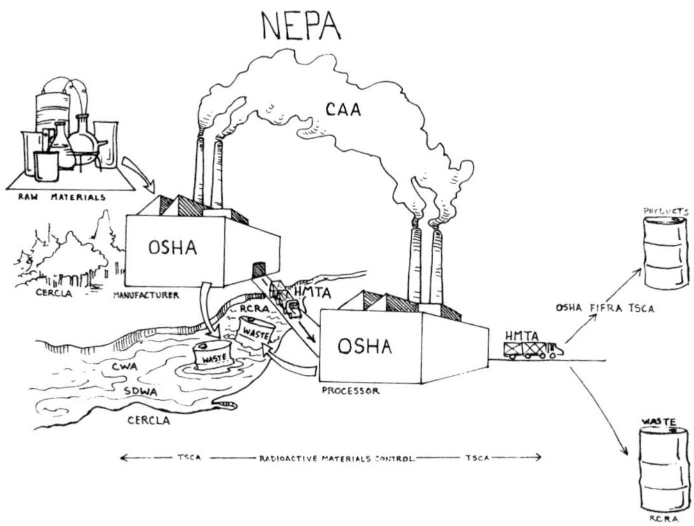

Figure 1.2. Diagram of various governmental regulatory agencies.

Background

FIFRA represents one of the longest duration examples of federal environmental regulation. The insecticide known as Paris Green was used as early as 1865 in this country to control the Colorado potato beetle. In 1910, Congress first set standards for ingredients in selected pesticides, including Paris Green and lead arsenate.

FIFRA was adopted in 1947 as a guideline for pesticide registration and labeling, due to a widespread increase in domestic use of synthetic organic pesticides following World War II. Theoretically, FIFRA had established three review points for a pesticide; however, as administered by the U.S. Department of Agriculture (USDA) there was actually no substantial safety review. Therefore, in the 1960s, the courts gave the USDA and subsequently EPA the discretion to ban pesticides based on a risk/benefit comparison. The supplementary Federal Environmental Pesticide Control Act of 1972 (FEPCA) of October 1972 and the amendments made in 1975, 1978, and 1980 further clarified the objectives of pesticides.

The need for regulation is apparently based on the characteristics and quantities of pesticides and herbicides. These substances are among the most toxic substances produced by the chemical industry and are manufactured in numerous and varied quantities. By 1981, EPA had registered 40,000 pesticide products made from one or more basic chemical compounds, and more than 1200 active pesticide ingredients.

Content

The regulations comprising FIFRA embody the following five goals:

1. Evaluation of risks using a registration system.
2. Exposure control via classification and certification.
3. Suspension, cancellation, or restriction of pesticides that pose a risk to the environment.
4. Enforcement through inspections, labeling notices, and regulation by state agencies.
5. Pesticides applications.

FIFRA requires that all pesticides be registered according to regulation prior to shipment, delivery, or sale in the U.S. The risks to the environment from pesticide use must be evaluated by EPA as part of the registration procedure. EPA has the authority to require manufacturers to submit relevant data deemed necessary to decide whether or not to register a pesticide. Evidence required for registration must demonstrate that the product will not cause unreasonable harm to humans, that it will perform its intended function without adverse effects, and that it will not result in harmful residues on food or feed.

In addition to standard registration processes, pesticides can be distributed and used under limited registrations or exemption mechanisms. These special case-by-case approvals include conditional registrations, state registrations for special local needs, emergency uses for specific problems, and experimental use permits.

Related to registration is control of pesticide labels, because such control is another key regulatory tool under FIFRA. It is a violation to sell a pesticide with a label that does not meet EPA standards. Pesticides classified as highly or moderately toxic must bear the words "Danger," "Poison," or "Warning" and/or the skull and crossbones symbol on the label. Others must bear the word "Caution." Further, it is a violation to use a pesticide in a manner that is inconsistent with directions specified on the label.

In registering a pesticide, EPA may classify the substance as being for general use or for restricted use. Most pesticides are classified for general use, which means that the product may be sold with no restrictions on who uses it. Pesticides classified for restricted use may be administered only by trained applicators. Because of these problems, both environmental and the industry agreed that EPA should give more flexibility than merely the choice between canceling or approving a pesticide. Congress therefore provided for the classification of pesticides in 1972 into general and restricted categories, with the latter group available only to Certified Applicators. There are several categories of applications, including private applicators and commercial applicators who use of supervise the application of pesticides on property other than their own. A pesticide label permitting use only "under the direct supervision of a Certified Applicator" means that the chemical is to be applied under the instructions and control of a Certified Applicator who, however, curiously is not required to be physically present when and where the pesticide is approved. These substances are usually extremely toxic or for some other reason present a special hazard.

EPA may suspend, cancel, or restrict a current pesticide registration if the substance is deemed to pose an unreasonable risk to humans or to the environment. Instead of such stringent measures, the agency more commonly deals with potentially hazardous pesticides through an extended special review process, termed "Rebuttal Presumption Against Registration" (RPAR). A manufacturer may respond to an RPAR by trying to rebut EPA's evidence of unreasonable risk. Based on EPA's final decision, an RPAR can result in cancellation of all or some registered restrictions.

The practice during the late 1980s was substituting a "Special Pesticide Review" for suspect chemicals to replace RPAR, which is admittedly an expensive and time-consuming procedure. By 1984, the backlog of RPAR reviews in progress had been reduced to nearly zero.

Enforcement

FIFRA is enforced by EPA through the Pesticides and Toxic Substances Enforcement Division. This division works in conjunction with the Office of Pesticide Programs in monitoring the conduct of pesticide enforcement and registration programs.

The Division makes decisions concerning suspension or cancellation of registrations and determines when additional enforcement actions are appropriate. In addition, the Division reviews scientific analyses of pesticide samples and makes recommendations to aid in achieving compliance, coordinates recall and seizure activities when necessary, and pursues legal action, both civil and criminal, when appropriate.

Individual states may have primary enforcement responsibility granted that the state has adequate pesticide use laws, does not permit any sale or use of pesticides prohibited by FIFRA, does not impose any labeling or packaging requirements different from those established by EPA, has adopted and is implementing effective enforcement procedures, and maintains records in accordance with the federal guidelines. Nearly all states have primary enforcement authority for FIFRA regulations. FIFRA allows the responsible agency to undertake the following actions in order to enforce the law:

- Register facilities involved in the manufacture, formulation, or distribution of pesticides.
- Inspect production facilities.
- Examine and test pesticides offered for sale.
- Impose fines or criminal penalties for violations.
- Stop sales of, seize, or recall products offered in violation of the regulations.

EPA may not require a state to have pesticide use laws and regulations more stringent than FIFRA. To date, states may, however, ban or limit use of specific pesticides and may require manufacturers to submit additional data to obtain state registration.

Relationship to Other Regulations

Spills of pesticides are subject to federal regulations in the Clean Water Act (CWA) covering discharges of toxic and hazardous materials to water, and pesticide manufacturers must apply for discharge permits if they release effluent into any regulated body of water. Pesticides in the atmosphere may be regulated by the section of the Clean Air Act (CAA) which pertains to hazardous air pollutants. The Resource Conservation and Recovery Act (RCRA) addresses the disposal of pesticides and wastes from pesticide manufacturing processes. The Department of Labor shares authority for the protection of workers from pesticide hazards through administration of the Occupational Safety and Health Act (OSH Act).

Current Status

Because some states (California, Pennsylvania, and Florida) have promulgated guidelines for pesticides which are more restrictive than federal regulations (ironically these states are the 1st, 2nd, and 3rd largest consumers of pesticides), there has been proposed legislation to limit a state's authority to gather data about a pesticide for state registration. However, there have been no changes to date. For the past 4 years, legislation introduced to amend FIFRA has fallen victim to the fact that pesticide law does not appear to be a high congressional priority. At least for the near future, other major environmental bills will probably block the passage of proposed amendments to FIFRA. These include a proposal to protect farm workers exposed to pesticides in the fields, more stringent standards on imported and exported chemicals, and standards for the amounts of pesticides used on fruits and vegetables for human consumption.

NATIONAL ENVIRONMENT POLICY ACT (NEPA)

Overview

NEPA defines a national environmental policy committed to use all practicable means to conduct *federal activities* in a way that will promote general welfare and be in harmony with the environmental consequences of their actions through preparation of an Environmental Impact Statement (EIS). The Act created the Council on Environmental Quality (CEQ) to issue standards for what is required to be in an EIS and what is not.

Background

NEPA was drafted to reflect a concern about the environmental consequences of the actions of federal agencies, and its principal objective was to encourage a productive and enjoyable harmony between man and his environment.

The Act defined the first environmental policy ever enacted by Congress and was signed into law by President Nixon on New Years Day, 1970.

NEPA does not directly address hazardous materials specifically and is relevant to the topic of hazardous materials management mainly because of its status as a historical precedent. For the most part, hazardous materials were not considered to be a public nuisance or an environmental health concern when NEPA was promulgated. For example, the 1970 edition of the annual report by the CEQ doesn't even mention hazardous wastes. However, NEPA represents a critical milestone in the general philosophy of environmental regulation and provides a foundation for virtually all environmental laws which followed its inception.

Upon NEPA's enactment on January 1, 1970 (almost a year prior to the creation of EPA), it was heralded by many to be the beginning of an environmentally oriented legislation decade. This Act is commonly known as the Magna Carta or "Bill of Rights" of environmental legislation.

NEPA has spawned extensive legislation. In fact, some experts feel that NEPA may have led to more lawsuits than all the other federal environmental laws combined. Ironically, the Supreme Court has found that NEPA creates no judicially enforceable substantive rights, but imposes only procedural requirements. However, the rights of those procedural requirements has been of such magnitude that virtually every federal agency has incorporated NEPA requirements into their routine procedures.

Content

NEPA is divided into two titles. The first title identifies environmental policy and goals and methods for accomplishing those goals. The second title creates the CEQ and defines its responsibility.

Probably the most important provision of NEPA is contained in Section 102(2)(c). It requires that a detailed statement (EIS) be prepared for every major federal action which significantly affects the quality of the human environment. The following must be addressed:

1. The environmental impact of the proposed action.
2. Any adverse environmental effects which cannot be avoided should the proposal be implemented.
3. Alternatives to the proposed action.
4. The relationship between local short-term activities and long-term enhancement of productivity of man's environment.
5. Any irreversible and irretrievable commitments of resources which would occur should the proposed action be implemented.

It must be noted that NEPA applies to federal agencies only. Only EISs are prepared by each of the responsible federal agencies. States, local agencies, and private parties may assist or be required to assist in this process. The analysis of the data and the conclusions must, however, be the work of the responsible federal agency.

The CEQ is charged with the issuance of standards on what is required to be addressed in an EIS and monitors progress toward achieving our national environmental goals. The CEQ is also charged with the following duties: assist and advise the President in the preparation of the Environmental Quality Report, gather environmental information, conduct studies on conditions and trends in environmental quality, and recommend national policies and legislation to protect the environment.

When violations of NEPA are alleged, any party that can show any evidence of resulting "injury in fact" has standing to sue.

Enforcement

NEPA applies to federal agencies only, and whether an EIS is required is a decision that is made by the responsible federal agency. The final draft of any EIS must be filed with EPA.

Many states, following the lead of the federal government, have enacted NEPA-like laws. However, these do not serve in lieu of the federal NEPA program.

Relationship to Other Regulations

Some statutes have excluded certain activities of specific agencies from NEPA coverage. EPA is exempt from many of the Act's requirements when engaged in regulatory activities. For example, EPA does not have to prepare an EIS prior to proposed environmental rule-making activities.

Current Status

Almost 23 years after the passage of NEPA, the concept of an EIS has become relatively noncontroversial. It is anticipated that as the reduction of NEPA's importance comes about, there will be an increase in the level of NEPA-related litigation.

ENVIRONMENTAL PROTECTION AGENCY (EPA)

The initiatives of environmental protection in the U.S. have passed to the federal government. NEPA consolidated all federal activities on air and water pollution, solid wastes, pesticides, noise, and environmental radiation in a new organizing action, the EPA. Many states followed suit by organizing all or most of the environmental protection work in an independent or autonomous unit. States follow the EPA leads as surrogates to keep out direct federal intervention and to qualify for various forms of monetary subsidies. When the subsidies go to local projects, the state agency is the control gate. An extensive and sometimes intensive bureaucratic process has evolved to move the authorities and mandates of Congress to EPA, sometimes directly and sometimes through the states to the point where the problem can be defined and solved.

EPA regulations, standards, and requirements are nationwide. Much has been accomplished. Environmental protection in governmental agencies is firmly institutionalized. It has greater strength and support than ever. All pollutant sources are to be controlled by this branch of government. Many streams and air sheds are showing improvement due to the supervision and enforcement of this agency. Ocean dumping and medical waste will be of revised concern to EPA in the 90s. Solid wastes are being managed much better due to the implementation of RCRA. New concerns for EPA will be the air loading of particulates, hydrocarbons, and carbon monoxide on a national scale, if not worldwide.

OCCUPATIONAL SAFETY AND HEALTH ACT (OSH ACT)

Overview

The purpose of the OSH Act is to assure safe and healthful employment conditions for all men and women working in the U.S. The Act addresses the proper management of hazardous materials in the working environment by setting standards which are intended to prevent injury and/or illness among workers. Employers are required to inform workers of the dangers posed by substances with which they work. An additional goal of the Act is the definition of a standard policy for regulating substances which cause cancer.

Background

A concern with the assurance of worker safety has long been a governmental issue, both at the state and federal level.

In 1877, the state of Massachusetts passed the first statute intended to insure worker safety. In 1910, in response to countless mining disasters, the U.S. Congress created the Bureau of Mines.

The current OSH Act was enacted in December of 1970, in the same month that EPA was created. To a great extent, the groundwork for the Act was laid as part of the "Great Society" programs of the late 1960s. A national health and safety program to protect workers was proposed by President Johnson in 1968, in response to numerous testimonies alleging that working conditions in U.S. industries were hazardous and costly in terms of time lost from work due to illnesses and injuries.

In 1968 legislative effort to address worker safety was unsuccessful, but the need for this type of federal law had been established. The Senate and House passed separate bills on worker safety in 1970. These were resolved into a compromise measure which became the OSH Act.

Business groups did obtain two provisions in the law as their price for support. First, industry insisted that states should be encouraged to assume primary responsibility thereby minimizing the role of the federal Occupational Health and Safety Administration (OSHA). Second, because of their distrust for the allegedly pro-union bias of the Department of Labor, responsibility for first-level adjudication of violations would be vested in an independent, three-member panel of judges in a separate Occupational Health and Safety Review Commission. Both of these provisions may be mistakes, and the first has openly been acknowledged as such by many industry leaders.

Congress did reject, however, an industry effort to separate the standard setting authority from the enforcement powers of the new organization. A special role of the National Institute for Occupational Safety and Health (NIOSH) was intended to assure it a primary position in the standard-setting process. Thus, the three main roles of OSHA are (1) setting the safety and health standards; (2) their enforcement through federal and state inspectors; and (3) public education consultation.

Content

The Act establishes conditions which call for the following actions:

1. Encouraging employers and employees to establish or perfect programs intended to reduce occupational hazards.
2. Providing that employees and employers have separate but dependent rights and responsibilities in relation to achieving safe and healthful working conditions.
3. Authorizing the Secretary of Labor to establish mandatory guidelines for occupational safety and health applicable to business affecting interstate commerce.
4. Creating an Occupational Safety and Health Review Commission for carrying out various functions defined by the act.
5. Augmenting previous advances made toward assuring occupational safety and health.
6. Providing means for research in the field of occupational safety and health.
7. Exploring ways of identifying occupational-related latent diseases.
8. Established medical criteria intended to assure (within reasonable limits) that no employee suffers diminished health, functional capacity, or life expect-ancy as a result of working conditions.

9. Providing for training programs to expand the field of occupational safety and health.
10. Providing for the development and enactment of occupational safety and health standards.
11. Providing an effective enforcement program which prohibits providing advanced warning to those responsible for areas to be inspected or investigated.
12. Providing grants to the states to encourage governments to expand, administer, and enforce state occupational safety and health laws.
13. Providing for appropriate reporting and record keeping procedures in regard to occupational safety and health.
14. Encouraging joint efforts between labor and management to reduce injuries and diseases that result from working conditions.

OSHA is charged with administering the provisions of the OSH Act, which is generally applicable to facilities with ten or more workers. Federal employees and workplaces regulated by other federal agencies are not exempt, which includes farms and the self-employed.

The major areas where OSHA regulations address hazardous materials are the following:

- Health standards
- Cancer policy
- Hazard communication standard
- Hazardous waste operations and training

Section 6 of the OSH Act grants authority to set safety and health standards. NIOSH was created to conduct research on safety and health issues and is charged with the development of criteria documents which are to be used as a basis for preparing new standards or for recommending more complete standards. These documents are prepared from a critical evaluation of all published information and data. Related federal agencies which may contribute data include the National Cancer Institute, National Heart, Lung and Blood Institute, National Institute of Environmental Health Sciences, National Center for Health Statistics, Center for Disease Control, and National Center for Toxicological Research. Also, NIOSH may perform specific health evaluation of problem/ exposed areas in a workplace, if an employee or employer requests it in writing to this agency.

Specific standards vary in form and requirements. In general, however, each addresses the following items:

- Exposure limits
- Labeling
- Protective equipment
- Control procedures
- Monitoring employee exposure

- Measuring employee exposure
- Medical examinations
- Access to records
- Training

Asbestos (1910.1001 and 1910.1101) was the first chemical covered by a specific final standard in 1972. By 1989, OSHA has also promulgated standards for the following substances:

- Coal tar pitch volatiles (1910.1002)
- 4-Nitrobiphenyl (1910.1003)
- Alpha-naththylamine (1910.1004)
- Methyl choromethyl ether (1910.1006)
- 3,3'Dichlorobenzidine (and its salts) (1910.1007)
- Bis chloromethyl ether (1910.1008)
- Beta-napthylamine (1910.1009)
- Benzidine (1910.1010)
- 4-Amino diphenyl (1910.1011)
- Ethylenimine (1910.1012)
- Beta-propiolactone (1910.1013)
- 2 Acetylaminofluorene (1910.1014)
- 4-Dimethyl aminoazobenzene (1910.1015)
- N-Nitrosodimethyl amine (1910.1016)
- Vinyl chloride (1910.1017)
- Inorganic arsenic (1910.1018)
- Lead (1910.1025)
- Benzene (1910.1038)
- Coke oven emissions (1910.1039)
- Cotton dust (1910.1043)
- 1, 2-dibromo-3-chloropropane (1910.1044)
- Acrylonitrile (1910.1045)
- Ethylene Oxide (1910.1047)
- Formaldehyde (1910.1048)

OSHA will be enforcing a new standard that was passed by Congress in 1989. It's new standard will be for blood-borne diseases, which include hepatitis B and acquired immune deficiency syndrome (AIDS). This standard will be among the first to be for a specific pathogen. In the future we will be looking for the specific standard on glycol ether or central nervous system diseases or neuropsychological disorders. Various provisions of the Act require record keeping and reporting, and many of the specific standards listed above contain additional requirements for training and biological monitoring. Employers must maintain the medical and chemical exposure records for at least 30 years and permit employees to see and copy these records upon written request. If the business should close down or file bankruptcy, these records are to be forwarded to NIOSH. Also, if the business is sold, the new owner will be responsible for the records and maintenance of them.

An OSHA cancer policy was promulgated in 1980, defining criteria to be used to identify, classify, and regulate chemicals with potential human carcinogenic effects. The implementation of the policy was placed on "hold" by the Reagan Administration. The policy is still in effect, but is not in use.

Of greater current impact is the Hazard Communication Standard, which was issued in 1983 and took effect in November 25, 1985. According to the rule, employers must furnish employees with instruction on the nature and effects of toxic substances with which they work in the workplace, either in written form, or in training programs. Instruction generally must include:

1. The chemical and common names of the substance (unless claimed as a trade secret by the manufacturer).
2. The location of the substance in the workplace.
3. Proper and safe handling practices.
4. First aid treatment and antidotes in case of overexposure.
5. The adverse health effects of the substance.
6. Appropriate emergency procedures.
7. Proper procedures for cleanup of leaks or spills.
8. Potential for flammability, explosion, and reactivity.
9. The rights of employees under this rule.

Most of this information is available from the Material Safety Data Sheet (MSDS), a document containing standardized information about the properties and hazards of listed toxic substances (MSDSs was known as the OSHA 20 Form in the 70s). Manufacturers, importers, and distributors of listed toxic substances are required to prepare and provide MSDSs to their direct purchasers. Originally this law was geared for the manufacturing industry, but now under the August 24, 1987, amended version it includes nonmanufacturing employees. Examples of nonmanufacturers would be hospitals, university laboratories, optical facilities, dentist offices, etc., anywhere chemicals are being used by an employee or the threat of an occupational disease would occur.

Enforcement

The OSH Act is enforced by the U.S. Department of Labor through procedures which are specified in the Act. The regulations encourage states to develop and operate their own job safety and health programs, such as Federal 7C1 programs. Approximately half of the states now have OSH Act enforcement authority. State laws such as those in New Jersey, New York, and California may be more stringent than federal regulations. Also, state laws remain in effect where no federal standards exist. If an investigation involving an employee's death in the workplace is conducted by OSHA and the investigation reveals some other employee or company responsibility (example: supervisor or manufacturer), OSHA has the right to refer their findings to the Department of Justice for Federal Criminal prosecution.

Relationship to Other Regulations

The OSH Act is related to several occupational-specific regulations including the Toxic Substances Control Act (TSCA) and FIFRA. Any labels required by OSHA may not conflict with the labeling requirements of the Hazardous Materials Transportation Act.

Current Status

Unlike many environmental regulations, the OSH Act has remained virtually unchanged, although new regulatory programs have been added since its inception. One of these additions, the cancer policy, has been essentially suspended. Another, the Hazard Communication standard, is being implemented in spite of court challenges, and its complete judicial review process will probably take several years to complete. On the horizon will be the issues of AIDs and other blood borne diseases for OSHA's control.

CLEAN AIR ACT (CAA)

Overview

The CAA was enacted to achieve air quality levels protective of public health and welfare and provides the regulatory framework for prevention and control of discharges into the air. The Act establishes national standards for air quality in terms of National Ambient Air Quality Standards (NAAQS), which are applicable to criteria pollutants. Hazardous pollutants are regulated by National Emission Standards for Hazardous Air Pollutants (NESHAP). The Act requires EPA to formulate plans to control pollutants through its own actions or through action by the states.

Background

The original CAA was passed in 1963, in response to a concern with the increase and widespread nature of air pollution brought about the urban and industrial development and the increasing use of motor vehicles. Certain environmental effects, such as reduced visibility, are readily apparent and, to a certain extent, easy to control; early air pollution control efforts focused on those problems through ordinances against black smoke and open burning. The most serious problems, however, are more insidious and require longer periods to become apparent, e.g., impaired breathing functions, cancer, lower crop yields, premature deterioration of paint and building materials. The Act was amended in the subsequent years of 1965 and 1967, with the most significant amendments being drafted in 1970 and 1977, with new provisions added to prevent significant deterioration of air quality, impose stringent requirements in areas that failed to attain national standards, restrict the use of dispersion techniques, and strengthen enforcement of the statutory provisions.

In 1970, Congress authorized the establishment of stringent uniform national ambient air quality standards to be determined by the newly created EPA.

Content

The CAA requires EPA to do the following:

- Identify air pollutants.
- Set national air quality standards.
- Formulate plans to control air pollutants.
- Set standards for sources of air pollution.
- Set standards limiting the discharge of hazardous substances into the air.

The Act itself is extremely complex and has come under considerable criticism because of this fact. At least three descriptive categories must be determined for each emission source in order to determine what constitutes regulatory compliance. The categories, as depicted in Figure 1.3, are type of source, location, and pollutant. Discussion of each follow.

Type of source is probably the easiest category to define. Mobile sources are vehicular, others are stationary, and whether a source is new or existing is defined by dates specified in the Act. Stationary sources are regulated by technology-based guidelines. (These depend on the location of the emission source, which is discussed below.) The control categories which are applicable to stationary sources are the following:

1. *New source performance standards.* Newly constructed facilities potentially emitting 100 tons/year.
2. *Control technique guidelines.* Existing plants that meet NAAQS.
3. *Reasonably available control technology.* Existing plants where NAAQS not met.
4. *Lowest achievable emission rates.* Mandated in nonattainment areas.
5. *Best available control technology.* Designed for use in prevention of significant deterioration (PSD) areas.

Location determines enforcement area and the agency in authority.

There are now more than 240 discrete enforcement areas, defined as one of the following categories of Attainment Area Classes:

- I — Virtually no emission increases allowed.
- II — Modest emission increases allowed.
- III — Considerable emission increases allowed.

The 1977 amendments introduced two additional location-related concepts. PSD addresses areas with exceptional air quality, and areas which do not meet national standards are addressed as nonattainment areas.

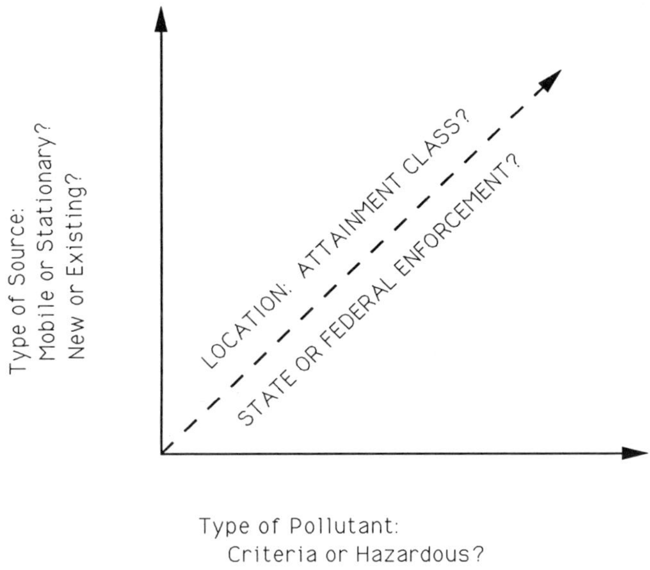

Figure 1.3. Regulatory elements of CAA.

The Act addresses two major classes of pollutants, which is the third descriptive category shown in Figure 1.3.

Criteria pollutants are regulated by primary standards designed to protect public health and by secondary standards designed to protect the public welfare. New Source Performance Standards (NSPS) pertain to new and modified stationary sources in 28 industrial categories if construction commenced after the standard was proposed. NAAQSs apply to the following commonly detected air pollutants: sulfur oxides, particulate matter, nitrogen oxides, carbon monoxide, photochemical oxidants (measured as ozone), nonmethane hydrocarbons, and lead. Characteristics of these seven common air pollutants are given in Table 1.1.

Hazardous pollutants are regulated by NESHAP, currently applicable to the following substances: asbestos, beryllium, mercury, and vinyl chloride. Hazardous pollutants differ from ambient in that they are usually localized and can be technically difficult and costly to control. Compliance requirements include compliance status information form, notification to agency, emissions testing, monitoring, and reporting.

The CAA grants EPA broad authority in the event of an emergency or imminent hazard, allowing the Administrator to bring action in a U.S. District Court to prevent or abate the emergency or hazard. Section 303 provides that in addition to bringing suit immediately to restrain the alleged pollution, the Administrator may issue orders to protect the health of threatened persons after consulting with state and local authorities.

Table 1.1. Characteristics of Major Air Pollutants

Pollutant	Characteristics	Health Effects	Other Effects
Sulfur dioxide	Corrosive	Associated with coughs, colds, asthma, bronchitis. Can aggravate heart disorders	Can cause acid rain
Nitrogen oxide	Highly reactive	Fatal at high concentrations	Can form ozone, a principal constituent of smog
Ozone	Pungent smelling faintly bluish	Irritates mucous membrane	Principal constituent of smog
Carbon monoxide	Colorless, odorless	Fatal at high levels, can cause headaches, dizziness, nausea, breathing difficulties	Impairs judgement and perception
Particulates	Dust, soot, smoke	Aggravates respiratory diseases	Aesthetic concerns
Lead	Mainly from auto emissions	Affects blood-forming, nervous and kidney systems	Young children are at high risk
Hydrocarbons	Mainly from auto emissions	No significant effects at low levels	Reacts with nitrogen oxides to form oxidants

Enforcement

The standards defined in the CAA are to be implemented through State Implementation Programs (SIPs). The Act provides considerable flexibility to the states in determining how best to meet its requirements. Where a state does not have authority, EPA enforces the CAA through its regional offices. Permit requirements vary among states and may vary from region to region even when administered by EPA.

Relationship to Other Regulations

The CAA must be interpreted in conjunction with RCRA where the incineration of hazardous waste is involved. Radionuclides are regulated under Section 112 of the Act. The Act bears a similarity to the CWA, in that both utilize technology-based guidelines. Pesticide emissions may also be regulated by the FIFRA.

Current Status

The CAA was amended in 1990, due to current trends involving toxic air pollutants. A more stringent air pollution bill has been introduced in the House, at least partially in response to the Bhopal chemical leak which claimed over 2000 lives in India. Among other measures, the bill would require EPA to list hazardous threshold levels for 85 chemical substances within a 2-year period. This bill was passed in 1986 as the Superfund Amendment and Reauthorization Act of 1986 (SARA) and listed 366 chemical substance hazardous threshold levels. The specific section that requires threshold levels is under the Emergency Planning and Community Right-to-Know. A more detailed explanation of SARA is under the SARA section of this chapter. Currently, provisions have been made under CAA to include Indoor Air Quality regulations. These regulations will address such troublesome problems as human supersensitivity to building materials and Sick Building or Tight Building Syndromes.

CLEAN WATER ACT (CWA)

Overview

The CWA controls discharges of effluent from point sources into waters of the U.S. The Act has five main elements: a permit program, a system of minimum national effluent standards for each industry, water quality standards, provisions for special programs such as toxic chemicals and oil spills, and a construction grant program for publicly owned treatment works (POTWs).

Background

Federal regulation of water quality dates back to the Refuse Act of 1899, which was designed to protect navigation. However, it additionally prohibited all discharges into navigable waters or tributaries thereof unless a permit was obtained in advance from the U.S. Army Corps of Engineers (COE). This provision provided a major milestone in federal efforts to regulate industrial aqueous discharges. By late 1971, it was evident that comprehensive water pollution control legislation was a prerequisite for a workable control program.

The present CWA is the successor to several water pollution control statutes that were passed in the 1950s and 1960s. For the most part, these laws were not effective in controlling water pollution. Major problems included

- Inability to determine with precision when a discharge violated applicable standards.
- Inapplicability of federal-state water quality standards to intrastate waters.
- Lack of state initiative in making load allocations required to set enforceable discharge standards.
- Cumbersome enforcement mechanisms and the requirement for state consent for federal enforcement.

Although a few states made the water quality approach work, it was clear by 1970 that an effective nationwide approach would require implementation of a permit program based on federal minimum "end-of-pipe" effluent criteria enforceable directly against the discharger.

In late 1972, Congress finally passed, over President Nixon's veto, a comprehensive recodification and revision of federal water pollution control law. The 1972 Act, Public Law 92-500, made the EPA responsible for setting effluent standards on an industry-by-industry basis and required such standards to be based on the capabilities and costs of pollution control technologies. EPA had to meet a stringent time-table for setting these standards.

The basic framework of the 1972 Act — the permit program, national effluent limitations, water quality standards, special provisions for oil spills and toxic substances, and a POTW construction grant program — proved reasonably sound and remains so today. Congress significantly amended the Act in 1977 in an effort to deal more effectively with toxic pollutants, and to resolve numerous definitional and policy issues raised by court and EPA decisions. The Act was revised again in 1978 to deal with spills of hazardous substances and in 1980 and 1981 to change certain aspects of the construction grant program for POTWs.

Content

The CWA regulations include standards for the following type of pollution discharges: direct discharges, indirect discharges, sources that spill oil or hazardous substances, and discharges of dredged or filled material.

It is convenient to associate each of the four major provisions of the CWA with a key acronym, as shown below:

- Direct discharges (NPEDS)
- Indirect discharges (POTW)
- Oil or hazardous substances (SPCC)
- Dredge and fill (COE)

Facilities that discharge directly into navigable waters must obtain a National Pollutant Elimination Discharge System (NPEDS) permit. Requirements are based on type of effluent, national technology-based guidelines, and state water quality standards. The type of controls required of direct discharges include the following, in ascending order of stringency: Best Practicable Technology (BPT), Best Conventional Technology (BCT), Best Available Technology (BAT), Best Engineering Judgment (BEJ), and Best Available Demonstrated Control Technology (BDCT).

Discharges into municipal sewers are indirect discharges and do not require a permit. Instead, industrial dischargers who discharge into POTWs must comply with pretreatment standards. General standards address fire or explosion hazards, corrosive discharges, viscous obstructions, sludge discharges, and heat sufficient to inhibit biological activities. Specific standards are set by industrial

Table 1.2. Industry Categories Used to Describe Standard Industrial Classification (SICs)

1. Adhesive and sealants
2. Aluminum forming
3. Asbestos manufacturing
4. Auto and other laundries
5. Battery manufacturing
6. Coal mining
7. Coil coating
8. Copper forming
9. Electric and electronic components
10. Electroplating
11. Explosives manufacturing
12. Ferroalloys
13. Foundries
14. Gum and wood chemicals
15. Inorganic chemicals manufacturing
16. Iron and steel manufacturing
17. Leather tanning and finishing
18. Mechanical products manufacturing
19. Nonferrous metals manufacturing
20. Ore mining
21. Organic chemicals manufacturing
22. Pesticides
23. Petroleum refining
24. Pharmaceutical preparations
25. Photographic equipment and supplies
26. Plastic and synthetic materials manufacturing
27. Plastic processing
28. Porcelain enamelling
29. Printing and publishing
30. Pulp and paperboard mills
31. Soap and detergent manufacturing
32. Steam electrical power plants
33. Textile mills
34. Timber products processing

Source: Environmental Policy Law, 8th Edition, pg. 333, Government Institutes, Inc.

categories, as described by Standard Industrial Classifications (SICs) (see Table 1.2). Pretreatment standards are generally intended to achieve BAT removal efficiencies.

Section 311 of the Act prohibits the discharge of hazardous chemicals and oil in quantities as may be harmful into waters of the U.S. It also requires certain industries to prepare Spill Prevention Control and Countermeasure (SPCC) plans to control oil pollution. Additionally, this section identifies about 300 substances as hazardous when spilled or discharged and establishes the minimum amount (reportable quantity) that when spilled must be reported to the National Response Center. Table 1.3 lists these substances and amounts.

The CWA regulates the discharge of dredged or fill material into waters of the U.S. The program includes all navigable waters and interstate waters and tributaries of both, all other waters in which use might affect interstate commerce, and wetlands adjacent to any of these waters. The program excludes ordinary agriculture, silviculture, and the maintenance of dams, levees, farm ponds, and forestry roads. The agency with primary authority for this portion of the Act is the COE.

The major portions of the Act that deal with hazardous materials are embodied in the above provisions and consist of water quality criteria and standards, effluent limitations and guidelines, and control of oil and hazardous substance discharges. (For the most part, dredge and fill regulations do not address hazardous substances.)

Section 504 of the CWA contains an imminent hazard provision allowing EPA to compel cleanup on sites creating imminent and substantial endangerment to the public health or the environment. The section is applicable to control of point sources that discharge pollutants into surface waters.

Enforcement

States administer the Act upon authorization by EPA, and states may set more stringent standards than those of the federal government. The dredge and fill portion of the Act (Section 404) is administered by the COE.

Relationship to Other Regulations

Section 307(a) may be used to set effluent limitations for some toxic pollutants which are also regulated by other legislation. The CWA may define reportable quantities of materials requiring spill notification under the Comprehensive Environmental Response, Compensation and Liability Act (CERCLA), and its provision for drawing up a National Contingency Plan (NCP) was revised and incorporated into the Superfund Law. The CWA is similar to the CAA in that it utilizes technology-based guidelines, and both Acts set national guidelines that the states must meet.

Current Status

Major revisions of the CWA were introduced during the late 1980s. Some of these revisions were the phasing out of current grant programs and the creation of new state-run loan funds for municipal sewage construction.

SAFE DRINKING WATER ACT (SDWA)

Overview

The overall intent of the SDWA is to ensure the quality of drinking water consumed by the public and to protect underground sources of water. The major provisions of the Act establish two broad programs: one relating to public water supply systems and the other relating to the underground injection of wastes.

Table 1.3. List of Reportable Quantities

Material	RQ in pounds	(kilograms)
Acetaldehyde	1000	(454)
Acetic Acid	1000	(454)
Acetic anhydride	1000	(454)
Acetone cyanohydrin	10	(4.54)
Acetyl bromide	5000	(2270)
Acetyl chloride	5000	(2270)
Acrolein	1	(0.454)
Acrylonitrile	100	(45.4)
Adipic acid	5000	(2270)
Aldrin	1	(0.454)
Allyl alcohol	100	(45.4)
Allyl chloride	1000	(454)
Aluminum sulfate	5000	(2270)
Ammonia	100	(45.4)
Ammonium acetate	5000	(2270)
Ammonium benzoate	5000	(2270)
Ammonium bicarbonate	5000	(2270)
Ammonium bichromate	1000	(454)
Ammonium bifluoride	5000	(2270)
Ammonium bisulfite	5000	(2270)
Ammonium carbamate	5000	(2270)
Ammonium carbonate	5000	(2270)
Ammonium chloride	5000	(2270)
Ammonium chromate	1000	(454)
Ammonium citrate	5000	(2270)
Ammonium fluoborate	5000	(2270)
Ammonium fluoride	5000	(2270)
Ammonium hydroxide	1000	(454)
Ammonium oxalate	5000	(2270)
Ammonium silicofluoride	1000	(454)
Ammonium sulfamate	5000	(2270)
Ammonium sulfide	5000	(2270)
Ammonium sulfite	5000	(2270)
Ammonium tartrate	5000	(2270)
Ammonium thiocyanate	5000	(2270)
Ammonium thiosulfate	5000	(2270)
Amyl acetate	1000	(454)
Aniline	1000	(454)
Antimony pentachloride	1000	(454)
Antimony potassium tartrate	1000	(454)
Antimony tribromide	1000	(454)
Antimony trichloride	1000	(454)
Antimony trioxide	5000	(2270)
Arsenic disulfide	5000	(2270)
Arsenic pentoxide	5000	(2270)
Arsenic trichloride	5000	(2270)
Arsenic trioxide	5000	(2270)
Arsenic trisulfide	5000	(2270)
Barium cyanide	10	(4.54)
Benzene	1000	(454)
Benzoic acid	5000	(2270)
Benzonitrile	1000	(454)
Benzoyl chloride	1000	(454)
Benzyl chloride	100	(45.4)
Beryllium chloride	5000	(2270)

Table 1.3. (continued)

Material	RQ in pounds	(kilograms)
Beryllium fluoride	5000	(2270)
Beryllium nitrate	5000	(2270)
Butyl acetate	5000	(2270)
n-Butyl phthalate	100	(45.4)
Butylamine	1000	(454)
Butyric acid	5000	(2270)
Cadmium acetate	100	(45.4)
Cadmium bromide	100	(45.4)
Cadmium chloride	100	(45.4)
Calcium arsenate	1000	(454)
Calcium arsenite	1000	(454)
Calcium carbide	5000	(2270)
Calcium chromate	1000	(454)
Calcium cyanide	10	(4.54)
Calcium dodecylbenzenesulfonate	1000	(454)
Calcium hypochlorite	100	(45.4)
Captan	10	(4.54)
Carbaryl	100	(45.4)
Carbofuran	10	(4.54)
Carbon disulfide	5000	(2270)
Carbon tetrachloride	5000	(2270)
Chlordane	1	(0.454)
Chlorine	10	(4.54)
Chlorobenzene	100	(45.4)
Chloroform	5000	(2270)
Chlorpyrifos	1	(0.454)
Chlorosulfonic acid	1000	(454)
Chromic acetate	1000	(454)
Chromic acid	1000	(454)
Chromic sulfate	1000	(454)
Chromous chloride	1000	(454)
Cobaltous bromide	1000	(454)
Cobaltous formate	1000	(454)
Cobaltous sulfamate	1000	(454)
Coumaphos	10	(4.54)
Cresol	1000	(454)
Crotonaldehyde	100	(45.4)
Cupric acetate	100	(45.4)
Cupric acetoarsenite	100	(45.4)
Cupric chloride	10	(4.54)
Cupric nitrate	100	(45.4)
Cupric oxalate	100	(45.4)
Cupric sulfate	10	(4.54)
Cupric sulfate ammoniated	100	(45.4)
Cupric tartrate	100	(45.4)
Cyanogen chloride	10	(4.54)
Cyclohexane	1000	(454)
2,4-D Acid	100	(45.4)
2,4-D Esters	100	(45.4)
DOT	1	(0.454)
Diazinon	1	(0.454)
Dicamba	1000	(454)
Dichlobenil	1000	(454)
Dichlone	1	(0.454)
Dichlorobenzene	100	(45.4)
Dichloropropane	5000	(2270)

Table 1.3. (continued)

Material	RQ in pounds	(kilograms)
Dichloropropene	5000	(2270)
Dichloropropene-Dichloropropane Mixture	5000	(2270)
2,2-Dichloropropionic acid	5000	(2270)
Dichlorvos	10	(4.54)
Dieldrin	1	(0.454)
Diethylamine	1000	(454)
Dinitrobenzene	1000	(454)
Dinitrophenol	1000	(454)
Dinitrotoluene	1000	(454)
Diquat	1000	(454)
Disulfoton	1	(0.454)
Diuron	100	(45.4)
Dodecylbenzenesulfonic acid	1000	(454)
Endosulfan	1	(0.454)
Endrin	1	(0.454)
Epichlorohydrin	1000	(454)
Ethion	10	(4.54)
Ethyl benzene	1000	(454)
Ethylenediamine	1000	(454)
Ethylene dibromide	1000	(454)
Ethylene dichloride	5000	(2270)
EDTA	5000	(2270)
Ferric ammonium citrate	1000	(454)
Ferric ammonium oxalate	1000	(454)
Ferric chloride	1000	(454)
Ferric fluoride	100	(45.4)
Ferric nitrate	1000	(454)
Ferric sulfate	1000	(454)
Ferrous ammonium sulfate	1000	(454)
Ferrous chloride	100	(45.4)
Ferrous sulfate	1000	(454)
Formaldehyde	1000	(454)
Formic acid	5000	(2270)
Fumaric acid	5000	(2270)
Furfural	1000	(454)
Guthion	1	(0.454)
Heptachlor	1	(0.454)
Hexachlorocyclopentadiene	1	(0.454)
Hydrochloric acid	5000	(2270)
Hydrofluoric acid	5000	(2270)
Hydrogen cyanide	10	(4.54)
Hydrogen sulfide	100	(45.4)
Isoprene	1000	(454)
Isopropanolamine-dodecylbenzenesulfonate	1000	(454)
Kelthane	5000	(2270)
Kapone	1	(0.454)
Lead acetate	5000	(2270)
Lead arsenate	5000	(2270)
Lead chloride	5000	(2270)
Lead fluoborate	5000	(2270)
Lead fluoride	1000	(454)
Lead iodide	5000	(2270)
Lead nitrate	5000	(2270)
Lead stearate	5000	(2270)
Lead sulfate	5000	(2270)
Lead sulfide	5000	(2270)

Table 1.3. (continued)

Material	RQ in pounds	(kilograms)
Lead thiocyanate	5000	(2270)
Lindane	1	(0.454)
Lithium chromate	1000	(454)
Malathion	10	(4.54)
Maleic acid	5000	(2270)
Maleic anhydride	5000	(2270)
Mercaptodimethur	100	(45.4)
Mercuric cyanide	1	(0.454)
Mercuric nitrate	10	(4.54)
Mercuric sulfate	10	(4.54)
Mercuric thiocyanate	10	(4.54)
Mercurous nitrate	10	(4.54)
Methoxychlor	1	(0.454)
Methyl mercaptan	100	(45.4)
Methyl methacrylate	5000	(2270)
Methyl parathion	100	(45.4)
Mevinpnos	1	(0.454)
Mesacarpate	1000	(454)
Monoethylamine	1000	(454)
Monomethylamine	1000	(454)
Naled	10	(4.54)
Naphthalene	5000	(2270)
Naphthenic acid	100	(45.4)
Nickel ammonium sulfate	5000	(2270)
Nickel chloride	5000	(2270)
Nickel hydroxide	1000	(454)
Nickel nitrate	5000	(2270)
Nickel sulfate	5000	(2270)
Nitric acid	1000	(454)
Nitrobenzene	1000	(454)
Nitrogen dioxide	1000	(454)
Nitrophenol	1000	(454)
Nitrotoluene	1000	(454)
Paraformaldehyde	1000	(454)
Parathion	1	(0.454)
Pentachlorophenol	10	(4.54)
Phenol	1000	(454)
Phosgene	5000	(2270)
Phosphoric acid	5000	(2270)
Phosphorus	1	(0.454)
Phosphorus oxychloride	5000	(2270)
Phosphorus pentasulfide	100	(45.4)
Phosphorus trichloride	5000	(2270)
Polychlorinated biphenyl	10	(4.54)
Potassium arsenate	1000	(454)
Potassium arsenite	1000	(454)
Potassium bichromate	1000	(454)
Potassium chromate	1000	(454)
Potassium cyanide	10	(4.54)
Potassium hydroxide	1000	(454)
Potassium permanganate	100	(45.4)
Propargite	10	(4.54)
Propionic acid	5000	(2270)
Propionic anhydride	5000	(2270)
Propylene oxide	5000	(2270)

Table 1.3. (continued)

Material	RQ in pounds	(kilograms)
Pyrethrin	1000	(454)
Quinoline	1000	(454)
Resorcinol	1000	(454)
Selenium oxide	1000	(454)
Silver nitrate	1	(0.454)
Sodium	1000	(454)
Sodium arsenate	1000	(454)
Sodium arsenite	1000	(454)
Sodium bichromate	1000	(454)
Sodium bifluoride	5000	(2270)
Sodium bisulfite	5000	(2270)
Sodium chromate	1000	(454)
Sodium cyanide	10	(4.54)
Sodium dodecylbenzenesulfonate	1000	(454)
Sodium fluoride	5000	(2270)
Sodium hydrosulfide	5000	(2270)
Sodium hydroxide	1000	(454)
Sodium hypochlorite	100	(45.4)
Sodium methylate	1000	(454)
Sodium nitrite	100	(45.4)
Sodium phosphate, dibasic	5000	(2270)
Sodium phosphate, tribasic	5000	(2270)
Sodium selenite	1000	(454)
Strontium chromate	1000	(454)
Strychnine	10	(4.54)
Styrene	1000	(454)
Sulfuric acid	1000	(454)
Sulfur monochloride	1000	(454)
2,4,5-T acid	100	(45.4)
2,4,5-T amines	100	(45.4)
2,4,5-T esters	100	(45.4)
2,4,5-T salts	100	(45.4)
2,4,5-TP acid	100	(45.4)
2,4,5-TP acid esters	100	(45.4)
TDE	1	(0.454)
Tetraethyl lead	100	(45.4)
Tetraethyl pyrophosphate	100	(45.4)
Thallium sulfate	1000	(454)
Toluene	1000	(454)
Toxaphene	1	(0.454)
Trichlorfon	1000	(454)
Trichloroethylene	1000	(454)
Trichlorophenol	10	(4.54)
Triethanolamine dodecylbenzenesulfonate	1000	(454)
Triethylamine	5000	(2270)
Trimethylamine	1000	(454)
Uranyl acetate	5000	(2270)
Uranyl nitrate	5000	(2270)
Vanadium pentoxide	1000	(454)
Vanadyl sulfate	1000	(454)
Vinyl acetate	1000	(454)
Vinylidene chloride	5000	(2270)
Xylene	1000	(454)
Xylenol	1000	(454)
Zinc acetate	1000	(454)

Table 1.3. (continued)

Material	RQ in pounds	(kilograms)
Zinc ammonium chloride	5000	(2270)
Zinc borate	1000	(454)
Zinc bromide	5000	(2270)
Zinc carbonate	1000	(454)
Zinc chloride	5000	(2270)
Zinc cyanide	10	(4.54)
Zinc fluoride	1000	(454)
Zinc formate	1000	(454)
Zinc hydrosulfite	1000	(454)
Zinc nitrate	5000	(2270)
Zinc phenosulfonate	5000	(2270)
Zinc phosphide	1000	(454)
Zinc silicofluoride	5000	(2270)
Zinc sulfate	1000	(454)
Zirconium nitrate	5000	(2270)
Zirconium potassium fluoride	5000	(2270)
Zirconium sulfate	5000	(2270)
Zirconium tetrachloride	5000	(2270)

Background

Passage of the CWA (see previous section) provided for the cleanup and protection of surface waters, but did not address the regulation of groundwater. The intent of the SDWA was to fill this regulatory gap and protect underground water sources.

By 1970 an effort had started in the federal government to review and revise the 1962 Drinking Water Standards. In 1971, EPA asked a coalition of the National Academy of Sciences and the National Academy of Engineering to revise the 1968 Water Quality Criteria. Investigations demonstrated that public confidence in the safety of drinking water supplies might in many instances be misplaced. The SDWA addressed this concern by allowing EPA to establish federal standards to control the levels of harmful contaminants in drinking water supplied by public water systems. In drafting the law, Congress also provided EPA with regulatory control of underground injection of substances into wells which might contaminate groundwater sources. The Act was signed by President Ford on December 16, 1974.

The SDWA, 42 U.S.C. 300f-300j-10, was passed in order to protect public water systems. The act requires the establishment of primary and secondary "maximum contaminant levels" (MCLs) of water delivered to the user. Primary standards relate to health, while secondary standards concern odor, taste, and other nonhealth problems. The EPA has authority to determine the applicable standards, while the states may apply to assume the implementation of the Act. The EPA has concentrated upon promulgating MCLs for organic toxic substances. A final rule has been published for the trihalomethanes, which result primarily from the interaction of chlorine (used as a disinfectant) with the natural organic substances present in raw water.

A 1982 report by the Government Accounting Office (GAO) stated that thousands of community water systems are presently contaminated by human waste, chemicals, and radioactivity in violation of federal drinking water standards. In addition, the GAO said that the public is not always properly warned when drinking water is unsafe. The report found that in 1980 alone, there were more than 146,000 violations of drinking water regulations by 28,000 of the nation's 65,000 community water systems, in most cases due to failure to test drinking water for contaminants.

Content

The major provisions of the Act as passed in 1974 included

1. The establishment of primary regulations for the protection of the public health.
2. The establishment of secondary regulations that are related to taste, odor, and appearance of drinking water.
3. The establishment of regulations to protect underground drinking water sources by the control of subsurface injection.
4. The initiation of research on health, economic, and technological problems related to drinking water supplies.
5. The initiation of a survey of rural water supplies.
6. The allocation of funds to states for improving their drinking water programs through technical assistance, training of personnel, and grant support.

The provisions of the Act that address the regulation of hazardous materials are found in Section 1412 (drinking water standards) and Section 1421 (underground injection).

The Act defines primary drinking water regulations applicable to public water supply systems, which consist of standards for maximum concentration of contaminants and standards for monitoring, analysis, reporting, and notification of users. Public systems are defined as those that provide piped water for human consumption to at least 15 service connections or 25 individuals daily. Standards vary for public systems depending on whether they are classified as community or noncommunity suppliers. Among the primary standards are MCLs for fluoride, arsenic, mercury, lead, nitrates, radioactivity, and various pesticides, inorganic, and organic chemicals.

The Act also establishes regulations to protect underground drinking water sources by the control of subsurface injection. There are five classes of wells defined and the requirements that have been promulgated for each class of well are set forth in the regulations. The classes and general requirements are as follows:

Class I	Description	Requirements
I	Injects hazardous waste beneath an underground source of water	Permit required to construct, must meet specified construction requirements
II	Operate in conjunction with oil and gas production operation, enhanced recovery, or storage	
III	Used in mining or recovery of nonfossil fuel	
IV	Injects hazardous materials	Specified requirements deferred to state
	above or into a substrate containing an underground source of drinking water	programs and RCRA
V	All injection wells not described above	Not currently regulated

The underground injection control program (UIC) is intended to be administered by the states. Class IV wells must be phased out of operation within six months after a site UIC program is approved, and no new Class IV wells will be permitted. Class IV wells are also addressed by the RCRA, which calls for their elimination through recently enacted amendments. Classes I, II, and III wells must be reevaluated and apply for a new state permit every 5 years.

Two other provisions of interest address "sole source aquifers" and imminent hazards. Sole source aquifers are underground water supplies that serve as the only source of drinking water for a specific area. Such aquifers may be subject to more stringent protective measures than those applicable to other groundwater sources.

Section 1431 addresses imminent hazards by authorizing EPA to commence "a civil action for appropriate relief, including a restraining order or permanent or temporary injunction" where it has been determined that a contaminant is or likely to enter a public water system, and that appropriate state and local authorities have not acted.

Enforcement

The Act establishes a joint federal-state system for assuring compliance with regulatory standards. States are intended to administer the Act subject to approval by EPA.

Relationship to Other Regulations

Regulation of Class I and IV wells is provided under the RCRA. Primary drinking water standards set maximum contaminant levels for radioactivity. The heavy metals which are included in primary drinking water standards are the basis for those specified for toxicity testing in RCRA.

Current Status

The SDWA was amended in 1986 to require EPA to regulate 83 contaminants in drinking water by 1989. The Act also requires EPA to establish a priority list of contaminants that may have adverse effects on the health of persons and which are known or anticipated to occur in public water systems. This type of evaluation process is illustrated in the California Constitutional Amendment called Proposition 65.

RADIOACTIVE MATERIALS REGULATIONS (RMR)

Overv iew

The possession, use, transfer, storage, and disposal of radioactive materials within the U.S. is regulated. This regulation is performed by such agencies as

1. U.S. Nuclear Regulatory Commission (NRC)
2. Agreement states
3. U.S. Department of Transportation (DOT)
4. U.S. Environmental Protection Agency (EPA)

Background

Each agency listed above retains authority over specific aspects associated with radioactivity. Most of the regulations established by the involved agencies pertaining to radioactive materials have evolved to the point where the overlaps and inconsistencies are minimized. NRC is the most dominant agency and historically has set the nuclear industry's operating guidelines and rules. DOT recognized the necessity for becoming more involved in the day-to-day decisions associated with the transport of radioactive material and promulgated the regulations found in 49 CFR 173.401.

Congress has historically found it difficult to agree on a comprehensive radioactive waste management policy. Congress enacted the Low-Level Radioactive Waste Policy Act in December of 1980, making states responsible for low-level radioactive waste. The individual states which have elected to administer the regulations set by NRC are known as agreement states. In 1982, Congress enacted the Nuclear Waste Policy Act, assigning responsibility for high-level radioactive waste to the federal government.

Content

For the most part, agreement states have adopted rules identical to the federal rules. Slight variations can exist, and in some states, such as California, New York, New Jersey, and New Hampshire, very significant changes are in place. Consequently, any use of radioactive materials in a particular location should be preceded with a review of the specific regulations.

NRC has retained regulatory control in certain areas: specifically those that involve weapons-grade nuclear material, high-level waste, commercial burial site, international shipments, and nuclear power plant operations to mention a few. NRC also retains the right to audit each agreement state's performance and can revoke the states' authorities if sufficient cause is demonstrated.

Agreement sites issue general licenses to facilities dealing with low or inconsequential quantities of radioactive materials. Specific licenses apply to definitely defined isotopes and quantities used under specific criteria. A broad license allows for more on-site flexibility than either a general or specific license and is administered by a committee.

Agreement states regulate the components of radiation protection plans. These include personnel monitoring, posting of areas, training, and maintenance of records and inventories. Waste disposal practices are also regulated by agreement sites, as are airborne and aqueous releases.

DOT has established the rules for proper packaging, description, labeling, placarding, manifesting, and transport of radioactive materials. In some instances, NRC and/or the site retain additional rules or authorities.

Shipping categories are the following: Radioactive Material, NOS; Radioactive Material, LSA; Radioactive Material, Limited Quantity. Packaging may be one of three allowable types:

1. Type A — meets the Mounds lab type of tests for drop tests, construction, water test, pressure test, etc.
2. Type B — cask type package capable of withstanding more severe abuse with no loss of containment.
3. Strong Tight Package — contains the materials under conditions normally incident in transport.

Labeling and placarding requirements are also regulated by DOT, as is the Radiation Shipping Record. This is a detailed manifest-type document that completely describes isotype, quantity, packaging, labeling, placarding, dose rates, chemical and physical characteristics, licenses of destination and shipper, and emergency numbers.

Various modes of shipment include rail, ship, air, and land vehicle. Associated with each of these are a separate set of specific rules.

The EPA regulations for radioactive materials are found in 40 CFR 190 subchapter F and pertain mainly to environmental contaminants from natural sources such as uranium mining and phosphate processing.

Enforcement

Final enforcement authority for radioactive materials control does not rest with one single agency. It is important, therefore, to be aware of the enormous diversity of agencies that may become involved in (1) the use of radioactive materials, (2) transportation, (3) spills, (4) over exposures, (5) licensing, etc. Each agency has its sphere of control and must be adhered to. Depending on the area of regulation, final enforcement may be the responsibility of either a federal or state agency.

Relationship to Other Regulations

Agencies concerned with public health, such as OSHA, seek to limit exposure of workers to radiation in the workplace. Radionucleotides are regulated under Section 112 of the CAA, and the primary standards defined in the SDWA set MCLs for radioactivity.

Current Status

One of the most significant current issues affecting NRC and all states is waste burial site identification. Regions have been established around groups of sites which have been charged with the creation of a compact. This compact must identify and establish the waste generated by those states. It is intended that these compacts be functioning and self-sufficient by 1986. Currently the compacts are still under fire for possible low-dose exposures to local communities, water sources, and food chains; only the 1990s will tell what will happen to them.

TOXIC SUBSTANCES CONTROL ACT (TSCA)

Overview

Toxic Substances Control Act of 1976 (TSCA) authorizes the Administrator of the EPA to protect health and the environment from harmful chemicals and mixtures. The TSCA regulates existing and new chemical substances and, therefore, primarily applies to ongoing chemical manufacturing operations and the products of these operations. TSCA grants EPA broad regulatory authority over most chemical substances. Pesticides, tobacco, foods, drugs, cosmetics, firearms, and regulated nuclear material are the main categories of chemicals which are not within the state's scope.

The goal of the Act is to reduce unreasonable environmental or human health risks from dangerous chemicals. The goal is to be achieved by controls on the manufacturer, distribution, use, and/or disposal of such dangerous chemicals.

EPA provides a toll-free number staffed by personnel who can answer TSCA-related questions. This number is 1-800-424-9065 or 1-202-554-1404.

Background

TSCA was drafted in response to repeated incidents involving adverse health, environmental, and economic damages inflicted by widely used substances such as polychlorinated biphenyls (PCBs), ketone, Aldrin-Dieldrin, chlorinated hydrocarbons, vinyl chloride, asbestos, etc.

National concern with widespread contamination by toxic compounds became evident in the 1960s, by the national arousal over the widespread contamination of food, water, and soil by certain highly toxic compounds of organic mercury. By 1972 the CAA and CWA had been promulgated to address some of these concerns, but there was no federal authority to require testing of the effects of various mercuric compounds or to regulate the multiple uses of mercury in industrial, commercial, and consumer products. However, both these acts dealt with substances only after they had entered the environment as wastes, and federal authority to regulate the complete cycle of chemicals from production to disposal was severely limited. TSCA was designed to fill this regulatory gap by granting EPA the authority to require testing before human or environmental exposure.

The task of drafting a federal toxics law was begun in 1971 and completed in 1976. TSCA was enacted on January 1, 1977, exactly 7 years to the day after the passage of the National Environmental Policy Act, and had been heralded by some as "the beginning of an environmental decade."

Content

TSCA regulates the safety of raw materials and, therefore, primarily applies to ongoing chemical manufacturing operations and the products of those operations. TSCA applies to any entity that manufactures, processes, or distributes in commerce any chemical substance or mixture. The terms "manufacturer" and "process" are defined very broadly by the Act. Manufacture includes importing and production, even production of a chemical as a by-product. Process includes incorporation of a chemical into an article by manufacturers of consumer goods or industrial products. Thus, TSCA can potentially apply to companies outside the chemical industry. However, pesticides, tobacco, foods, drugs, cosmetics, firearms, and regulated nuclear material are not within the statute's scope.

TSCA has two main regulatory goals. First is the acquisition of sufficient data to identify and evaluate potential chemical hazards. The second goal is to reduce these hazards by controlling the production, use, distribution, and disposal of such substances as deemed necessary. These goals are to be achieved through four major activities.

1. Screening of new chemicals.
2. Testing of chemicals identified as potential hazards.

3. Gathering of information on existing chemicals.
4. Controlling chemicals proven to pose a hazard.

Virtually all new chemicals must undergo review prior to manufacture. Premanufacturing notices (PMNs) must be sent to EPA by firms intending to produce a new chemical or to put an existing chemical to a new use. Under Section 4 of the Act, testing can be required under either of the following conditions:

1. If a chemical may pose an unreasonable risk to health or the environment and there are insufficient test data on the substance to evaluate its effects.
2. If a chemical produced in large volumes could enter the environment in substantial quantities or be in contact with humans in significant amounts and there are insufficient test data on the substance to evaluate its effects.

Section 8 of TSCA gives EPA the authority to require the collection and submittal of certain information by chemical manufacturers and processors. Control measures granted to EPA include

1. Limiting allowed uses of chemicals.
2. Requiring special labeling.
3. Requiring special use instructions.
4. Requiring special record keeping.
5. Setting disposal standards.
6. Requiring quality control procedures in chemical manufacturing.

TSCA record keeping requirements include

1. Reporting of production and exposure data on specified chemicals.
2. Retention of reports of significant adverse reactions (records must be kept for 30 years).
3. Submission of health and safety studies related to certain listed chemicals.
4. Notice of substantial risk by chemical manufacturers, processors, or users who become aware of a substantial threat from a chemical are required to notify EPA.

Section 6 provides authority for the control or banning of substances that are found to pose an unreasonable risk to health or the environment. Allowable regulatory actions available range from labeling requirements to complete prohibition of PCBs, chlorofluorocarbons (CFCs), asbestos in schools, and disposal of 2,3,7,8-tetrachlorodibenzo-*p*-dioxin (TCDD) wastes are regulated under this provision.

Specific standards for the manufacture, use, and disposal of PCBs in concentration greater than 50 ppm include the following:

1. May only be used as part of closed systems.
2. Containers with PCBs must be clearly marked.
3. PCBs in concentrations of 500 ppm or greater must be disposed of through incineration.
4. Regulated concentrations less than 500 ppm may be buried in secure landfills or burned in a high-efficiency boiler.

TSCA deals with CFCs by basically outlawing most aerosol uses, and with asbestos by requiring inspection for exposure of students to airborne asbestos fibers. EPA must be given a 60-day notification prior to any disposal of TCDD wastes. Similar to other environmental laws, TSCA contains an imminent hazard provision. EPA is authorized to take "appropriate action in relation to a chemical substance or mixture which presents an imminent and unreasonable risk of serious or widespread injury to health or the environment."

Enforcement

TSCA is enforced by EPA through the Office of Toxic Substances, and authorized enforcement actions include inspections, imposition of penalties, and seizure of substances manufactured in violation of the Act. Unlike many other federal environmental regulations, TSCA has no provision which authorizes state agencies to undertake program enforcement.

Relationship to Other Regulations

EPA is not authorized to proceed under TSCA to control a risk that can be adequately dealt with by actions taken under another statute, and EPA's exercise of authority under TSCA does not constitute a preemption of jurisdiction under the OSH Act. In addition to TSCA, Section 112 of the CAA and Section 307 of the CWA also specifically address toxic substances.

Current Status

In November 1984, EPA set forth general exemption standards for small manufacturers of chemical substances. Manufacturers that qualify under these standards are exempt from many of the reporting and record keeping rules.

It appears unlikely that TSCA-related legislation will be passed, due to its perceived low priority in relation to other environmental bills. However, TSCA grants widespread potential regulatory powers to EPA, and the importance of TSCA is likely to expand in the future. TSCA is currently gathering information to pass a law for industry to monitor the mutagenic and neurological disorders in employees exposed to 1,1,1-trichloroethane. Opposition is strong from industry and the outcome will be determined in 1991.

RESOURCE CONSERVATION AND RECOVERY ACT (RCRA)

Overview

The management of the hazardous waste which is being generated in large quantities by our modern, post-World War II economy has come to be recognized as a major environmental problem over the past 10 years. In October 1986, Congress enacted the Resource Conservation and Recovery Act (RCRA), the first comprehensive federal effort to deal with the problem of solid waste generally and hazardous waste specifically.

RCRA regulates the handling of hazardous waste at currently operating or future facilities and is intended to provide for the environmentally sound disposal of waste materials. Thus, it addresses a gap left by the CAA and CWA, which only require that industry remove hazardous substances from air emissions and water discharges (see Table 1.2).

RCRA establishes the five major elements characterizing the federal approach to hazardous waste management:

1. Classification of hazardous waste.
2. Cradle-to-grave manifest system.
3. Standards for generators, transporters, and facilities which treat, store, or dispose of hazardous waste.
4. Enforcement through a permitting program.
5. Authorization of state programs to operate in lieu of the federal program.

It does not address what has come to be recognized as the equally serious problem of abandoned and inactive sites. Legislation establishing remedies and allocating responsibilities for correcting problems at those sites is contained in the Comprehensive Environmental Response, Compensation, and Liability Act of 1980, commonly known as Superfund, and is discussed in the CERCLA section of this chapter.

EPA staffs a toll-free line to answer RCRA-related questions. The number is 1-800-424-9346.

Background

RCRA traces its beginnings to the Solid Waste Disposal Act, which was passed in 1965. In 1970 the Act was amended to require an investigation of hazardous waste management activities in the U.S. Ironically, a concern with hazardous waste disposal practices was fueled by passage of the CAA and CWA. Sludge and other debris generated by air and water pollution control equipment required disposal guidelines.

RCRA was passed on October 21, 1976, replacing the Solid Waste Disposal Act. Development of the congressionally mandated regulatory program was subject to various delays. The first hazardous waste rules were issued in February of

1980, beginning what has become one of the largest and most controversial regulatory programs.

Content

RCRA has two subtitles which create significant regulatory programs. *Subtitle C* authorizes a comprehensive federal program to regulate hazardous wastes from generation to ultimate disposal. This concept is known as cradle-to-grave regulation. *Subtitle D* addresses the disposal of nonhazardous solid waste.

This section of Chapter 1 will deal mainly with Subtitle C, which is an extensive topic by itself. In 1984, hazardous waste regulations occupied over 250 pages of the *Code of Federal Regulations* (CFR). Additionally, significant amendments extending the scope of the original Act were promulgated in November of 1984.

Part 261 of RCRA defines the criteria for determining whether a waste is hazardous and, therefore, subject to regulation. In order to be regulated, a waste must be both a solid waste and a hazardous waste. The actual physical state of the waste may mean little according to the Act: certain liquids, sludges, or contained gases are solids by RCRA definitions. Substances generally deemed not to be solid wastes include domestic sewage, CWA-regulated point discharges, irrigation return flows, radioactive material which is regulated by other agencies, and *in situ* mining wastes.

Figure 1.4 presents a general schematic decision tree for determining if a solid waste is hazardous. A waste is generally subject to regulation if it exhibits any of the four characteristics of a hazardous waste (ignitability, corrosivity, reactivity, or EP toxicity) or if it is specifically listed in the Act as being a hazardous waste. Exceptions include industrial wastewater discharges, nuclear materials, fly ash, mining overburden, drilling fluids, wastes generated by small quantity generators, and wastes generated by activities associated with recycling, reuse, or reclamation of waste materials.

The Act defines standards which are applicable to generators and transporters of hazardous waste. Generator, transporter, and treatment, storage, or disposal (TSD) facility requirements are summarized in Table 1.4. Generators must assure that wastes are shipped in proper containers, are accurately labeled, and are accompanied with proper placards for use by the transporter. Generators must also comply with applicable storage standards for containers and tanks, conduct proper operating, maintenance, and inspection procedures, and prepare a contingency plan to be followed in the event of an emergency.

In addition, the owner or operator must do the following:

1. Analyze wastes.
2. Provide security.
3. Conduct inspections.
4. Provide employee training.
5. Maintain emergency equipment.
6. Formulate an emergency response plan.

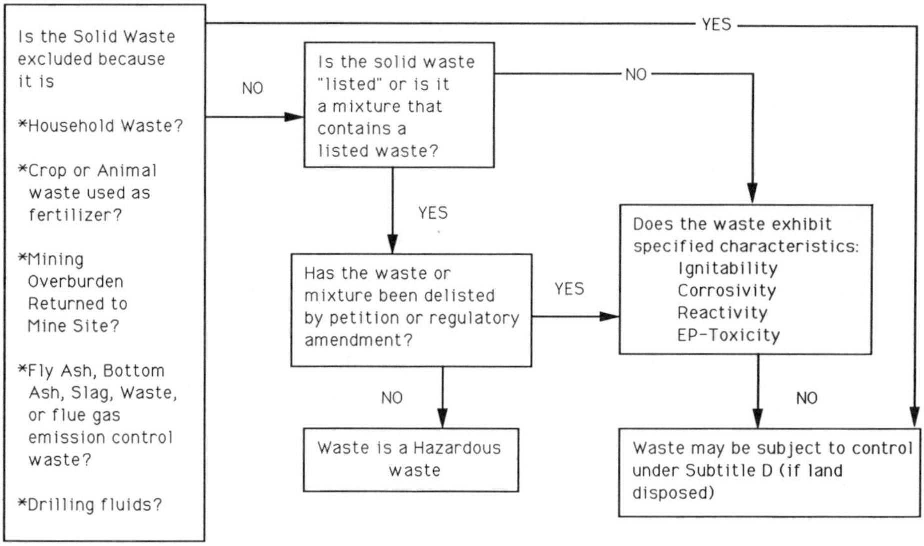

Figure 1.4. Determination of hazardous waste.

Table 1.4A. General Requirements of TSDF Facilities[a]

Notify EPA
Obtain identification number
Utilize manifest system
Observe proper packing system
Ship to TSDF
Prepare annual report
Store waste less than 90 days

[a] Transporters must comply with DOT regulations and the hazardous waste manifest systems, which are discussed in Chapter 9. General requirements for transporters are shown in Table 1.4B.

Table 1.4B. Transporter Requirements[a]

File notice with EPA
Use manifest
Delivery to TSDF
Record retention (3 years)
Discharges: take appropriate action and report to DOT

[a] All facilities which treat, store, or dispose of hazardous waste (TSD facilities) must eventually comply with specified standards. Requirements are dependent upon the type of TSD facility and whether the facility is currently under operation or is defined as a new facility. Interim status standards apply to facilities in operation or under construction on November 19, 1980, before a final permit obtained. Standards for TSD facilities can be categorized as general or specific. General requirements applicable to virtually all TSD facilities are listed in Table 1.4C.

Table 1.4C. TSD Facility General Requirements

Notify EPA
Obtain interim status (if applicable)
Observe facility standards
Institute record keeping
Prepare annual and other reports
Obtain permit
Formulate closure plan

Specific standards apply to designated categories of TSD facilities. The types which are regulated include containers, tanks, surface impoundments, waste piles, land treatment units, landfills, incinerators, thermal treatment units, chemical, physical, and biological treatment units, and underground injection wells. RCRA-mandated facility-specific standards have been extremely controversial and subject to change.

TSD facilities must have a written plan to provide for closing the site at the end of the life of the facility. There must be a guarantee that financial resources will be available. This guarantee may be in the form of a trust fund, a security bond, letter of credit, or insurance policy.

All TSD facilities are required to obtain an RCRA permit. Permits consist of two parts: Part A of the application requires a minimal amount of general information, while Part B is a complex engineering report, requiring extensive site-specific information and substantial documentation. Permit conditions are usually determined on a case-by-case basis. Permit amendments are usually determined on a case-by-case basis, too. Permit amendments, modification procedures, and rules applicable to expired permits are addressed in the Act.

RCRA contains an imminent hazard provision. It states that EPA may take action against persons whose actions, past or present, present an imminent or substantial endangerment to health or to the environment.

The 1984 amendments to RCRA substantially broadened the scope of the Act. Addressed in the amendments are restrictions on land disposal, small-quantity generators, burning and blending of wastes, underground storage tanks, interim status facilities, inspections, and citizen suits. The new law adds 72 provisions in all and directs EPA to carry out 58 of them in the next 2 years. The law, which is currently under EPA interpretation, has been, and continues to be, the subject of much speculation. See Table 1.5 comparison of RCRA and TSCA regulations.

Enforcement

RCRA is intended to be administered at the state level, upon final authorization by EPA. To receive final authorization, a state program must be "equivalent to and consistent with" the federal program and must provide for adequate enforcement. Sites may regulate more materials than are regulated by EPA and may impose more stringent environmental standards.

Table 1.5. Comparison of RCRA and TSCA Regulations Governing PCBs

RCRA	TSCA
Waste Generators	
Must notify EPA and obtain ID number	No notification requirements for generators
May not store waste more than 90 days without a permit. (Eventually storage over 90 days must be to accumulate for proper treatment, recovery, or disposal)	PCBs may be stored up to 30 days with minimal controls. Storage beyond one year is strictly prohibited
Must complete manifest form for waste shipment off site	Manifest not required, but frequently used anyway
Must submit exception reports to EPA concerning lost shipments	Exception reports are not required. However, generators are responsible for proving that PCB wastes are properly disposed
Must keep detailed records of waste shipments for at least three years, and submit biannual report on waste generation	Detailed records on waste generation must be maintained for 5 years. An annual summary must be prepared and retained on site
Requirements apply to any facility generating more than 100 kg of waste per month (or 1 kg/month of acutely hazardous waste)	Applies to any generator of PCBs
Transporters	
Shipments must be accompanied by manifest	No comparable requirement under TSCA. Some Department of Transportation regulations apply
Waste may only be taken to permitted or interim status treatment, storage, or disposal facility	PCB generators responsible for proving proper disposal, such as certification that material was properly destroyed
Land Disposal Facilities	
Construction of facilities in floodplains is strictly prohibited	TSCA regulations contain similar requirements for land disposal facilities
A groundwater protection plan, including monitoring wells, is required	TSCA regulations contain similar requirements for land disposal facilities
New facilities must have double liners with leachate collection and treatment	

Source: EPA

RCRA enforcement provisions are numerous and substantial and include the definition of "knowing endangerment," which is a felony.

Subtitle D of RCRA is at the opposite end of the regulatory spectrum from the TSCA which provides only for federal enforcement. Other than publishing criteria for sanitary landfills and an inventory of open dumps, EPA has very little to do with the regulation of nonhazardous solid waste disposal. Generally, nonhazardous waste disposal programs are conducted by site agencies without federal involvement.

Relationship to Other Regulations

The treatment, disposal, and transportation of hazardous wastes involved in the cleanup of the Superfund site generally requires compliance with RCRA-specified standards, but does not require an RCRA permit. Hazardous wastes are considered a subset of hazardous materials and as such are subject to DOT requirements for packaging, marking, labeling, placarding, spill reporting, and shipping paper preparation. RCRA currently regulates Class I and IV type underground injection wells as defined by the SDWA.

Current Status

RCRA has been added to by amendments approximately 11 times between 1976 and 1986. The most extensive amendments occurred in 1984 (known as the Hazardous and Solid Waste Amendments). The amendments establish such major regulations as land (disposal) ban of numerous (on a congressionally established time schedule) underground storage tank laws, waste minimization, used oil laws, et al. Additional amendments were passed in 1986, increasing enforcement inspections, monitoring, and corrective actions. As a result of the concern of proper disposal of biological and medical waste, Congress passed an amendment to the Solid Waste Disposal Act which would require the Administrator of the EPA to promulgate regulations on the management of infectious waste. The relatively new Act is entitled as the "Medical Waste Tracking Act of 1988." A copy of this Act can be found in the Appendix section of this book.

HAZARDOUS MATERIALS TRANSPORTATION REGULATIONS (HMTR)

Overview

DOT regulations establish a labeling system applicable to hazardous materials, require the display of placards on land carriers transporting hazardous cargo, and mandate the use of a uniform manifest in the shipment of hazardous waste. The laws also identify stringent specifications which specify the type of packages and containers that can be used to ship hazardous materials. If shipments are to be made by public carrier, i.e., air, passenger rail, or water, additional restrictions apply.

Background

DOT regulates transporters of hazardous materials to ensure safe and proper delivery to their authorized destination. The first regulations pertaining to the transport of hazardous materials were drafted soon after the Civil War, to control the transport of explosives by train. Current transportation laws are mainly based on the Hazardous Materials Transportation Act (HMTA), which was passed in 1975.

EPA promulgated standards for hazardous waste transporters on May 19, 1980, and coordinated these standards with HMTA. DOT views hazardous wastes a subset of hazardous materials; thus, a hazardous waste will always be regarded as a hazardous material subject to DOT regulation. RCRA sites that EPA's hazardous waste regulations must be consistent with the DOT regulations, which are found in 49 CFR 172, 173, 178, and 179.

Content

Identifying a hazardous material in accordance with DOT regulations involves marking, labeling, and placarding.

Marking is generally written information that identifies the package contents. In order to correctly mark a shipment, the shipping name must be determined by selecting an accurate technical shipping name from the tables in the DOT regulations. If there is not an applicable shipping name, identification must be used that corresponds to the specific hazard class of the material being shipped. Table 1.6 lists common classifications in descending order of priority. In selecting appropriate markings by this system, the highest priority hazard ranking must be used to describe a material if more than one classification is applicable.

If a hazardous waste container has a capacity of 110 gal or less, it must be marked with the following information at a minimum:

1. Proper shipping name and identification number.
2. Name and address of cosignee and consignor (except in certain situations).
3. Warning.
4. Generator's name and address.
5. Manifest document number.
6. No abbreviations may be used except for ORM.

For containers greater than 110 gal, requirements vary, and the regulations should be consulted for specific situations.

A label is a sticker of specified design which goes on the package and pictorially indicates the hazard class of the contents. DOT requires labels according to the classification of the hazardous material. A descriptive summary of the labeling system is as follows:

Table 1.6. DOT Hazard Classes (Descending Priority)[a]

1. Radioactive material
2. Poison A
3. Flammable gas
4. Nonflammable gas
5. Flammable liquid
6. Oxidizer
7. Flammable solid
8. Corrosive material (liquid)
9. Poison B
10. Corrosive material
11. Irritating material
12. Combustible material
13. ORM-B
14. ORM-A
15. Combustible liquid
16. ORM-E

[a] EPA's hazardous waste characteristics do not directly correspond to DOT's hazard classes, as shown in Table 1.7. DOT created the Other Regulated Material (ORM) hazard class to accommodate those hazardous wastes which are regulated by EPA, but which do not otherwise fall into one of the previously developed hazard classes.

Substance	Label
Explosives	Orange with black lettering
Flammable gas	Red with black lettering
Flammable liquid	Red with black lettering
Nonflammable gas	Green with black lettering
Flammable solid	Red and white vertically striped
Corrosive	Diamond shaped, white top, black bottom
Oxidizing substance	Yellow with black lettering
Poisons	Black on white skull and crossbone
Radioactive	Magenta propeller on yellow

If a package qualifies for DOT's small quantity exclusion listed in the regulations, it does not have to be labeled.

A placard is similar to a label in that it pictorially illustrates the hazard class of material being shipped. DOT requires the display of placards on land carriers that transport hazardous cargo. The placards are intended to serve as a notice of the potential danger of the cargo before labels on packages are observed or before any shipping papers are read. Placards are intended to convey the same basic identification information as DOT-required labels.

The generator of a hazardous waste is required to provide the transporter with proper placards identifying the material being shipped. DOT's regulations specify that any quantity of a waste having a hazard class rating must be placarded (see Table 1.7).

Shipping paper requirements for hazardous materials are specified by DOT regulations. Hazardous waste shipments must be accompanied by a uniform hazardous waste manifest. The uniform hazardous waste manifest is an EPA

Table 1.7. Comparison of EPA's Hazardous Waste Characteristics With DOT's Hazardous Classes

EPA Characteristic	DOT Hazard Class
Ignitability	Flammable liquid Combustible liquid Flammable solid Flammable gas Oxidizer
Corrosivity	Corrosive metal
Reactivity	Explosive Flammable solid Organic peroxide Poison B (sulfides, cyanides, etc.)
EP toxicity	Poison B ORM-A ORM-B

document, and DOT authorizes its use as a shipping paper (Figure 1.5). A checklist that indicates the originator's requirements is shown in Table 1.8.

Enforcement

Primary regulatory responsibility rests with DOT. Although Section 3003 of RCRA directs EPA to regulate transporters of hazardous waste, EPA's regulations incorporate and rely on the requirements already established by DOT under the HMTA. The uniform manifest for hazardous waste transport was developed jointly by EPA and DOT. States and municipalities cannot enforce standards inconsistent with or more stringent than federal DOT regulations.

Relationship to Other Regulations

Transporter requirements are established by the RCRA for certain instances involving spills, mixing of wastes, and storage of wastes. If the transporter selects the disposal site, he may be strictly and jointly liable with the generator and owner/operator, vehicle, and financial responsibility requirements under the CERCLA. CERCLA defines substances and reportable quantities which, if released into the environment by a transporter, must be reported to the National Response Center.

Current Status

The provisions of HMTA and related regulations have remained substantially unchanged since promulgation of final rules. One of the most recent developments

Please print or type *(Form designed for use on elite (12-pitch) typewriter.)* *Form Approved. OMB No. 2050-0039. Expires 9-...*

UNIFORM HAZARDOUS WASTE MANIFEST

1. Generator's US EPA ID No. Manifest Document No.
2. Page 1 of
Information in the shaded areas is not required by Federal law

3. Generator's Name and Mailing Address
4. Generator's Phone ( )
5. Transporter 1 Company Name 6. US EPA ID Number
7. Transporter 2 Company Name 8. US EPA ID Number
9. Designated Facility Name and Site Address 10. US EPA ID Number

A. State Manifest Document Number
B. State Generator's ID
C. State Transporter's ID
D. Transporter's Phone
E. State Transporter's ID
F. Transporter's Phone
G. State Facility's ID
H. Facility's Phone

GENERATOR

11. US DOT Description *(Including Proper Shipping Name, Hazard Class, and ID Number)*	12. Containers No.	Type	13. Total Quantity	14. Unit Wt/Vol	I. Waste No
a.					
b.					
c.					
d.					

J. Additional Descriptions for Materials Listed Above

K. Handling Codes for Wastes Listed Above

15. Special Handling Instructions and Additional Information

16. **GENERATOR'S CERTIFICATION:** I hereby declare that the contents of this consignment are fully and accurately described above by proper shipping name and are classified, packed, marked, and labeled, and are in all respects in proper condition for transport by highway according to applicable international and national government regulations.

If I am a large quantity generator, I certify that I have a program in place to reduce the volume and toxicity of waste generated to the degree I have determined to be economically practicable and that I have selected the practicable method of treatment, storage, or disposal currently available to me which minimizes the present and future threat to human health and the environment, **OR**, if I am a small quantity generator, I have made a good faith effort to minimize my waste generation and select the best waste management method that is available to me and that I can afford.

Printed/Typed Name	Signature	*Month Day Year*

TRANSPORTER

17. Transporter 1 Acknowledgement of Receipt of Materials

Printed/Typed Name	Signature	*Month Day Year*

18. Transporter 2 Acknowledgement of Receipt of Materials

Printed/Typed Name	Signature	*Month Day Year*

FACILITY

19. Discrepancy Indication Space

20. Facility Owner or Operator: Certification of receipt of hazardous materials covered by this manifest except as noted in Item 19.

Printed/Typed Name	Signature	*Month Day Year*

EPA Form 8700-22 (Rev. 9-86) Previous editions are obsolete. Reorder No. 860 Lab Safety Supply Co.

ORIGINAL-RETURN TO GENERATOR

Figure 1.5. Uniform hazardous waste manifest EPA Form 8700-2 (Rev. 9-86).(Compliments of Lab Safety Supply Co.)

Table 1.8. Hazardous Waste Pickup Checklist

Manifest

1. Is manifest printed on EPA form 8700-22 (rev. 4-85)? Compare manifest's section 16 to pickup ticket.
2. "Manifest Document Number" must be five digits, and must be identical to "State Manifest Number".
3. Generator name, address, phone number.
4. Federal EPA identification number.
5. Transporter #1:____________________. (Chemical Hauler), E:A# ______________. phone number ____________________.
6. Transporter #2: Name, EPA#, and phone number must match pickup ticket exactly.
7. Section 9: Designated facility: ____. address of facility, EPA# _________________________. phone number ___________.
8. Proper DOT shipping name: If material has an N.O.S. shipping name (for example: Waste Flammable Liquid N.O.S.) then it must be followed by a description of those constituents which comprise the hazard (for example: Waste Flammable Liquid N.O.S. (Acetone/Toluol Mixture)). Description must include proper hazard class, UN/NA number, and EPA Waste Code number.
9. Total quantity of waste: either pounds or gallons are acceptable. Be consistent!!! The proper abbreviation for "pounds" is "P"; the proper abbreviation for gallons is "G".
10. Number and type of containers: the proper designation for metal drums is "DM".
11. Section K: Handling code must be as follows: T63.
12. Generators signature and date.
13. Typos must be completely obliterated. NO STRIKEOVERS OR CROSSOUTS!!!

Packaging

1. No OPEN HEAD DRUMS!!!
2. Is container completely, securely sealed? All bungs must have gaskets. Bungs must be tight. NO leaks!!!
3. Proper DOT shipping name on container.
4. "WASTE" precedes proper DOT shipping name.
5. Generator's name and address on container.
6. Manifest number on container.
7. Applicable DOT Hazard Warning label.
8. Date accumulation began.
9. All other marks, labels, trademarks, etc., must be obliterated.

was the requirement for hazardous waste transporters to utilize a uniform manifest. This requirement became effective on September 20, 1984.

COMPREHENSIVE ENVIRONMENTAL RESPONSE, COMPENSATION AND LIABILITY ACT (CERCLA)

Overview

CERCLA (1980) grants the federal government the authority to undertake activities under the directives of a National Contingency Plan in order to cleanup dangerous inactive disposal sites and emergency spill situations. Included is the authority to carry out investigations, testing, and monitoring of disposal sites, in addition to implementing actual remedial measures. Financial support for these actions is provided by the Hazardous Substance Response Trust Fund (commonly termed the "Superfund"). Also, financial responsibility for cleanup is placed on companies or individuals determined to be responsible for contamination. The responsibility for such cleanup action may also extend to generators and transporters whose hazardous waste was shipped to the site from which there is a release. If the responsible parties do not voluntarily take appropriate removal and remedial action, EPA has the authority to order them to do so. If they still do not respond, EPA can use the federal funds to then seek recovery of its expenses from the responsible parties.

Background

The other regulations discussed in this text have a prospective focus. They deal with future and existing activities, not past occurrences. They are not intended to provide remedies for situations where environmental damage has already occurred and responsible parties are unidentifiable, unavailable, or lacking in funds. CERCLA was enacted to remedy such situations.

The impetus for CERCLA largely consisted of the 1978 discovery of hazardous chemicals seeping into homes at Love Canal, New York. The resulting legislation was admittedly a compromise bill, enacted literally during the last hours of the 96th Congress. The final law, signed by President Carter on December 11, 1980, eliminated many of the provisions in earlier drafts and limited funding mechanisms to a life of 5 years unless reauthorized. Many legislators felt such measures were acceptable in the face of a perceived urgent need for legislation.

One of the first Superfund expenditures was for the investigation and cleanup of Love Canal.

Content

The major provisions of CERCLA address the following topics: response and remedial actions, National Contingency Plan (NCP), liability, financing remedial actions, and notification requirements.

CERCLA grants authority for EPA to provide required remedial actions whenever any hazardous substance is released or there is a substantial threat of such a release. The scope of this authority is vast, since the key terms contained in the Act have extremely broad statutory definitions. For example, "hazardous substance" includes any material so designated by the CWA, CAA, or RCRA. The Act authorizes EPA to take the following actions: dispose of removed material, conduct monitoring programs, limit site access, evacuate affected individuals, and undertake studies, including monitoring, testing, and surveys.

The NCP contained in the CWA was revised by CERCLA. It includes

1. Methods for identifying facilities at which hazardous substances have been disposed.
2. Methods for evaluating and remedying releases of hazardous substances and analysis of relative cost.
3. Methods, criteria, and equipment for determining the appropriate extent of cleanup.
4. Determination of federal, state, and local roles in carrying out the plan.
5. Means of assuring the cost effectiveness of remedial actions.
6. Determining priorities among releases or threatened releases for the purposes of taking effective remedial action.

The NCP also defines criteria for determining priorities, and based upon these criteria EPA has developed a Hazard Ranking System (HRS). The HRS takes into account the following considerations:

1. The population at risk.
2. The hazard potential of the hazardous substances involved.
3. The potential for contamination of drinking water supplies.
4. The potential for direct human contact.
5. The potential for destruction of sensitive ecosystems.

The NCP requires a determination of whether the no-action alternative may be appropriate, because in some circumstances a remedial response action may cause a greater environmental or health danger than no action.

The NCP is divided into eight subparts:

- Subpart A — Introduction, scope, and definitions.
- Subpart B — Responsibilities, including federal, state, local, and private.
- Subpart C — Organization, coordination, and communication.
- Subpart D — Regional and local contingency plans.
- Subpart E — Oil spill response.
- Subpart F — Hazardous substances response.
- Subpart G — Trustees for natural resources.
- Subpart H — Use of dispersants and other chemicals.

Under CERCLA, EPA is authorized to collect damages from parties that are determined to be responsible for the contamination. There are three classes of potential responsible parties: (1) owners, operators, generators, (2) transporters, and (3) tenants using hazardous substances.

Liability extends to

1. All removal and remedial costs incurred by the federal or state governments consistent with the NCP.
2. Any other necessary response cost incurred that is consistent with the NCP.
3. Damages for injury to national resources controlled by any governmental entity.
4. All costs incurred from the cleanup and restoration of the environment.

However, CERCLA imposes no liability for federally permitted releases or for personal injury resulting from releases of hazardous substances.

In addition, Title II of CERCLA provides for a Hazardous Substance Response Trust Fund, which is financed by taxes on crude oil, certain petroleum products, chemical feedstocks, and appropriations from general revenues. This fund is commonly known as "Superfund." The Collection of the taxes commenced on April 1, 1981, and would terminate on September 30, 1985. The reauthorization of CERCLA was made when Congress passed the Superfund Amendments and Reauthorization Act (SARA) of 1986, on October 17, 1986, by President Reagan to the Second Session of the 99th Congress of the U.S. in Washington, DC. This new Act is commonly known as SARA.

The notification requirements of CERCLA address both spills and disposal sites. Sections 102 and 103 require the reporting of the release of hazardous substances into the environment in cases where the release is not allowed in accordance with a permit. A federally permitted release is any discharge or emission authorized by permits issued under the CWA, SDWA, CAA, RCRA, Marine Protection, Research and Sanctuaries Act (ocean dumping), or the Atomic Energy Act. All persons presently or previously owning facilities at which hazardous substances were handled, and all transporters to such sites, are required to notify EPA of their actions if releases occur or are determined to have occurred.

No notification is required for

1. Releases that have been reported to the NRC under Subtitle C or RCRA.
2. The application of registered pesticides in accordance with EPA requirements.
3. Stable, continuing releases that have been previously reported or issued from a previously reported site.

The authority granted to the federal government for remedial actions under CERCLA far surpasses the imminent hazard provisions of the RCRA and other environmental statutes. Therefore, CERCLA is the logical vehicle for federal suits involving imminent hazards to the environment. However, the conservative practice is to base complaints in lawsuits on a variety of alleged violations, citing various regulations and violations and, thus, providing the broadest foundation on which to seek legal redress.

Enforcement

CERCLA is enforced by EPA. CERCLA-mandated cleanups may be coordinated with state and local governments.

Relationship to Other Regulations

CERCLA mandated revisions in the NCP, which was first enacted by the CWA. Reportable quantities requiring notification of spills is 1 lb (see Table 1.3), unless EPA regulations in CERCLA or Section 311 of the CWA specify a higher quantity. Treatment, transportation, and disposal of hazardous wastes at sites regulated by CERCLA must meet standards in the RCRA. Any substance designated as hazardous under CERCLA must bear the same designation under the HMTA.

Current Status

There are two factors supporting passage of 1986 amendments extending CERCLA. There was a real and tangible need for reauthorization, because the law's taxing mechanism for generating hazardous waste site cleanups was to expire in October. Also favoring reauthorization is the fact that industry, Congress, environmentalists, and the Reagan Administration have all gone on record in support of extending the law.

Legislation under consideration in 1986 consisted of bills in the houses of Congress. The Senate reauthorization bill calls for a 5-year, $7.5 billion program, and the House bill called for a 5-year, $10 billion plus program. The bill was passed in December 1986. The bill was signed by President Reagan and is known as the Superfund Amendments and Reauthorization Act of 1986 or SARA. The amendments increased the authority of EPA in many respects and required certain actions in modifying the NCP.

29 CFR 1910.120 HAZARDOUS WASTE OPERATIONS AND EMERGENCY RESPONSE (HWOER)

Introduction

In the past decade, industry, government, and the general public have become increasingly aware of the need to respond to the hazardous waste problem, which has grown steadily over the past 40 years. In 1980, Congress passed CERCLA — the Superfund Law — to provide for liability, compensation, cleanup, and emergency response for hazardous substances released into the environment and the cleanup of inactive waste disposal sites or hazardous materials operations. Under 29 CFR 1910.120 Hazardous Waste and Site Operations a precedent was initiated to require proper employee training for any one who is exposed to hazardous waste, hazardous substances, or any substances that cause ill health effects to be trained within 30 days of employment.

History

1910.120 was outlined as a guidance document for individuals/managers responsible for occupational safety and health programs at inactive hazardous waste sites. It assumes a basic knowledge of science and experience in occupational safety and health by the employee. Introduction of the regulations-specific topics are to expand the employee's knowledge. It was a product of a four-agency committee (NIOSH, OSHA, the U.S. Coast Guard (USCG) and the EPA) mandated by CERCLA section 301(f) to study the problem of protecting the safety and health of workers at hazardous waste sites, and by CERCLA section 111 (c)(6) to develop a program to protect the health and safety of employees involved in the response to hazardous substances, releases, removals, or remedial actions.

Since this regulation covers provisions for medical, record keeping, and training, the training section will be elaborated upon. A copy of the final rule as recorded in the *Federal Register*, Vol. 51, No. 244, Friday, December 19, 1986 pg. 45663–45675, appears in the appendix of this text.

Training Requirements

All employees (such as equipment operators and general laborers) exposed to hazardous substances, health hazards, or safety hazards for 30 or more days shall be thoroughly trained in the following:

1. Names of personnel and alternates responsible for site safety and health.
2. Safety, health, and other hazards present on the site.
3. Use of personal protective equipment (PPE).
4. Work practices by which the employee can minimize risks from hazards.
5. Safe use of engineering controls and equipment on the site.
6. Medical surveillance requirements including recognition of symptoms and signs which might indicate overexposure to hazards.
7. Implementation of a health and safety plan.
8. Decontamination of equipment, etc.

All employees at the time of job assignment shall receive a minimum of 40 hr of initial instruction off the site (classroom), and a minimum of 3 days of actual field experience under the direct supervision of a trained, experienced supervisor. Workers who may be exposed to unique or special hazards (such as ticks, snakes,

AIDS, etc.) shall be provided additional training. The level of training provided shall be consistent with the employee's job function and responsibilities.

In addition, on-site management and supervisors directly responsible for or who supervise employees engaged in hazardous waste operations shall receive training as provided above (40 hr) and at least 8 additional hours of specialized training on managing such operations at the time of job assignment.

All trainers of 1910.120 items shall have received a level of training higher than and including the subject matter of the level of instruction that they are providing. It must be noted that employees shall not participate in field activities until they have been trained to a level required by their job function and responsibility.

Employees and supervisors that have received and successfully completed the training and field experience specified in paragraphs (e)(1), (e)(2), and (e)(3) of the regulation shall be certified by their instructor as having completed the necessary training. Any person who has not been so certified or meets the requirements of paragraph (e)(1) of 1910.120 shall be prohibited from engaging in hazardous waste operations after March 16, 1987.

Employees (firemen, health inspectors, environmentalists, industrial hygienists, drillers, etc.) who are responsible for responding to hazardous emergency situations that may expose them to hazardous substances shall be trained in how to respond to expected emergencies.

Employees specified in paragraph (e)(1) and managers specified in paragraph (e)(3) of 1910.120 shall receive 8 hr of refresher training on the items specified in paragraph (e)(1) of this section and other relevant topics.

Employees who can show by an employee's work experience and/or training that the employee has had initial training equivalent to that training required in paragraphs (e)(1), (e)(2), and (e)(3) of 1910.120 shall be considered as meeting the initial training requirements of those paragraphs. Equivalent training includes the training that existing employees might have already received from actual, on-site experience.

Finally, off-site emergency response training for handling emergency responses involving hazardous substances shall be conducted on a monthly basis and shall be at least 24 hr annually. The training shall include, as a minimum, recognition of hazards, selection, care, and use of PPE and safe operating procedures to be used at the incident scene.

For more details of training and other sections of the 1910.120 standard, see the Appendix section of the book.

Current Status

The 1910.120 regulation is in its second year. OSHA is in the process of inspecting the various sections of the law. To date, no one knows what the total fines will be paid by violators and how many there will be. Also, the interpretation of this law and the quality of the training will be the test to reduce uncalled-for exposures at hazardous waste operations and emergency response situations.

SARA TITLE III: EMERGENCY PLANNING AND COMMUNITY RIGHT-TO-KNOW

Overview

On October 17, 1986, the Superfund Amendments and Reauthorization Act of 1986 (SARA) was enacted into law. One part of the new SARA provisions is Title III: The Emergency Planning and Community Right-to-Know Act of 1986. Title III establishes requirements for federal, state, and local governments and industry regarding emergency planning and "community right-to-know" reporting on hazardous and toxic chemicals. This legislation will build upon EPA's Chemical Emergency Preparedness Program (CEPP) and numerous state and local programs aimed to helping communities better meet their responsibilities in regard to potential chemical emergencies. The community right-to-know provisions of Title III will help to increase the public's knowledge and access to information on the presence of hazardous chemicals in their communities and releases of these chemicals into the environment.

The amendment was written to include four specific titles:

- Title I: Provisions Relating Primarily to Response and Liability (Sec. 101–127).
- Title II: Miscellaneous Provisions (Sec. 201–213).
- Title III: Emergency Planning and Community Right-to-Know (Sec. 300–330).
- Title IV: Radon Gas and Indoor Air Quality Research (Sec. 401–405).

Specifically, this section of Chapter 1 will be concentrating on Title III, because of its major impact on government, industry, academia, and the public.

Title III has four major sections: Emergency Planning (§ 301–303), Emergency Notification (§ 304), Community Right-to-Know reporting requirements (§ 311, 312), and Toxic Chemical Release Reporting-Emissions Inventory (§ 313).

Background

SARA is the acronym used for the extended amendment to CERCLA. SARA stands for the Superfund Amendments and Reauthorization Act of 1986. (See CERCLA section.)

Content

Each of the four major sections will be discussed in detail in the following paragraphs.

§ 301–303: Emergency Planning

The emergency planning sections were designed to develop state and local government's emergency response and preparedness capabilities through better coordination and planning, especially within the local community.

Title III required that the Governor of each state designate a state emergency response commission by April 17, 1987. If a state commission is not designated, the Governor will operate as the commission until the Governor makes such a designation. While existing state organizations can be designated as the state emergency response commission, the commission can have broad-based representation. Public agencies and departments concerned with issues relating to the environment, natural resources, emergency services, public health, occupational safety, and transportation all have important roles in Title III activities. Various public and private sector groups and associations with interest and expertise in Title III issues also can be included in the state commission.

In addition, the state commission should have designated local emergency planning districts before July 17, 1987, and appoint local emergency planning committees within 1 month after a district is designated. The state commission is responsible for supervising and coordinating the activities of the local emergency planning committee, for establishing procedures for receiving and processing public requests for information collected under other sections of Title III, and for reviewing local emergency plans.

This local emergency planning committee must include elected state and local officials, police, fire, civil defense, public health officials, environmental, hospital and transportation officials, as well as representatives of facilities subject to the emergency planning requirements, community groups, and the media. Before September 17, 1987, each facility subject to the emergency planning requirements must designate a representative to participate in the planning process. The local committee must establish rules, give public notice of its activities, and establish procedures for handling public requests for information.

The local committee's primary responsibility will be to develop an emergency response plan before October 17, 1988, and update it each year thereafter. In developing this plan, the local committee will evaluate available resources in preparing for responding to a potential chemical accident. The plan must include

1. Identification of facilities and extremely hazardous substances transportation routes.
2. Emergency response procedures, on-site and off-site.
3. Designation of a community coordinator and facility coordinator(s) to implement the plan.
4. Emergency notification procedures.
5. Methods for determining the occurrence of a release and the probable affected area and population.
6. Description of community and industry emergency equipment and facilities and the identity of persons responsible for them.

7. Evacuation plans.
8. Description and schedules of a training program for emergency response personnel.
9. Methods and schedules for exercising emergency response plans.

Those planning activities of the local committees and facilities would be *focused on, but not limited to,* the 402 extremely hazardous substances published in the November 17, 1986, *Federal Register.* The list included the threshold planning quantities for each substance. EPA can revise the list and threshold planning quantities based on the toxicity, reactivity, volatility, dispersability, combustibility, or flammability of a substance.

Any facility that produces, uses, or stores any of the listed chemicals in a quantity greater than its threshold planning quantity is subject to the emergency planning requirements. In addition, the state commission or the Governor can designate additional facilities, after public comment, to be subject to these requirements. By May 17, 1987, covered facilities must notify the state commission that they are subject to these requirements. If a facility begins to produce, use, or store any of the extremely hazardous substances in threshold quantity amounts, it must notify the state commission within 60 days.

Each state commission must notify EPA of all facilities subject to the emergency planning requirements, including facilities designated by the state commission or the Governor.

§ 304: Emergency Notification

Facilities must immediately notify the local emergency planning committee and the state emergency response commission if there is a release of a listed hazardous substance that exceeds the reportable quantity for that substance. Substances subject to this requirement are substances on the list of 402 extremely hazardous substances as published in *Federal Register* on November 17, 1986, and substances subject to the emergency notification requirements under CERCLA Section 103(a).

The initial notification can be by telephone, radio, or in person. Emergency notification requirements involving transportation incidents can be satisfied by dialing 911, or in the absence of a 911 emergency number, call the operator.

This emergency notification needs to include

1. The chemical name.
2. An indication of whether the substance is extremely hazardous.
3. An estimate of the quantity released into the environment.
4. The time and duration of the release.
5. The medium into which the release occurred.
6. Any known or anticipated acute or chronic health risks associated with the emergency and, where appropriate, advice regarding medical attention necessary for exposed individuals.
7. Proper precautions, such as evacuation.
8. Name and telephone number of contact person.

Section 304 also requires the follow-up written emergency notice after the release. The follow-up notice or notices shall include

1. Update information included in the initial notice.
2. Provide information on
   a. Actual response actions taken.
   b. Any known or anticipated data or chronic health risks associated with the release.
   c. Advice regarding medical attention necessary for exposed individuals.

Until state commission and local committees are formed, releases should be reported to appropriate state and local officials.

§ 311–312: Community Right-to-Know Reporting Requirements

There are two "community right-to-know" reporting requirements which apply primarily to manufacturers and importers. Section 311 requires that facilities which must prepare or have available Material Safety Data Sheets MSDSs under the OSHA regulations to submit either copies of its MSDs or a list of MSDS chemicals to

1. The local emergency planning committee.
2. The state emergency response commission.
3. The local fire department.

If the facility owner or operator chooses to submit a list of MSDS chemicals, the list must include the chemical name of each substance and any hazardous component as provided on the MSDS. This list must be organized in categories of health and physical hazards as set forth in OSHA regulations unless modified by EPA.

If a list is submitted, the facility must submit the MSDS for any chemical on the list upon the request of the local planning committee. Under Section 311, EPA may establish threshold quantities for hazardous chemicals below which no facility must report.

The initial submission of the MSDSs or list is required no later than October 17, 1987, or 3 months after the facility is required to prepare or have available an MSDS under OSHA regulations. A revised MSDS must be provided to update MSDSs which were originally submitted if significant new information regarding a chemical is discovered.

The reporting requirement of Section 312 involves submission of an emergency and hazardous chemical inventory form to the local emergency planning committee, the state emergency response commission and the local fire department. The hazardous chemicals covered by Section 312 are the same for facilities which are required to submit MSDS or the list for Section 311 (see Table 1.9).

Under Section 312, EPA may also establish threshold quantities for hazardous chemicals below which no facility must be subject to this requirement.

Table 1.9. Key Dates to Remember Under SARA Title III

November 17, 1986	EPA published List of Extremely Hazardous Substances and Planning Threshold Quantities in *Federal Register* (§302, 303, 304)
November 17, 1986	EPA initiates comprehensive review of emergency systems (§305(b))
January 27, 1987	Format for Emergency Inventory Forms and reporting requirements published in *Federal Register* (§311, 312)
March 17, 1987	National Response Team publishes guidance for preparation and implementation of emergency plans (§303(f))
April 17, 1987	State governors appoint state emergency response commissions (§301(a))
May 17, 1987	Facilities subject to Section 302 planning requirements notify state emergency response commission (§302(c))
	Interim report on emergency system review due to Congress (§305(b))
June 1, 1987	EPA publishes toxic chemical release (i.e., emissions inventory) form (§313(g))
July 17, 1987	State emergency response commission designates emergency planning districts (§301(b))
August 17, 1987 (or 30 days after designation of districts, whichever is sooner)	State emergency response commission appoints members of local emergency planning committees (§301(c))
September 17, 1987 (or 30 days after committee is formed, whichever is earlier)	Facility notifies local planning committee of selection of a facility representative (§303(d)(1))
October 17, 1987	MSDS or list of MSDS chemicals is submitted to State commission, local committee, and local fire department (§311(d))
March 1, 1988 (and annually thereafter)	Facilities submit their emergency inventory forms to state commission, local committee, and local fire department (§312(a)(2))
April 17, 1988	Final report on emergency systems study due to Congress (§305(b))
July 1, 1988 (and annually thereafter)	Covered facilities submit initial toxic chemical forms to EPA and designated state officials (§313(a))
October 17, 1988	Local emergency planning committees complete preparation of an emergency plan (§303(a))
June 30, 1991	Comptroller general report to Congress on toxic chemical release information collection, use, and availability (§313(k))
October 17, 1991	EPA report to Congress on Mass Balance Study (§313(1))

The inventory form incorporates a two-tier approach. Under Tier I, facilities must submit the following aggregate information for each applicable OSHA category of health and physical hazard:

1. An estimate (in ranges) of the maximum amount of chemicals for each category present at the facility at any time during the preceding calendar year.
2. An estimate (in ranges) of the average daily amount of chemicals in each category.
3. The general location of hazardous chemicals in each category.

Upon request of a local committee, state commission, or local fire department, the facility must provide the following Tier II information for each substance subject to the request:

1. The chemical name or the common name as indicated on the MSDS.
2. An estimate (in ranges) of the maximum amount of the chemical present at any time during the preceding calendar year.
3. A brief description of the manner of storage of the chemical.
4. The location of the chemical at the facility.
5. An indication of whether the owner elects to withhold location information from disclosure to the public.

The public may also request Tier II information from the state commission and the local committee. The information submitted by facilities under Sections 311 and 312 must generally be made available to the public by local and state governments during normal working hours.

EPA published a uniform format for the inventory forms on January 27, 1987. Tier I information shall be submitted on or before March 1, 1988, and annually thereafter on March 1.

§ 313: Toxic Chemical Release Reporting

Section 313 of Title III requires EPA to establish an inventory of toxic chemical emissions from certain facilities. Facilities subject to this reporting requirement are required to complete a toxic chemical release form for specified chemicals. The form must be submitted to EPA and those state officials designated by the Governor on or before July 1, 1988, and annually thereafter on July 1, reflecting releases during each preceding calendar year.

EPA is required to publish a format for the Toxic Chemical Release form which was released in January 1988. The following information must be included:

1. The name, location, and type of business.
2. Whether the chemical is manufactured, processed, or otherwise used and the general categories of use of the chemical.
3. An estimate (in ranges) of the maximum amounts of the toxic chemical present in the facility at any time during the preceding year.

4. Waste treatment/disposal methods and efficiency of methods for each waste stream.
5. Quantity of the chemical entering each environmental medium annually.
6. A certification by a senior official that the report is complete and accurate.

EPA must establish and maintain a national toxic chemical inventory based on the data submitted. This information must be computer accessible on a national data base.

In addition to the requirements for the emissions inventory in Section 313, EPA will arrange for a mass balance study to be carried out by the National Academy of Sciences using information collected from states that conduct a mass balance-oriented annual quantity toxic chemical release program. Mass balance is the accounting of the total quantity of substances brought into a facility vs the amount that is shipped out. The difference is an indication of the amount released into the environment. A report of this study must be submitted by EPA to Congress no later than October 17, 1991.

Other Title III Provisions

Section 322 of Title III addresses trade secrets and applies to emergency planning, community right-to-know, and toxic chemical release reporting. Any person may withhold the specific chemical identity of a hazardous chemical for specific reasons. Even if the chemical identity is withheld, the generic class or category of the chemical must be provided. The withholder must show each of the following:

1. The information has not been disclosed to any other person other than a member of the local planning committee, a government official, an employee of such person or someone bound by a confidentiality agreement, that measures have been taken to protect the confidentiality, and that the withholder intends to continue to take such measures.
2. The information is not required to be disclosed to the public under any other federal or state law.
3. The information is likely to cause substantial harm to the competitive position of the person.
4. The chemical identity is not readily discovered through reverse engineering.

However, even if chemical identity information can be legally withheld from the public, Section 323 provides for disclosure under certain circumstances to health professionals who need the information for diagnostic purposes or from local health officials who need the information for assessment activities. In these cases, the person receiving the information must be willing to sign a confidentiality agreement with the facility.

Information claimed as trade secret and substantiation for that claim must be submitted to EPA. This includes information that otherwise would be submitted only to state or local officials, such as the emergency and hazardous material inventory (312). People may challenge trade secret claims by petitioning EPA, which must then review the claim and rule on its validity.

Table 1.10. Summary for Chemical Lists for Title III

List of extremely hazardous substances (FR 11/17/86) [402 substances]	302: Emergency planning 304: Emergency notification	Facilities with more than established planning of these substances must notify the state commission. Initial focus for preparation of emergency plans by local emergency planning committees. Certain releases of these substances trigger Section 304 notification to state commission and local committees.
Substances requiring notification under Section 103(a) of CERCLA [717 substances]	304: Emergency notification	Certain releases of these substances trigger Section notification to state commission and local committees as well as Section 103(a) requirement for National Response Center notification.
Hazardous chemicals considered physical or health hazards under OSHA's Hazard Communication Standard [This is a performance standard; there is no specific list of chemicals]	304: Emergency notification 311: Material safety data sheets	Identifies facilities subject to emergency notification requirements. MSDS or list of MSDS chemicals provided by covered facilities to state commission, local committee, and local fire department. Covered facilities provide site-specific information on chemicals to state commission, local committees, and local fire departments.
Toxic chemicals identified as of concern by states of New Jersey and Maryland [329 chemicals/chemical categories]	313: Toxic chemical release reporting	These chemicals are reported on an emissions inventory to inform government officials and the public about releases of toxic chemicals in the environment.

EPA must publish regulations governing trade secret claims. The regulations were published on October 15, 1987, and cover the process for submission of claims, petitions for disclosure, and a review process for these petitions.

Under Section 305, EPA is required to review emergency systems for monitoring, detecting, and preventing releases of extremely hazardous substances at representative facilities that produce, use, or store these substances. EPA will report interim findings to Congress no later than May 17, 1987 and issue a final report of findings and recommendations to Congress by April 17, 1988 (see Tables 1.9, 1.10, and 1.11).

The report must include EPA's findings regarding each of the following:

1. Status of current technological capabilities to (1) monitor, detect, and prevent significant releases of extremely hazardous substances; (2) determine the mag-

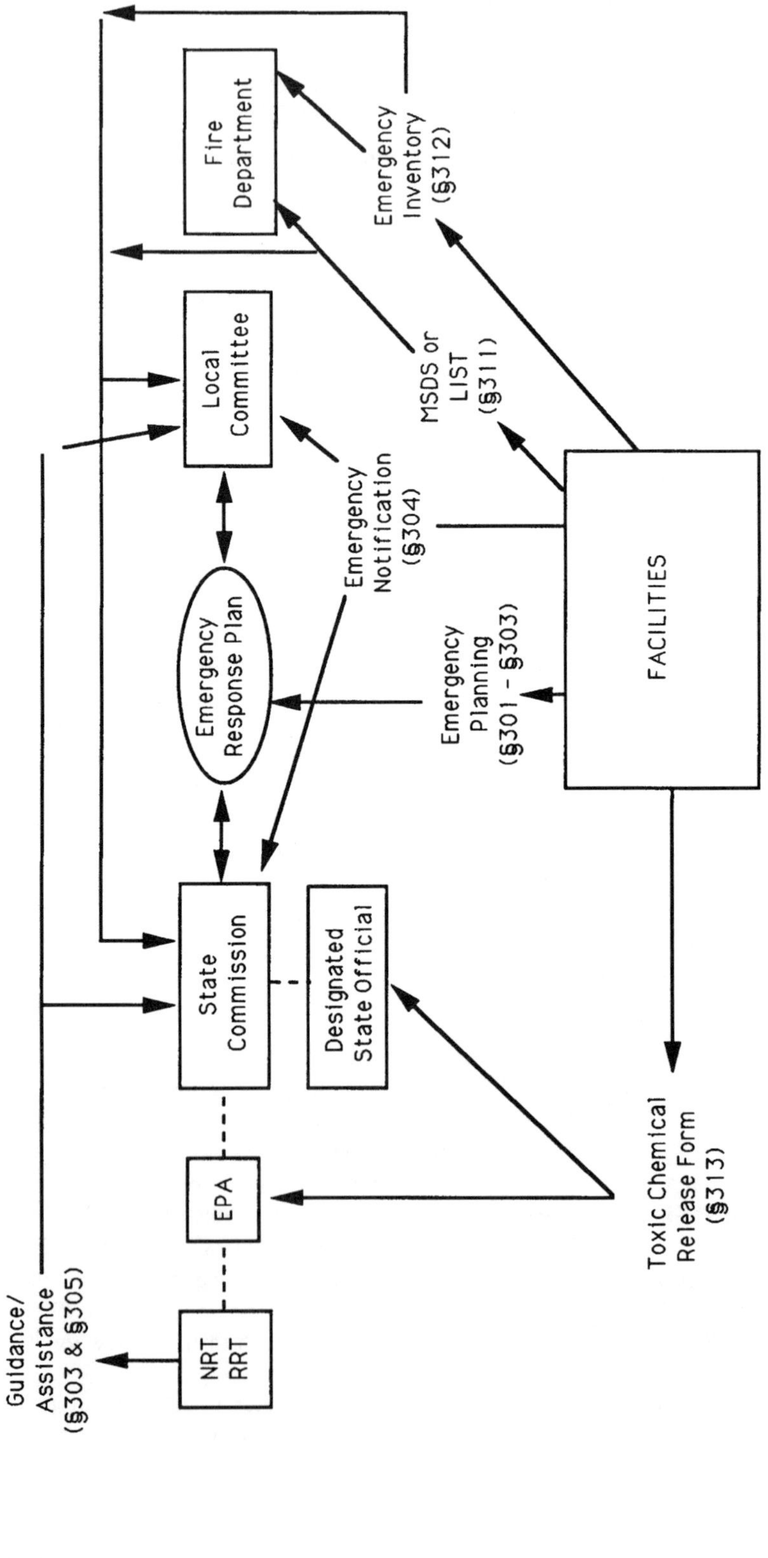

Table 1.11. Title III-Major Information Flow Requirements

nitude and direction of the hazard posed by each release; (3) identify specific substances; (4) provide data on specific chemical composition of such releases; and (5) determine relative concentration of the constituent substances.

2. Status of public emergency alert devices or systems for effective public warning of accidental releases of extremely hazardous substances into any media.
3. The technical and economic feasibility of establishing, maintaining, and operating alert systems for detecting releases.

The report must also include EPA's recommendations for

1. Initiatives to support the development of new or improved technologies or systems that would assist the timely monitoring, detection, and prevention of releases of extremely hazardous substances.
2. Improving devices or systems for effectively alerting the public in the event of an accidental release.

Relationship to Other Regulations

SARA, as stated earlier, is an amendment of CERCLA still regulated by EPA. The interesting thing to be observed in the future will be the role of state and local government enforcement of this law and the role of the public citizen.

Current Status

This law has now been in existence for 3 years, and the enforcement side of EPA is taking action. From this action, the key area of concern in the future will be the legal tort system for third-party suits for noncompliance or criminal charges of negligence upon owners/executives of major corporations who take a laid-back attitude in complying with Title III of SARA. Even after the second year of existence, many of the original filers of Tier I and II requirements did not realize they had to resubmit the forms every year until notified otherwise by EPA or their local emergency coordinator.

PROPOSITION 65 OF CALIFORNIA

Overview

The California Safe Drinking Water and Toxic Enforcement Act of 1986, otherwise known as Proposition 65, requires the state's health and welfare agency to publish a list of chemicals that its scientific advisers have determined cause cancer and reproductive harm. These chemicals are banned from discharge into the state's water supply. This law will set the precedent for other states to follow where the voter has been advised by environmental activists.

Table 1.12. Representative Reproduction Toxicants Under California's Proposition 65 Regulation

Aminopterin
Chlorocyclizine hydrochloride
1,2-Dibromo-3 chloropropane (DBCP)
Diethylstilbestrol (DES)
Diphenylhydantoin
Ethyl alcohol in alcoholic beverages
Ethylene oxide
Etretinate
Isotretinoin
Lead
Methyl mercury
Thalidomide
Valproate
Warfarin

Background

Proposition 65 is the first state law that is more stringent than the SDWA or CWA. The chemicals that are banned from discharge into the state's water supply must be identified for the public. The responsibility of this action is laid upon the manufacturer of those chemicals and products containing them. The manufacturers must warn consumers, workers, and the public exposed to them of their health hazards.

The 136 chemicals on the carcinogen list of October 31, 1987, ranged from arsenic to vinyl chloride. They include such reproductive toxicants as ethyl alcohol (as in alcoholic beverages), lead, and the sterilizing agent ethylene oxide (see Table 1.12). Public supporters of this law eventually like to incorporate wholesale existing lists of hazardous substances, such as the U.S. Public Health Service's National Toxicology Program's list of 145 chemicals. Some 230 chemicals are to be aimed for adoption to the list.

Enforcement

Approximately 1 year after listing of the carcinogens, the ban on a chemical's discharge into water suppliers and warning requirements will go into effect. For reproductive toxins it will be 20 months. California's Health and Welfare Agency will be responsible for the implementation of this controversial law. An interesting item of this law is to accept a determination by a federal or state agency that a carcinogenic chemical or a product containing it is safe in meeting the requirement to prove "*no significant risk.*"

Such chemicals and products will not have to be accompanied by warning labels or other hazard notices. The regulation is termed "interim," however, because the Health and Welfare Agency reserves the right to relist these chemicals and products if evidence of "significant risk" is subsequently found.

The acceptance of federal or state agency approval does not apply to chemicals and products containing them that are listed as reproductive toxicants. Manufacturers and marketers of such chemicals and products must either warn about their hazard or be prepared to prove that they represent "no significant risk."

"Bounty Hunter Provision"

Under Proposition 65, private citizens and organizations are allowed to bring lawsuits against *any* manufacturer that fails to warn about a product's dangers if a state or local prosecutor fails to act. Thus, the ability to prove "no significant risk" becomes an important part of this "Bounty Hunter Provision."

The suing citizens and groups receive 25% of any fine (up to $2500/day) imposed on the manufacturer. The prospect of repeated lawsuits in different parts of California, with heavy costs of legal defense and expert witness fees, will only raise the cost of these products. The burden of proof is set upon the manufacturer, not the state or the plaintiff. The fact is the product must carry "no significant risk" to stay on California's shelves. Also, draft regulations have proposed specific warnings for environmental exposure. Environmental exposures would include all exposures which are not consumer products exposures or occupational exposures.

Implementation and Exemptions

The state's Health and Welfare Agency has proposed the following warning methods:

- A sign could be posted in the area where environmental exposures occur (such as a field where pesticides containing listed chemicals are being sprayed).
- A warning notice could be mailed or delivered to each person in the area where the exposure occurs. The notice must be given at least once a quarter.
- A warning could appear in newspapers or on a broadcast in the exposure area at least once every quarter.

Examples of notices that the agency considers meet the requirements of the regulation are:

"WARNING. This area contains a chemical known to the State of California to cause cancer."

and

"WARNING. This area contains a chemical known to the State of California to cause birth defects or other reproductive harm."

Similar warning messages are required for consumer-product exposure and workplace exposure. The regulation accepts the use of MSDSs to convey the warning to workers exposed to the listed chemicals and products containing them.

Under the law, a manufacturer can obtain an exemption from the warning requirements if its product containing a listed chemical "poses no significant risk assuming lifetime exposure at the level in question for substances known to the state to cause cancer, and that the exposure will have no observable effect assuming exposure at 1000 times the level in question for substances known to the state to cause reproductive toxicity."

Current Trend

Presently other states are considering avenues to protect and enact their own lawsuits targeting protection of citizens against harmful exposures. In Massachusetts, a bill is pending that in its current form would ban or severely restrict the sale and use of thousands of chemicals. Included are limits on the amounts of such chemicals kept in inventory at locations in the state. Taxes on such chemicals are intended to act as an economic disincentive.

California's Proposition 65 and similar laws being considered in other states show the growing role states play in environmental protection. Thus, until the federal agencies in charge of the various environmental regulations "totally enforce" them, it will be the role of the individual state to do so.

REFERENCES

FIFRA

Carson, R. Silent Spring (New York, 1961).

PL 95-516, 86 Stat. 973, 21 October 1972.

PL 94-140, 28 November 1975.

PL 95-396, 92 Stat. 819, 30 September 1978.

PL 96-539, 94 Stat. 3194, 17 December 1980.

FIFRA SS 3(d), 7 U.S.C SS 136 a(d).

Miller, M. L., Attorney Reid and Priest. "Federal Regulation of Pesticides," in Environmental Law Handbook, 8th ed. (Rockville, MD: Government Institutes, 1985), pp. 404–405.

Environmental Statutes (Rockville, MD: Government Institutes, 1985), pp. 191–230.

NEPA

Environmental Statutes (Rockville, MD: Government Institutes, 1985) chap. 5: NEPA pp. 383–408, PL 91–190, 42 U.S.C. SS 4321–4347.

National Environmental Policy Act, Timothy A. Vanderver, Jr., Attorney Patton, Boggs, and Blow, Washington, DC, pp. 370–397.

Schoenbaum, T. J., *Environmental Policy Law, Cases, Readings, and Text,* 1985 ed., University Caseboole Series (Mineola, NY: Foundation Press, 1985), chap. 2.

EPA

40 CFR Series EPA, Government Printing Office, Washington DC (1988).

OSHA

Public Law 91-596, 91st Congress S.2193, December 29, 1970, 84 Stat. 1590.

Environmental Law Handbook Occupational Safety and Health Act, by Marshall Lee Miller, Attorney, Reid and Priest, Washington, DC (Rockville, MD: Government Institutes, 1985), pp. 336–337.

29 CFR 1910 General Industry Standards, July 1, 1988, U.S. Government Printing Office, Washington DC.

CAA

Clean Air Act, Environmental Statutes, 1985 ed. (Rockville, MD: Government Institutes, 1985), pp. 1–190.

Air Pollution Control, G. William Frich, Attorney Lathrop, Koentz, Righter, Clagett, and Norquist, Kansas City, MO and Lydia N. Wegman, Attorney Suidler, Berlin and Strelow, Washington, DC, *Environmental Law Handbook*, 8th ed. (Rockville, MD: Government Institutes, 1985), pp. 213–263.

Schoenbaum, T. J., *Environmental Policy Law, Cases, Readings, and Text*, 1985 ed., University Caseboole Series (Mineola, NY: Foundation Press, (1985), pp. 831–993.

CWA

Ibid. Environmental Statutes, Federal Water Pollution Control Act, (1985), pp. 231–366.

Ibid, *Environmental Law Handbook*, 8th ed., J. Gordon Airbuckle and Timothy A. Vanderver, Jr., Attorneys Patton, Boggs, and Blow, Washington DC, (1985), pp. 265–270.

33 U.S.C. SS 1251 et seq.

Ibid. Environmental Policy Law, pp. 612–657.

SDWA

Ibid. Environmental Statutes, 1985 ed. (1985) pp. 577–618.

Ibid. Environmental Policy Law.

Ibid. Environmental Policy Law, pp. 684–704.

40 C.F.R. Parts 141–143 (1981).

Eleventh Annual Report of the Council on Environmental Quality 86 (1980), Government Printing Office, Washington, DC (1980).

New Orleans Times-Picayune, March 7, 1982, SS 1, at 10, col. 1.

RMR

40 CFR 190 Subchapter F, Government Printing Office, Washington, DC (1988).

TSCA

Druley, R. M. and Ordway, G. L. *The Toxic Substances Control Act* (Washington, DC: Bureau of National Affairs, Inc., 1976), pp. 1–9.

Ibid. Environmental Policy Law, 8th ed., Marshall Lee Miller, Attorney Reid and Priest, Washington, DC (1985) pp. 142–146.

RCRA

Ibid. Environmental Statutes, 1985 ed., (1985), pp. 451–562.

Ibid. Environmental Policy Law, Ridgeway M. Hall, Jr., Nancy S. Bryson, Attorneys Crowell and Moring, Washington, DC (1985), pp. 61–108.

42 USC 9601 et seq. (1981), Citations throughout this chapter are to sections of the act, rather than to the U.S. Code.

42 USC Section 9601 et seq.

HMTR

49 CFR Parts 178 to 199 Department of Transportation Hazardous Materials Transportation Regulations, U.S. Government Printing Office, Washington, DC (1989).

CERCLA

Ibid. Environmental Statutes, 1985 ed.

Ibid. Environmental Policy Law, 8th ed., Ridgeway M. Hall, Jr., Nancy S. Bryson, Attorneys Crowell and Moring, Washington, DC (1985), pp. 109–119.

47 CFR 31180, 16 July 1982.

1910.120

NIOSH/OSHA/USCA/EPA. "Occupational Safety and Health Guidance Manual for Hazardous Waste Site Activities," U.S. Department of Health and Human Services, Document #85-115, U.S. Government Printing Office, Washington, DC (October 1985).

"Hazardous Waste Operations and Emergency Response," *Fed. Regist.* 51(244): 45663–45675, December 19 (1986).

OSHA, U.S. Department of Labor. "Training Requirements in OSHA," Standards and Training Guidelines, OSHA 2254 (Revised), U.S. Government Printing Office, Washington, DC (1987), pp. 26 and 27.

SARA

EPA Document, Community Right to Know, U.S. Government Printing Office, Washington, DC (1987).

EPA Document 20S-0002, When All Else Fails! Enforcement of the Emergency Planning and Community Right-to-Know Act, U.S. Government Printing Office, Washington, DC (1990).

PROPOSITION 65

1. Goerth, C. R. "Hazards Warnings Mandated by California's Proposition 65," *Occupational Safety and Health Magazine*, (March 1988).

CHAPTER 2

Chemical Properties of Hazardous Materials

INTRODUCTION

The terms chemicals and allied products are defined as the overall title for the products of the entire industry. Three general classes of products are commonly identified with the classification: (1) basic chemicals, such as acids, alkalis, salts, and organic chemicals; (2) chemical products to be used in further manufacture, as, for example, synthetic fibers, plastic materials, dry colors, and pigments; and (3) finished chemical products for ultimate consumption, such as drugs, cosmetics, soaps, and detergents, or material or supplies for other industries, such as paints, fertilizer, pesticides, and explosives.

Each one of the various chemical properties of the hazardous materials will be discussed in this chapter.

OVERVIEW

The number of chemical substances that man can synthesize from natural products is virtually unlimited. No one discussion can adequately address the properties of all hazardous materials. Therefore, it may be necessary to use a questionnaire, as shown in Table 2.1. It is possible, however, to consider the broad categories that describe the manner in which a chemical or chemical mixture may be hazardous:

1. The material presents a fire danger. It is explosive, flammable, or undergoes a reaction which releases heat.
2. The material is a cryogen. It presents a hazard due to its extremely cold physical state.

Table 2.1. Typical Form for Getting More Information from a Supplier

HAZARDOUS MATERIALS INFORMATION SHEET

(Please complete all applicable sections.)

1. Product Name, Number, Synonym ____________ Chemical Formula ____________
2. Manufacturer's Name ______________________________
3. Manufacturer's Address ______________________________
4. Chemical and Physical Properties: a. Molecular Wt. _____ b. Boiling Point ______°C
   c. Melting Point ________°C d. Specific Gravity (water=1) or Bulk Density ______°C
   @ ________°C e. Vapor Density (air=1) ______ f. Vapor Pressure (mmHg) ______
   @ ___________°C: __________ @ __________°C: __________ @ __________°C
   g. Solubility ______________________________
   h. pH/conc. ____________ i. Index of Refraction ____________ @ ____________°C
   j. Corrosive action on materials (e.g., aluminum, carbon steel, copper, rubber, plastics, etc.) ______________________________
   k. Does the material decompose when exposed to air? water? heat? strong oxidizers? possible products? ______________________________
   l. Does the material generate heat through polymerization or condensation? _______
   m. Composition (give chemical names of components, information will be treated as confidential)

COMPOUND	PERCENT	COMPOUND	PERCENT

NOTE: Please be specific. For example, it is important to know whether an alcohol is methanol: an aromatic hydrocarbon is benzene: a chlorinated material is carbon tetrachloride.

5. Flammability and Explosive Properties: a. Flash Point, F. Closed Cup ____________
   Open Cup __________ If flash point changes during evaporation give date ________
   b. Explosive limits (% by vol. in air): LOWER ____________ UPPER ___________
   c. Susceptibility to spontaneous heating: YES _____________ NO ____________
   d. Fire point, F ____________________ Auto-ignition temp. F ______________
   e. What products might be formed in the event of fire or abnormal temperatures? ______________________________
   f. Suitable extinguishing agents ______________________________
6. Procedures in Case of Container Breakage or Leakage ____________________
7. Transportation and Storage Requirements ____________________
8. Physiological Properties (give animal tested, observation time, dosage value and range, dilution medium, etc.):
   a. Acute oral toxicity ______________________________
   b. Acute local effects on eyes ______________________________
   c. Acute local effects on skin. Primary irritant? ____________________
      Sensitizer? ______________________________
   d. Acute inhalation toxicity (vapor, mist, fume, dust. Indicate effects of concentration and ime.) ______________________________
   e. Chronic effects ______________________________
   f. Warning properties (odor; irritation of eyes, nose, throat) ______________
   g. Threshold limit value (estimate, if not on current list of ACGIH) ___________
9. First Aid Treatment
   a. Skin contact ______________________________
   b. Eye contact ______________________________
   c. Inhalation ______________________________
   d. Antidote and treatment in case of swallowing ____________________
10. Recommended Pre-placement or Periodic Medical Examination (health standards, clinical tests, frequency, etc.) ______________________________

Table 2.1. (Continued)

11. Precautions for Normal Condition of Use ____________________
12. Recommended Personal Protective Equipment ____________________
13. Suggested Method for Air Analysis ____________________
14. Pertinent Literature References ____________________
15. Information Furnished by NAME ____________________ Date __________
    TITLE ____________________
    COMPANY ____________________
    ADDRESS ____________________

(If more space is needed for comment, please attach an additional sheet. Please attach product information data sheets or other publications related to the safe handling and use of this material.)

3. The material is toxic. Depending on the concentration, duration, and type of exposure, the material may cause sickness or even death.
4. The material is radioactive. Exposure induces reactions in human tissues which can cause cell death.

Tables 2.2 and 2.3 depict representative chemical formulas and elements.

PHYSICAL PROPERTIES OF HAZARDOUS MATERIALS

Physical State

The physical state for material (gas, liquid, or solid) has a definite relationship to the potential danger posed by that material. A material that is dangerous in one state can be relatively innocuous in another. There are specific terms to describe the various physical state of a material. As defined by the American National Standards Institute, these are as follows.

Dust

Dust is solid particles generated by handling, crushing, grinding, rapid impact, detonation, and decrepitation of organic or inorganic materials, such as rock, ore, metal, coal, wood, and grain. Dust does not tend to flocculate except under electrostatic forces; they do not diffuse in air, but settle under the influence of gravity.

Dust is a term used in industry to describe airborne solid particles that range in size from 0.1 to 25 μm (1 μm = 1/10,000 cm = 1/25,000 in.) (μm is the abbreviation for micron). Dusts above 5 μm in size usually will not remain airborne long enough to present an inhalation problem.

Dust may enter the air from various sources. It may be dispersed when a dusty material is handled, such as when lead oxide is dumped into a mixture or a product is dusted with talc. When solid materials are reduced to small sizes in processes such as grinding, crushing, blasting, shaking, and drilling, the mechanical action of the grinding or shaking device supplies a source of energy to disperse the dust formed.

Table 2.2. Chemical Formulas

Acetic acid — HOAc
Aluminum chloride — $A1C1_3$
Aluminum — Al
Ammonia — NH_3
Ammonium nitrate — NH_4NO_3
Ammonium hydroxide — NH_4OH
Ammonium — NH_4
Benzene — C_6H_6
Boron trifluoride — BF_3
Boron oxide — B_2O_3
Bromine trifluoride — BrF_3
Bromine gas — Br_2
Cadmium oxide — CdO
Cadmium hydroxide — $Cd(OH)_2$
Calcium oxychloride — $Ca(OCl)_2$
Calcium oxides — CaO_1
Calcium cyanide — $Ca(CN)_2$
Calcium chloride — $CaC1_2$
Carbon tetrachloride — $CC1_4$
Carbon monoxide — CO
Carbon dioxide — CO_2
Carbon bisulfide — CS_2
Cesium oxide — Cs_2O
Chlorine trifluoride — ClF_3
Chlorine oxide — $C1O_2$
Chlorine gas — C1
Chloroform — $CHC1_3$
Chromium trioxide — CrO_3
Chromium oxide — Cr_2O_3
Cobalt oxides — CoO_1
Copper iron sulfide — $CuFeS_2$
Cyanide — CN
Dioxin — 2,3,7,8 - TCDD
Ethanol — EtOH
Fluorine oxide — F_2O_2
Fluorine gas — F_2
Formaldehyde — HCHO
Hydriodic acid — HI
Hydrochloric acid — HC1
Hydrofluoric acid — HF
Hydrogen peroxide — H_2O_2 or HOOH
Hydrogen sulfide — H_2S
Hydrogen gas — H_2
Hypochlorous acid — HOC1
Iodine heptafluoride — IF_7
Iron oxide — Fe_2O_3
Lithium hydride — LiH
Lithium hydroxide — LiOH
Magnesia — M_gO
Magnesium ethyl — $Mg(C_2H_5)_2$
Mercuric fluoride — HgF_2
Methanol — CH_3OH
Nitric acid — HNO_3
Nitrogen oxides — NO_2
Nitrogen fluoride — NF_3
Nitrogen oxide — N_2O_4
Nitrosyl chloride — NOC1

Table 2.2. (Continued)

Osmium tetroxide — OsO_4
Oxides of phosphorus — PO_1
Oxygen fluoride — OF_2
Oxygen gas — O_2
Ozone — O_3
Phosgene — $COC1_1$
Phosphorus pentoxide — P_2O_5
Phosphorus trioxide — P_2O_3
Phosphorus trichloride — $PC1_3$
Potassium carbide — KHC
Potassium chlorate — $KC1O_3$
Potassium chromate — K_2CrO_4
Potassium hydroxide — KOH
Rubidium carbide — Rb_2C_2
Silica — SiO_2
Silver oxide — AG_2O
Sodium hydroxide — NaOH
Sodium carbide — Na_2C_2
Sodium nitrate — $NaNO_3$
Sodium oxybromide — NaOBr
Sodium oxychloride — NaOC1
Sodium oxide — Na_2O
Sodium nitride — NaN_3
Sodium peroxide — Na_2O_2
Sodium potassium alloy — NaK
Sodium perchlorate — $NaC1O_3$
Sulfur chloride — $SC1_2$
Sulfur dioxide — SO_2
Sulfur oxides — SO_1
Sulfuric acid — H_2SO_4
Sulfurous acid — H_2SO_3
Tellurium oxide — TeO_2
Thallium nitrate — $T1(NO_3)_3$
Thallous oxide — $T1_2O$
Thiosulfuric acid — $H_2S_2O_3$
Vanadium oxides — VO_1
Zinc oxide — ZnO
Zinc dichromate — $ZnCr_2O_7$
Zinc chloride — $ZnC1_2$
Zinc chromate — $ZnCrO_4$

Fumes

Fumes are solid particles generated by condensation from the gaseous state, generally after volatilization from molten metals. This physical change is often accompanied by a chemical reaction, such as oxidation. Fumes flocculate and sometimes coalesce.

A fume is formed when a volatilized solid, such as metal, condenses in cool air. The solid particles that make up a fume are extremely fine, usually less than 1.0 μm. In most cases, the hot material reacts with the air to form an oxide. Examples are lead oxide fumes from smelting, and iron oxide fumes from arc welding. A fume can also be formed when a material such as magnesium metal is burned or when welding or gas cutting is done on galvanized metal. Metal fumes can be generated from soldering, too.

Table 2.3. Elements

Element	Symbol	Element	Symbol
Actinium	Ac	Mendelevium	Md
Aluminum	Al	Mercury	Hg
Americium	Am	Molybdenum	Mo
Antimony	Sb	Neodymium	Nd
Argon	Ar	Neon	Ne
Arsenic	As	Neptunium	Np
Astatine	At	Nickel	Ni
Barium	Ba	Niobium	Nb
Berkelium	Bk	Nitrogen	N
Beryllium	Be	Nobelium	No
Bismuth	Bi	Osmium	Os
Boron	B	Oxygen	O
Bromine	Br	Palladium	Pd
Cadmium	Cd	Phosphorus	P
Calcium	Ca	Platinum	Pt
California	Cf	Plutonium	Pu
Carbon	C	Polonium	Po
Cerium	Ce	Potassium	K
Cesium	Cs	Praseodymium	Pr
Chlorine	Cl	Promethium	Pm
Cobalt	Co	Radium	Ra
Columbium	Cb	Radon	Rn
Copper	Cu	Rhenium	Re
Curium	Cm	Rhodium	Rh
Dysprosium	Dy	Rubidium	Rb
Einsteinium	Es	Ruthenium	Ru
Erbium	Er	Samarium	Sm
Europium	Eu	Scandium	Sc
Fermium	Fm	Selenium	Se
Fluorine	F	Silicon	Si
Francium	Fr	Silver	Ag
Gadolinium	Gd	Sodium	Na
Gallium	Ga	Strontium	Sr
Germanium	Ge	Sulfur	S
Gold	Au	Tantalum	Ta
Hafnium	Hf	Technetium	Tc
Helium	He	Tellurium	Te
Holmium	Ho	Terbium	Tb
Hydrogen	H	Thallium	Ti
Indium	In	Thorium	Th
Iodine	I	Thulium	Tm
Iridium	Ir	Tin	Sn
Iron	Fe	Titanium	Ti
Krypton	Dr	Tungsten (Wolfram)	W
Lanthanum	La	Uranium	U
Lawrencium	Lr	Vanadium	V
Lead	Pb	Xenon	Xe
Lithium	Li	Ytterbium	Yb
Lutetium	Lu	Yttrium	Y
Magnesium	Mg	Zinc	Zn
Manganese	Mn	Zirconium	Zr

Gases and vapors are not fumes, even if newspaper columnists often (and incorrectly) call them that.

Smoke

Smoke is carbon or soot particles less than 0.1 µm in size which result from the incomplete combustion of carbonaceous materials such as coal or oil. Smoke generally contains droplets as well as dry particles. Tobacco, for instance, produces a wet smoke composed of minute tarry droplets. The size of the particles contained in tobacco smoke is about 0.25 µm.

Aerosols

Aerosols are liquid droplets (mist and fog) or solid particles dispersed in air (fumes, dust, and smoke) that are of fine enough particle size to remain so dispersed for a period of time.

Mists

Mists are suspended liquid droplets generated by condensation from the gaseous to the liquid state or by breaking up a liquid into a dispersed state, such as by splashing, foaming, or atomizing. A mist is formed when a finely divided liquid is suspended in the atmosphere. Examples are the oil mist produced during cutting and grinding operations, acid mists from electroplating, acid or alkali mists from pickling operations, paint spray mist from spraying operations, and the condensation of water vapor to form a fog or rain.

Gases

Gases are normally formless fluids that occupy the space or enclosure and can be changed to the liquid or solid state only by the combined effect of increased or decreased temperature. Gases can diffuse into the atmosphere. Examples are welding gases, internal combustion engine exhaust gases, and air.

Vapors

Vapors are the gaseous form of a substance which are normally in the solid or liquid state (at room temperature and pressure). The vapor can be changed back to solid or liquid state either by increasing the pressure or decreasing the temperature alone. Vapors also diffuse. Evaporation is the process by which a liquid is changed into the vapor state and mixed with the surrounding atmosphere. Solvents with low boiling points will volatilize readily.

In addition to the definitions concerning concentrations of matter, which find daily usage in the vocabulary of chemical concentration, other terms used to describe degree of exposure are

1. ppm — Parts of vapor or gas per million parts of contaminated air by volume.
2. mppcf — Millions of particles of a particulate per cubic foot of air.

3. mg/m^3 — Milligrams of a substance per cubic meter of air.
4. µg/L — Micrograms of a substance per liter of solution.

DENSITY OF GASES AND LIQUIDS

The density of a gas in relation to the density of air is termed the vapor density. It is calculated as follows:

$$\text{Vapor density} = \frac{\text{Weight of gas or vapor}}{\text{Weight of an equal volume of air}}$$

The relative density of a liquid is termed the specific gravity. It is calculated as follows:

$$\text{Specific gravity} = \frac{\text{Weight of liquid}}{\text{Weight of an equal volume of water}}$$

The vapor density or specific gravity is an important piece of information when dealing with environmental pollutants. It will affect the way the material travels through the environment, the kinds of hazards it poses, and the safety precautions that must be taken to afford protection against the material's hazardous effects. Refer to the *Merck Index* or similar technical references for vapor density or specific gravity of the material in question.

Vapor Density

Hazards of Materials Lighter than Air

If the vapor density is less than 1, the material is lighter than air and will rise and dissipate in the air if unconfined. The material could collect in high spots such as ceilings if confined. One example is hydrogen or helium.

Hazards of Materials Heavier than Air

If the vapor density is greater than 1, the material is heavier than air and will sink. The material will linger near the ground and collect in low spots. Examples are carbon dioxide (can displace oxygen), chlorine (toxic), gasoline vapors (explosive), and trichloroethylene (can displace air).

Specific Gravity

Hazards of Materials Lighter than (but Insoluble in) Water

If specific gravity is less than 1, the material will float on water.

If the material enters the environment, it will sink in porous materials, such as soil or sand, until it encounters water (i.e., a groundwater aquifer), where it will then float.

Hazards of Materials Heavier than Water

If the specific gravity is greater than 1, the material will sink in water. This is true of materials like carbon disulfide, virtually all chlorinated solvents, and polychlorinated biphenyls.

If the material enters the environment, it will sink in porous material such as soil or sand, then sink unaffected by water until it reaches a nonporous surface.

FIRE DANGERS

An understanding of numerous fire hazards may be gained from understanding the fundamentals of chemistry and physics, which describe the properties of matter. Although the phenomena of burning materials is a familiar one, it is not always obvious to the casual observer that liquids and solids do not actually burn as the liquid or solid state of matter. Instead, they give off vapors that can only ignite when a combustible mixture with air has been obtained. Several fire hazard-related definitions are based on vapor ignition:

1. Lower explosive limit: the minimum concentration of gas or vapor in air below which a substance does not burn when exposed to an ignition source.
2. Upper explosive limit:* the maximum concentration of gas or vapor in air above which ignition does not occur.
3. Flammable range: the numerical difference between the upper and lower explosive limits.
4. Flash point: the minimum liquid temperature at which a spark or flame will cause an instantaneous flash in the vapor space above the liquid.
5. Fire point: the temperature at which the liquid gives off enough vapor to continue to burn when lighted.

Spontaneous Combustion is another term commonly associated with the subject of fire danger. It is synonymous with autoignition temperatures, the temperature at which ignition occurs without the introduction of an ignition source.

Many hazardous materials have fire danger-associated properties. These categories of hazardous materials include water-reactive materials, oxidizing agents, organic compounds, polymers, explosives, and corrosive materials. A brief discussion of these categories follows (Figure 2.1) that illustrates the relationship of vapor air mixtures and flammability.

Water-Reactive Materials may be hydroscopic, which means that their physical characteristics are significantly altered by the effects of water vapor. For example, a container of concentrated sulfuric acid may overflow when exposed to the water vapor present in the air in a classroom or laboratory. Other water-reactive mate-

* Also known as lower and upper flammability limit.

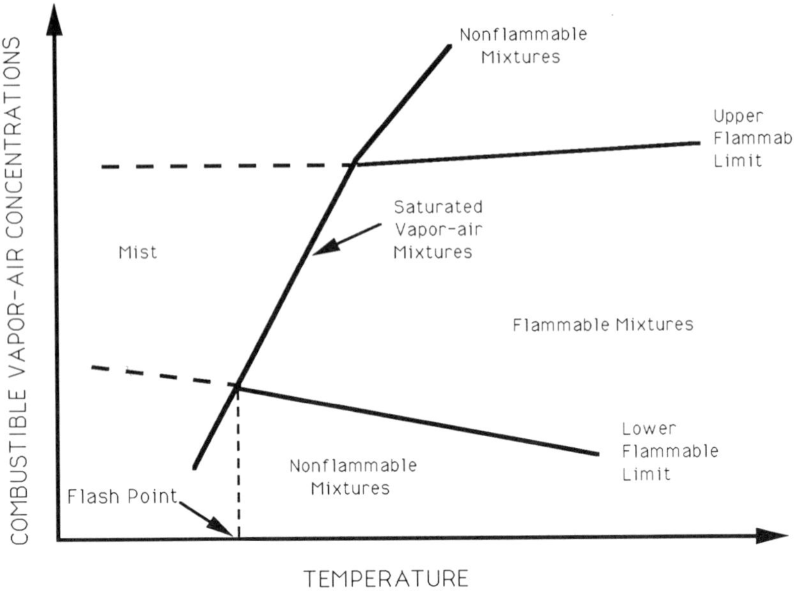

Figure 2.1. Vapor air mixture diagram. The vapor air mixture diagram shows relation between upper and lower flammable (explosive) limits, flammable and nonflammable regions, flash point, and vapor pressure curve. Liquid droplets exist to the left of the saturated vapor/air mixture curve. Actually this diagram shows what happens to a vapor/air mixture as concentrations and temperature vary. Adapted from Van Dolan, et al., *Fire Technol.* May (1965), by permission.

rials may undergo hydrolysis, which is the decomposition or alteration of a chemical substance by water. Products of hydrolysis are often flammable. (In addition, they may be toxic or corrosive.) Water can chemically react with some burning substances, resulting in greater fire intensity than the original combustion. Because of this property of certain water-reactive materials, precaution must be taken in using water to extinguish fires.

A substance that gains electrons in a chemical reaction is an oxidizing agent, this includes fluorine, oxygen, chlorine, and sulfur. Hazardous features of oxidizing agents include the following: they are chemically reactive, their reactions usually generate heat, and they may propagate fires by the release of oxygen.

Organic Compounds are associated with a number of potential hazards. Most are extremely flammable and are insoluble in water. Therefore, water may be ineffective as a fire extinguisher. Many organic compounds burn with very sooty flames, thus impeding the effects of fire fighters. In addition to fire danger-related hazards, numerous organic compounds have associated health hazards.

Polymers, which are generally organic compounds, commonly comprise resins, plastics, and fibers. Burning polymers often melt as they burn, and the resulting dripping material can spread fires. The surface of burning polymers usually chars, generating large amounts of smoke. Their vapors are extremely toxic in a few compounds. An example would be the polymer of nylon or polymethane.

An Explosive is a chemical compound or mixture of compounds which suddenly undergoes a very rapid chemical transformation, with the simultaneous production of large quantities of heat and gases. DOT classifies explosives as Class A, B, or C. Class A explosives are materials, such as nitroglycerine, that are capable of detonation by means of a spark, flame, or shock. Classes B and C explosives are materials that ordinarily explode under extreme conditions of temperatures.

A Corrosive Material is one that undergoes a chemical process which converts minerals and metals into unwanted products. It is also known to eat away flesh. Acids and bases are common examples of corrosive materials. Acids can react with metals, gases, metallic oxides, and carbonates. Bases can react with acids, metals, and salts. A substance may be corrosive in any physical state: as a liquid, a solid, or a gas. Corrosive materials can cause ignition with organic materials. (Example, sulfuric acid and nitric have this vapor.) Additionally, they may emit toxic gases. Under RCRA, the definition of a corrosive material or mixture is a substance that has a pH less than 2.5 or greater than 12.5. The pH of water (neutral of liquid) is 7. See Glossary for terms and definitions of acids and bases.

A special type of fire danger is posed by the phenomenon called *BLEVE* (pronounced bleh' vee). BLEVE is an acronym that stands for boiling liquid expanding vapor explosion, which occurs when tanks containing flammable liquids under pressure are exposed to fire and then explode. The combination of the weakened structure of the tank and the buildup of internal pressure from the gas results in an instantaneous release and ignition of the vapor. The danger resulting from a BLEVE is immense. People have been killed and dismembered by fragments of exploded tanks, which travel several hundred feet at great velocities. BLEVES are often unexpected, as they are likely to occur 10 to 30 min after the tanks have been exposed to flames.

The following is a partial list of common compressed gases that may result in a BLEVE when exposed to external heat:

- Propane
- Butane
- Butene
- Ethylene
- Hydrogen sulfide
- Isobutane
- Isobutylene
- Propylene
- Vinyl chloride
- Methane

Recommended responses to a potential BLEVE are first to evacuate the area, and second, to cool the containers if it is possible to do so with minimal risk.

Fire Extinguishers

The most important property of a fire extinguishment is nonflammability. Most substances used as fire extinguishers are already oxidized and are thus incapable of further reactions with oxygen.

Water is the most commonly known and widely used fire extinguishment. However, in certain cases, the application of water to burning material may actually intensify the fire. Also, water which is applied to a fire may run off to areas where it is incapable of contacting the burning material. If water is judged to be an appropriate fire extinguishment, its use can be enhanced by employing it as a spray, fog, foam, or gel. Other nonwater-based common fire extinguishments include carbon dioxide and halogenated hydrocarbons. Figure 2.2 depicts appropriate extinguishments for various types of fires.

CRYOGENS

Cryogens are gases that must be cooled to less than –150°F to achieve liquefication. The handling and storage of such materials requires special attention. Cryogens have a high expansion rate upon vaporization, which may cause the rupture of their container. They have an ability to liquify other gases. This property can result in an explosion if air is caused to solidify and block the venting tubes of a storage container. Additionally, cryogens have a potential to damage living tissue, as contact with these materials in their liquid state will cause frostbite and deep skin cell destruction. Table 2.4 lists the boiling temperatures of four common cryogenic liquids.

OXIDIZERS

These materials can cause ignition combustion or detonation of organic materials, powdered metals, and other reducing agents. They are used as chlorinating and bleaching agents, fertilizer, pyrotechnic mixtures, and rock fuels. Some common oxidizing agents are hypochlorites, chlorates, and perchlorates. In general a list of various incompatible chemicals are listed in Table 2.5.

MSDSs

The acronym MSDS stands for the words Material Safety Data Sheets. An MSDS is a written or printed paper(s) concerning a hazardous chemical that is prepared in accordance with paragraph (b) of the OSHA Hazard Communication

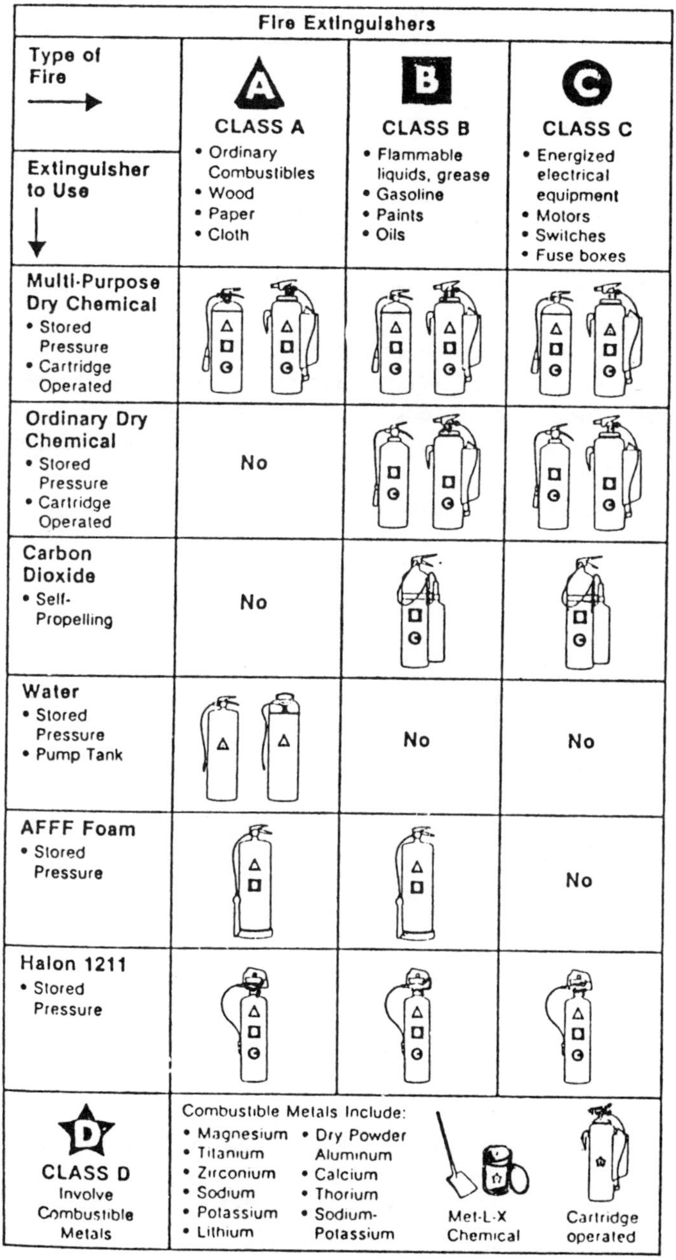

Figure 2.2. Fire extinguishers — various types. Reprinted from *Safety Engineering Standards*, by permission. Industrial Indemnity is the sole owner of the rights granted, and Industrial Indemnity controls these rights.

Table 2.4. Boiling Temperature of Various Cryogenic Liquids

Liquid	Temperature
Nitrogen	–320°F
Argon	–302°F
Helium	–452°F
Carbon dioxide	–102°F

Standard 1910.1200. Prior to this standard, MSDSs used to be known in industry as the OSHA 20 Form.

MSDSs are to be developed or obtained by chemical manufacturers and importers for each hazardous chemical they produce or import. Employees shall have an MSDS for each hazardous chemical which they use.

Each MSDS shall contain at a minimum the following information:

1. Manufacturer's name, address, and emergency phone number.
2. Hazardous ingredients/identify information.
3. Physical/chemical characteristics.
4. Fire and explosion hazard data.
5. Reactivity data.
6. Health hazard data.
7. Precaution for safe handling and use.
8. Control measures.
9. Disposal methods.

Examples of various hazardous materials that are element compounds are illustrated in Tables 2.2 and 2.3, respectively. Table 2.6 illustrates the proper procedures for storage of hazardous materials.

TOXICITY

Toxicity of chemicals will be discussed in this chapter only from a chemical vs toxicological perspective. A detailed discussion of the science of toxicology is in Chapter 3.

Pharmacology is the study of the nature, properties, and effects that chemicals have on living systems. Toxicology is a branch of pharmacology, and toxicity relates to a harmful effect on the body by physical contact, ingestion, or inhalation.

Paracelsus, the founder of modern pharmacology, who died in 1541, wrote "Poison is in everything, and no thing is without poison. The dosage makes it either a poison or a remedy" This is an apt description of how the toxicity of a substance is directly related to its concentration. Toxicity actually depends upon four factors:

1. The quantity of the material.
2. The rate and extent of absorption into the bloodstream.

Table 2.5. Incompatible Chemicals

The following is a partial list of chemicals that are incompatible with each other. Reactions can take place which may liberate poisonous or flammable gases, cause explosions by contact or by their reaction products, or may ignite spontaneously. An occasional review of this list will be a reminder of these hazards and may suggest other incompatibles that may be encountered in a rubber or plastics laboratory. Substances in the right hand column should be stored and handled so they cannot possibly accidentally contact corresponding substances in the left hand column.

Alkaline and alkaline earth	Carbon dioxide, carbon tetrachloride, and other metals, such as sodium and chlorinated hydrocarbons (also prohibit water, potassium, cesium, lithium, foam, and dry chemical on fire involving these magnesium, calcium, aluminum metals)
Acetic acid	Chromic acid, nitric acid,hydroxyl-containing compounds, ethylene glycol, perchloric acid, peroxides, and permagnates
Acetone	Concentrated nitric and sulfuric acid mixtures
Acetylene	Chlorine, bromine, copper, silver, fluorine, and mercury
Ammonia (anhyd.)	Mercury, chlorine, calcium hypochlorite, Iodine, bromine, and hydrogen fluoride
Ammonium nitrate	Acids, metal powders, flammable liquids, chlorates, nitrites, sulfur, finely divided organics, or combustibles
Aniline	Nitric acid, hydrogen peroxide
Bromine	Ammonia, acetylene, butadiene, butane and other petroleum gases, hydrogen, sodium carbide, turpentine, benzene, and finely divided metals
Calcium carbide	Water (see also acetylene)
Carbon, activated	Calcium hypochlorite
Copper	Acetylene, hydrogen peroxide
Chlorates	Ammonium slats, acids, metal powders, sulfur, finely divided organics, or combustibles
Chromic acid	Acetic acid, naphthalene, camphor, glycerine, turpentine, alcohol, and other flammable liquids
Chlorine	Ammonia, acetylene, butadiene, butane and other petroleum gases, hydrogen, sodium carbine, turpentine, benzene, and finely divided metals
Chlorine dioxide	Ammonia, methane, phosphine, and hydrogen sulfide
Cumene hydroperoxide	Acids (organic or mineral)
Fluorine	Isolate from everything

Table 2.5. (continued)

Hydrocyanic acid	Nitric acid, alkalis
Hydrogen peroxide	Copper, chromium, iron, most metals or their salts, any flammable liquid, combustible materials, aniline, nitromethane
Hydrofluoric acid, anhyd. (hydrogen fluoride)	Ammonia, aqueous and anhydrous
Hydrogen sulfide	Fuming nitric acid, oxidizing gases
Hydrocarbons (benzene, butane, propane, gasoline, turpentine, etc.)	Fluorine, chlorine, bromine, chromic acid, sodium peroxide
Iodine	Acetylene, ammonia (anhyd. or aqueous), hydrogen
Mercury	Acetylene, fulminic acid, ammonia
Nitric acid (conc.)	Acetic acid, aniline, chromic acid, hydrocyanic acid, hydrogen sulfide, flammable liquids, flammable gases, and nitritable substances
Oxygen	Oils, grease, hydrogen, flammable liquids, solids, or gases
Oxalic acid	Silver, mercury
Perchloric acid	Acetic anhydride, bismuth and its alloys, alcohol, paper, wood
Phosphorus (white)	Air, oxygen
Potassium chlorate	Acids (see also chlorates)
Potassium perchlorates	Acids (see also perchlorates)
Potassium permanganage	Glycerine, ethylene glycol, benzaldehyde, sulfuric acid
Silver	Acetylene, oxalic acid, tartaric acid, fulminic acid, ammonium compounds
Sodium	See alkaline metals (above)
Sodium nitrite	Ammonium nitrate and other ammonium salts
Sodium peroxide	Any oxidizable substance, such as ethanol, methanol, glacial acetic acid, acetic anhydride, benzaldehyde, carbon disulfide, glycerine, ethylene glycol, ethyl acetate, methyl acetate, and furfural
Sulfuric acid	Chlorates, perchlorates, permanganates
Toluene diisocyanate and other isocyanates	Avoid contact with strong alkalis, such as caustic soda, to prevent uncontrollable polymerizations

Table 2.5. (continued)

Trichlorethylene	Reacts with strong alkalis to form gases that ignite spontaneously
Zirconium	Prohibit water, carbon tetrachloride, foam, and dry chemical on zirconium fires

Table 2.6. Proper Procedures for Storage of Hazardous Materials

Face labels out
Keep stacks straight and aligned
Check for location accuracy
Don't stack too high
Check for loose closures
Place into proper location as soon as possible
Don't block exits or emergency equipment
Report all spills or leaks immediately

3. The rate and extent of metabolization into by-products.
4. The rate and extent of excretion.

If there is no exposure, regardless of the relative toxicity of a hazardous material, there can be no harm. Exposure occurs when a substance is absorbed through the skin, ingested, or inhaled. The respiratory system is the most potentially hazardous route by which a substance can enter the body, since gases are absorbed into the bloodstream almost immediately upon inhalation. Gases considered to be hazardous include the following categories:

- Anesthetic: Material capable of causing a marked loss of muscular powers.
- Asphyxiant: Material that is capable of arresting respiration.
- Irritant: Material which produces local irritation or inflammation of eyes, skin, or respiratory membranes.

Smoke is a gas mixture and particulate matter that commonly acts both as an asphyxiant and an irritant. Smoke poses a hazard in the following ways: it obscures vision, interferes with normal respiration, aids in the absorption of toxic gases, and is itself carcinogenic. As of October 6, 1987, it was discovered by scientists that smoke interferes with a gene in DNA to induce cancer.

Although toxicity is not a physical constant and, therefore, is difficult to measure accurately, many efforts have been made to quantify toxicity. Units by which toxicity is measured include lethal dose, 50% kill (LD_{50}), and lethal concentration, 50% kill (LC_{50}). Both are measurements of the amount of a substance which results in death for 50% of test specimens after a specific duration of exposure. The lethal dose is expressed in units of milligrams per kilogram, and lethal concentration is usually expressed in parts per million (ppm) by volume. The threshold limit value (TLV) is the upper limit of a toxicant

Table 2.7. Difficulties in Identifying Chronic Effects of Hazardous Materials

Chemical mixtures are involved
Duration and concentration of exposure must be determined
Small populations at risk make statistical analysis difficult
Psychophysiological complaints may mask quantifiable effects

concentration to which an average, healthy person may be exposed all day, every day, without suffering adverse effects. For 8 hr/day the TLV is usually expressed as parts per million for gases in air and as milligrams or micrograms per cubic meter for fumes and mists in air. The term TLV is used by the American Conference of the Governmental Industrial Hygiene Association. The term permissible exposure limit (PEL) is the 8-hr exposure valve used by OSHA.

EVALUATING TOXICITY

One of the earliest documentations of health effects of hazardous materials was by a contemporary of Erasmus, who published a study of the diseases of metal miners in Central Europe in 1556. During the industrial revolution and continuing to the present, toxic effects in humans are considered to be substantiated when a significant number of individuals develop symptoms of a disease related to a particular environment. The Appendix lists some confirmed and suspected occupational carcinogens and the organs or tissues that are their targets.

Exposure to hazardous materials is usually categorized as either acute or chronic. Acute refers to short-term, often one-time exposure, and chronic refers to long term, repeated exposure. Chronic effects are more difficult to quantify than acute effects for reasons listed in Table 2.7.

Psychophysiological complaints include those symptoms that are often impossible to attribute to a single physical cause: depression, irritability, dizziness, nausea, weakness, fatigue, insomnia, and numbness. Such stress-related illnesses may be difficult to separate from exposure results, especially where stress effects remain after actual exposure.

If psychophysical complaints are from an exposure to chemical(s), then evaluation using various biological monitoring tests for target organ systems (specific antibody tests), amino acids, enzymes, metabolic pathways of the liver and the specific chemical(s)/(sub)metabolites must be considered. The results of these types of tests will show the various scientists and health care professionals the combined effects of exposure.

Chemical interaction can occur at an individual cell and its organelles. This later concept may be the true mechanism of psychophysical complaint(s) of individuals who have been poisoned at a cellular level, thus planting the seedling of disease in mankind.

Deleterious effects on health from hazardous materials have been documented, and in many cases the nature of the physiological problems is consistent with the

Table 2.8. Examples of Various Levels of Ionizing Radiation

<200 mrem/hr	In a container
<10 mrem/hr	1 m away
<2 mrem/hr	Safe
Acute ≥700 rem if person survives (unlikely)	
300–500 rem	Survival (possible)
100–250 rem	Survival (probable)
General public 500 mrem and a radiation worker 5000 mrem.	

exposure effects from known contaminants. Long-term health problems potentially related to hazardous material exposure have not been adequately studied. A complete understanding of effects undoubtedly will involve a range of physiological systems, including genetic predisposition.

RADIOACTIVE MATERIALS

Exposure to radioactive materials causes changes in body chemistry. This, in turn, can lead to the ultimate destruction of an entire organ or death.

Radiation exposure causes a physical change that is generally associated with aging and increases the mortality rate of an exposed population. Various cancers are results of radiation exposure, as are genetic mutations. Table 2.8 lists examples of various levels of ionizing radiation exposure limits.

The term "dose" is often used in the sense of the exposure dose, expressed in roentgens, which is a measure of the total amount of ionization that the quantity of radiation could produce in air. The biological dose, given in rems, is a measure of the biological damage to living tissue from the radiation exposure.

QUALITY ASSURANCE/QUALITY CONTROL

Any discussion of the science and technology of hazardous materials involves analyses, so that the materials in question can be accurately identified and quantified. Obviously, there must be some mechanism or protocol to assure the validity of analyses. Such programs are called quality assurance and quality control (QA/QC). Chapter 12 will go into more detail on this subject matter.

Quality assurance is the testing and inspection of all or a portion of the final product to ensure that the desired quality level of product reaches the consumer. Quality control is the inspection, analysis, and action applied to a portion of the product or operation to estimate overall quality of the product and determine what, if any, changes must be made to achieve or maintain the required level of quality.

For the purposes of this discussion, the terms "product" or "operation" may be assumed to refer mainly to the analytical procedures used to identify and quantify hazardous materials. However, it is important to note that QA/QC procedures may be applied to any product or operation, ranging from the preparation of permit applications to conducting a test burn for a hazardous waste incinerator.

Spikes and dupes are often used to evaluate the quality of analytical procedures. A spike is a sample to which a known amount of a specific substance has been added. A dupe is one of two or more samples which contain the same materials in identical amounts. With spikes, the accuracy of the analytical results can be compared to the "right" answer. The precision of analysis can be evaluated by determining how closely the results for duplicated samples (dupes) match.

Analyses to evaluate laboratory procedures are often conducted "blind" by a facility with "third-party" status. This means that the individuals who conduct the analyses have no information about the content of the samples and, therefore, have no preconceived ideas about desired results.

In addition to analytical procedures, good quality assurance for environmental measurements includes assessments of the following:

1. Precision
2. Accuracy
3. Detection limits of analytical methods
4. Data sampling procedures
5. Sample handling and storage
6. Calibration
7. Documentation
8. Evaluation of appropriateness of measurement method

Unlike storage, the terms "treatment" and "disposal" refer almost exclusively to hazardous wastes (as opposed to hazardous materials). The long-term effects of these new technologies and some variations of conventional technologies have not yet been determined. Monitoring is an important aspect of all management processes to ensure that the desired result is achieved. (See Chapter 10 for a detailed account of various monitoring equipment.)

There are four major methods of treating hazardous wastes. These are

1. Physical treatment
2. Chemical treatment
3. Biological treatment
4. Incineration

REFERENCES

1. American Conference of Governmental Industrial Hygienists. "TLVs(R): Threshold Limit Values and Biological Exposure Indices for 1988–1989" (Cincinnati, OH: ACGIH, 1988), p. 4.

2. OSHA-3091. "Safety and Health Guide for the Chemical Industry," U.S. Department of Labor, U.S. Government Printing Office, Washington, DC (1986).
3. OSHA-3111. "Hazard Communication Guidelines for Compliance," U.S. Department of Labor, U.S. Government Printing Office, Washington, DC (1988).
4. OSHA-3084 (Revised). "Chemical Hazard Communication," U.S. Department of Labor, U.S. Government Printing Office, Washington, DC (1988).
5. Schumake, A. "Hazardous Materials Management," (Gainesville, FL: University of Florida, 1985), chap. 3.
6. Phoenix Safety Company, *Phoenix Hazardous Material Training Manual*, (Phoenixville, PA: PSC, 1986), chap. 2.
7. Olishifski, J. B. *Fundamentals of Industrial Hygiene*, (Chicago: National Safety Council, 1976), pp. 13–15.

CHAPTER 3

Toxicology

INTRODUCTION

The constant introduction to new industrial processes, operations, and techniques involving new chemicals makes a knowledge of toxicology important to the hazardous materials handler who is frequently exposed to hazardous materials in his/her daily work duties.

Toxicology is basically the study of poisons, in particular, industrial poisons, which are raw and used chemical materials. This chapter will discuss common terms used in this field of science and their association with recognized agencies that develop concentration level criteria.

Life of modern man has been greatly improved by the developments of chemicals such as pharmaceuticals, cosmetics, food additives, building materials, housewares, pesticides, industrial chemicals, etc. Through the use of chemicals, we can increase food production, cure diseases, build more efficient houses, brighten our offices and homes, make ourselves more attractive, and send men to the moon. As has been said, "without chemicals, life wouldn't be the same." Every corner of the world, every part of our life has benefited from chemicals. At the same time, we must be cautious to insure that our own existence is not endangered by the uncontrolled and our expanded use of chemicals, which the masses know so little about.

Following decades of reckless and unconscious handling of chemicals, which have resulted in several disastrous incidences of pollution (example: Bhopal, Wolburn, and Chernobyl), man has recognized the need for better control of the present use and the future development of chemicals. This need has turned into a demand from the public to have chemicals tested and retested before they reach the consumers. Industry in general and toxicologists in particular have been challenged with the job of setting safety levels for all chemicals — an almost

Table 3.1. Types of Toxicologist

Type	Focus of Work
Clinical toxicologist	Concerned with the effects of chemical (drug) poisoning and the treatment of poisoned people
Descriptive toxicologist	Concerned directly with the toxicity testing of chemicals
Environmental toxicologist	Concerned with the ultimate environmental fate of chemicals and their impacts upon the biological ecosystems and human population
Forensic toxicologist	Concerned with applying techniques of analytical chemistry to answer medical/legal questions about harmful effects of chemicals
Industrial toxicologist	Concerned with disorders produced in individuals who have been exposed to harmful materials during the course of their employment
Mechanistic or biochemical	Concerned with elucidating the biochemical mechanisms by which chemicals exert toxic effects
Regulatory toxicologist	Concerned with assessing descriptive data with regard to the risk involved in the marketing of chemicals and their legal uses

Reprinted from *Industrial Toxicology: Safety and Health Applications in the Workplace*, P. L. Williams and J. L. Burson, Eds., (New York: Von Nostrand, 1985), by permission.

impossible task often compounded by sensitive political issues for both government and industry. Before making a judgment, the toxicologist needs the knowledge of the chemical's primary and cumulative toxicity, its no observed effect level (NOEL), and its mutagenic, teratogenic, and carcinogenic potential. (Some of these data can be obtained from animal studies conducted on a short-, intermediate-, or long-term basis.)

WHAT IS TOXICOLOGY?

Toxicology is the study of chemical or physical agents that produce adverse responses in the biologic systems with which they interact. These responses span a broad physiological spectrum, ranging from something relatively minor, such as irritation or tearing, to a more serious response such as liver or kidney damage, or even permanent disabilities such as lung diseases, kidney failure, or cancer. Because these responses occur within such a broad range, toxicologists have defined a number of specialties (see Table 3.1). Every toxicologist, however, performs one or both of the two basic functions of toxicology, which are (1) to examine the nature of the adverse effects produced by chemicals or physical agents, and (2) to assess the probability of their occurrence.

While a great variety of toxicities can result from chemical exposure in the workplace, these toxicities can be classified into a few general categories. For

Table 3.2. Combined Tabulation of Toxicity Classes

Commonly Used Terms	LD Single Oral Dose for Rats (g/kg)	Vapor Exposure Causing 2 to 4 Deaths in 6-Rat Group (ppm)	LD Skin for Rabbits (g/kg)	Probable Lethal Dose for Man
Extremely toxic	0.001 or less	Less than 10	0.005 or less	Taste (1 grain)
Highly toxic	0.001 to 0.05	10 to 100	0.005 to 0.043	1 tsp (4 cc)
Moderately toxic	0.05 to 0.5	100 to 1,000	0.044 to 0.340	1 oz (30 g)
Slightly toxic	0.5 to 5.0	1,000 to 10,000	0.35 to 2.81	1 pint (250 g)
Practically nontoxic	5.0 to 15.0	10,000 to 100,000	2.82 to 22.6	1 qt
Relatively harmless	15.0	100,000	22.6	1 qt

Reprinted from *Handbook of Toxicology*, Vol. 1, W. S. Spector, Ed. (Philadelphia: W. B. Saunders Co., 1956), by permission.

example, it is usually useful to separate adverse responses by the temporal duration of their effects:

1. *Acute toxicity*: Generally has a sudden onset for a short period of time (that is, a reversible effect), approximately 24 hr (see Table 3.2).
2. *Chronic toxicity*: Marked by long or permanent duration, constant or continuous (that is, a permanent or irreversible effect). Approximately 48 to 72 hr plus. (Can be defined as *over* 24 hr, too.)

Likewise, toxicants may be distinguished by their general sites of action:

1. *Local toxicity*: Occurs at the site of application or exposure, between the toxicant and the biological system.
2. *Systemic toxicity*: Requires absorption of the toxicant within the body and then distribution of the toxicant via the bloodstream to susceptible organ(s), which are the sites of action.

Toxicities are also categorized by the response they produce or the organ they effect. Thus, acute or chronic problems might be traceable to hepatotoxins (liver poisons), nephrotoxins (kidney poisons), or neurotoxins (poisons to the nervous system), as well as to immunotoxins (poisons of the immune system), teratogens (poisons causing birth defects), allergen sensitizers (agents that stimulate the production of natural bodily substances that can lead to harmful, allergic-like reactions), or carcinogens (toxicants that cause malignant tumors or cancers).

Thus, to describe, evaluate, and predict the occurrence of an industrially related toxicity, one should understand the aspects of physiology, biology, biochemistry, pathology, mathematics, and other scientific disciplines. Sometimes just being a good *detective* will not hurt.

ROUTE AND SITE OF EXPOSURE

The major routes by which toxic agents gain access to the body are through the gastrointestinal tract (ingestion), the skin (topical), and through parenteral administration. Toxic agents generally exhibit the greatest potency and produce the most rapid response when given by the intravenous route, and an approximate descending order of effectiveness for other routes would be inhalation, intraperitoneal, subcutaneous, intramuscular, intradermal, oral, and topical. With oral exposure the onset of symptoms and the severity of effects are usually influenced by the stomach contents, since absorption tends to be more rapid and more complete from an empty stomach. The vehicle and other formulation factors can markedly affect and alter the absorption following ingestion, inhalation, or topical exposure, and this effect may also occur with parenteral administration. Similarly, the site of administration can influence the toxicity of agents given parenterally. For example, an agent that is detoxified in the liver would be expected to be less toxic when given via portal circulation than when given via systemic circulation.

Industrial exposure to toxic agents occurs most frequently by inhalation and topical exposure, and the accidental and suicidal poisoning occurs most frequently through oral ingestion. Comparison of the lethal dose of an agent by different routes of exposure often provides useful information concerning the absorption of the agent. For many agents the lethal dose for topical exposure is about ten times the lethal dose for oral administration, which in turn is about ten times the lethal dose with intravenous administration. In those situations where the lethal dose for oral or topical administration is close to the lethal dose for intravenous administration, it usually means that the toxic agent is absorbed rapidly and easily. Conversely, in those cases where the lethal dose by the dermal route is several orders of magnitude higher than the oral lethal dose, it is likely that the skin can be expected to provide an effective barrier to poisoning by the agent. Toxic effects by a route of exposure can also be influenced by the agent and vehicle to which the system is exposed and the rate at which exposure occurs. Blood level studies are often needed to clarify the role of these and other factors as well as exposure route differences. Today, even blood samples can show antibody sensitization to various chemicals, such as formaldehyde or total chemical low-level exposure such as in xenobiotics. One must remember that in industry, multiple routes of exposure can occur, too. Examples would be breathing and skin absorption exposure occurring simultaneously.

LENGTH AND FREQUENCY OF EXPOSURE

Toxic effects may be produced by acute and/or chronic exposure (short term) which are defined as a single exposure or multiple exposure occurring within a short time (24 hr or less). For many agents, the toxic effects of acute exposure are quite different from those produced by chronic exposure (wrist drop from exposure to lead or loss of tooth enamel from hydrochloric acid). Acute exposure to agents that are rapidly absorbed are likely to produce immediate toxic effects, but acute exposure can also produce delayed toxicity that may or may not be similar to the toxic effects of chronic exposure. This is in the application of xenobiotics. Chronic exposure to a toxic agent may produce some immediate or acute effects with each administration in addition to the long-term, low-level, or chronic effects of the agent. In characterization of the toxicity of a specific chemical agent, it is evident that information is needed not only for the singledose (acute) and long-term (chronic) effects, but also for exposures to intermediate duration. Conventionally, such exposures are referred to as short-term (a week or so) and subchronic (usually 3 months) in toxicity testing programs.

The other time-related factor that is important in the temporal characterization of exposure is the frequency of administration. In general, fractionation of the dose reduces the effect. A single dose of a test substance that produces an immediate severe effect might produce less than half of the effect when given in two divided doses and no effect when given in ten doses over a period of several hours or days. Such fraction effects occur when metabolism or excretion occurs between successive doses or when the injury produced by each administration is partially or fully reversed prior to the next administration. A diagrammatic representation of these concepts is presented in Figure 3.1. It is evident that with any type of multiple dose (multiposal) the production of toxic effect is not only influenced by the frequency of administration, but may, in fact, be totally dependent on frequency rather than duration of exposure. Chronic toxic effects occur, therefore, when the agent accumulates in the biologic system (absorption exceeds metabolism and/or excretion) or when an agent produces irreversible toxic effects or when there is insufficient time for the system to recover from the toxic effect within the exposure frequency interval. When the rate of elimination is less than the rate of absorption, the toxic agent usually does not accumulate indefinitely, but reaches a steady state where the rate of elimination equals the rate of administration (see Figure 3.1).

VARIABLES INFLUENCING ACUTE AND CHRONIC EFFECTS

The following variables should be considered when evaluating acute and chronic toxic effects.

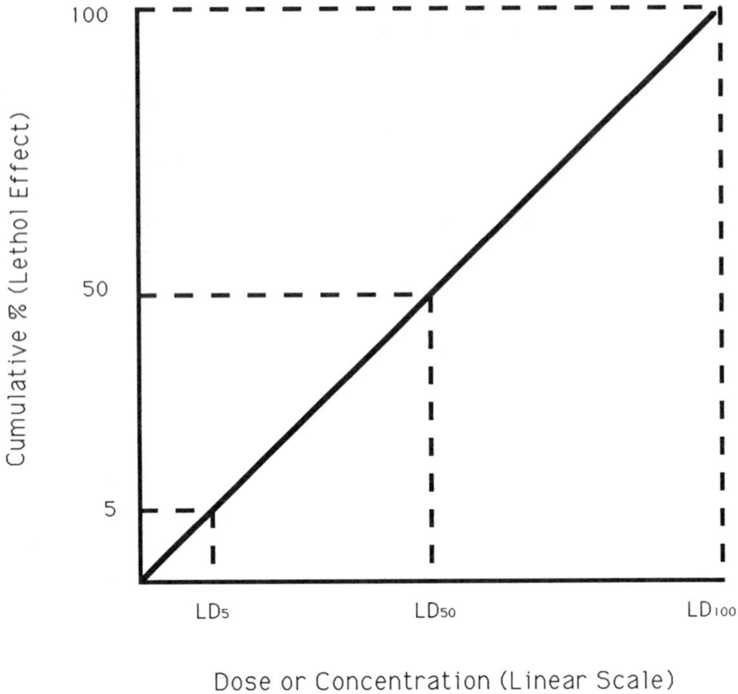

Figure 3.1. Dose/response curves for a chemical agent administered to a uniform population of test animals. The slope of the curve can be considered an index of the margin of safety of the material.

Route of Exposure

How a substance enters the body determines how much of it enters (rate of absorption) and which organs are exposed to the largest concentration of the substance. Both of these variables may in turn determine rates of metabolism and excretion. For example, the amount of chemical that is orally toxic may not be toxic, or may be more toxic, when applied to the skin. An example would be fluoride. In drinking water it has no toxic effects at 2 mg/L, but at 4 mg/L it destroys tooth enamel and at 10 mg/L causes osteitis-like symptoms.

Sex

Gender characteristics may affect the toxicity of some substances. Woman have a larger percent of fat in their total body weight than men; women also have different susceptibilities to reproduction system disorders or teratogenic effects. Some cancers and disease states are sex linked. Large sex-linked differences may also be present in animal data.

Age

Older people have differences in their musculature and metabolism that change the disposition of chemicals within the body and, therefore, the levels required to induce toxicity. At the other end of the spectrum, children have higher respiration rates and different organ susceptibilities (they are less sensitive to central nervous system [CNS] stimulants and more sensitive to CNS depressants), differences in the metabolism and elimination of chemicals, and many other biological characteristics that distinguish them from adults in the consideration of risk or chemical hazards.

EFFECTS OF CHEMICAL INTERACTION (SYNERGISM, POTENTIATION, AND ANTAGONISM)

Synergists are chemicals whose combined toxicity causes a greater additive effect. Potentiation occurs when a chemical that does not produce a specific toxicity nevertheless increases the toxicity caused by another chemical. Antagonists are chemicals that, when combined, diminish each other's measured effects.

MODES OF CHEMICAL INTERACTION

There are four ways by which chemical interactions can be increased or decreased.

1. *Functional*: Both chemicals affect the same physiologic function.
2. *Chemical*: The chemical interaction between the two compounds affects the toxicity of one of the chemicals.
3. *Dispositional*: The absorption, metabolism, distribution, or excretion of one of the chemicals is altered by the second chemical.
4. *Receptor-mediated*: When two chemicals bind to the same tissue receptor, the second chemical, which differs in activity, competes for the receptor and thereby alters the effect produced by the first chemical.

GENETIC MAKEUP

Unfortunately or fortunately we are not born physiologically equal; this provides both advantages and disadvantages. For example, people deficient in glucose-6-phosphatedehydrogenase (G6PD deficiency) are anemic. On the other hand, some individuals metabolize such drugs as isoniazid faster than other people and are therefore less susceptible to their serious side effects.

FACTORS INFLUENCING THE TOXICITY OF A CHEMICAL

Some additional factors influencing the toxicity of a chemical in question are listed below.

1. Physical and chemical properties of toxic substances may affect absorption or alter the probability of exposure. Chemical composition (salt, free base, anion, etc.), presence of impurities (may alter absorption or toxicity), and stability of product in the mixture are examples.
2. Exposure conditions may affect the amount of toxicant an individual is exposed to. Concentration, amount, type of exposure (skin, oral, inhalation, etc.), and duration (acute or chronic) are examples.
3. Heterogeneous makeup of persons exposed may alter the toxic response. Genetic status, immunologic status, nutritional status, hormonal status, age, sex, body type, health, and concurrent disease are examples.
4. Environmental conditions may heighten or lessen the effects of exposure or the toxicity. How the toxic substance is carried (air, water, soil, food), presence of additional chemicals (synergism, antagonism), temperature of chemicals (synergism, antagonism), temperature and air pressure (volatility, ventilation, etc.), safety protection equipment, and methods for handling toxic chemicals, medical facilities, and training are examples.

PHYSICAL CLASSIFICATION OF TOXIC AGENTS

This type of classification is an attempt to base the discussion of toxic agents on the form in which they are present in the air.

Gases and Vapors

In common industrial hygiene usage, the term "gas" is usually applied to a substance that is in the gaseous phase of a material, which is ordinarily a solid or a liquid at room temperature and pressure. When considering this fact with toxicity of a gas or vapor, the solubility of the material is irritant gas. Its solubility in aqueous media will determine the amount of material that reaches the lung and, hence, its site of action. A highly soluble gas, such as ammonia, is taken up readily by the mucous membranes of the nose and upper respiratory tract. Sensory response to irritation in these areas provides the individual with warning of the presence of an irritant gas. On the other hand, a relatively insoluble gas such as nitrogen dioxide is not scrubbed out by the upper respiratory tract, but penetrates readily to the lung. Amounts sufficient to lead to pulmonary edema and death may be inhaled by an individual who is not at the time aware of the hazard. The solubility coefficient of a gas or vapor in blood is one of the factors determining rate of uptake and saturation of the body. With a very soluble gas, saturation of the body is slow, is largely dependent upon ventilation of the lungs, and has only slightly influenced changes in circulation. In the case of slightly soluble gas, saturation is rapid, depends chiefly on the rate of circulation, and is little influenced by the rate of breathing. If the vapor has a high fat solubility, it tends to accumulate in the fatty tissues it reaches carried in the blood. Since fatty tissue often has a meager blood supply, complete saturation of the fatty tissue may take a longer period. It is also of importance whether the vapor or gas is one which is readily metabolized. Conversion of metabolite would tend to lower the concentration in the blood and shift the equilibrium towards increased uptake. It is also of importance whether such metabolic products are toxic.

Aerosols

An aerosol is composed of solid or liquid particles of microscopic size dispersed in a gaseous medium (for our purposes, air). Special terms are used for indicating certain types of particles. Some of these are "dust," a dispersion of solid particles usually resulting from the fracture of larger masses of material such as in drilling, crushing, or grinding operations; "mist," a dispersion of liquid particles, many of which are visible; "fog," visible aerosols of a liquid formed by condensation of vaporized materials: "smoke," aerosols resulting from incomplete combustion that consist mainly of carbon and other combustible materials. The toxic response to an aerosol depends obviously on the nature of the material, which may have a target organ the respiratory system or may be a systemic toxic agent acting elsewhere in the body. In either case the toxic potential of a given material dispersed as an aerosol is only partially described by a statement of the concentration of the material dispersed as an aerosol is only partially described by a statement of the concentration of the material in terms of weight per unit volume or number of particles per unit volume. For a proper assessment of the toxic hazard, it is necessary to have information also on the particle size distribution of the material. Understanding of this fact has led to the development of instruments that sample only particles in the respirable size range. The particle size of an aerosol is the key factor in determining its site of deposition in the respiratory tract and, as a sequel to this, the clearance mechanisms which apply to the removal of any aerosol from the atmosphere, as well as upon anatomical and physiological factors such as the geometry of the lungs and the respiratory cycle. The interrelationship of these factors has been examined both theoretically and experimentally. The monograph by Hatch and Gross, "Pulmonary Deposition and Retention of Inhaled Aerosols," gives an excellent discussion of the subject and should be required reading for anyone entering the field of environmental toxicology. The most recent theoretic treatment is that of the Task Force on Lung Dynamics, which also gives an excellent summary of past work.

Physiological and Chemical Classification

Toxic compounds may be classified according to their chemical nature and physical makeup.

Such classification attempts to frame the discussion of toxic materials according to their biological action.

Irritants

The basis of classifying these materials is their ability to cause inflammation of mucos membranes with which they come in contact. While many irritants are strong acids or alkalis familiar as corrosive to nonliving things such as lab coats or bench tops, bear in mind that inflammation is the reaction of a living tissue and is distinct from chemical corrosion. The inflammation of tissue results from concentrations far below those needed to produce corrosion. As was indicated earlier in discussing gases and vapors, solubility is an important factor in determining the site of irritant action in the respiratory tract. Highly soluble materials such as ammonia, alkaline dusts and mists, hydrogen chloride, and hydrogen fluoride affect mainly the upper respiratory tract. Other materials of intermediate solubility such as the halogens, ozone, diethyl or dimethyl sulfate, and phosphorus chlorides affect both the upper respiratory tract and the pulmonary tissue. Insoluble materials such as nitrogen dioxide, arsenic trichloride, or phosgene affect primarily the lung. There are exceptions to the statement that solubility serves to indicate site of action. One such is ethyl ether and other insoluble compounds that are readily absorbed unaltered from the alveoli and hence do not accumulate in that area. In the upper respiratory passages and bronchi where the material is not readily absorbed, it can accumulate in concentrations sufficient to produce irritation. Another exception is in materials such as bromobenzyl cyanide which is a vapor from a liquid boiling well above room temperature. It is taken up by the eyes and skin as a mist. In initial action, then, it is a powerful lachrymator and upper respiratory irritant, rather than producing a primarily alveolar reaction, as would be predicted from its low solubility.

Irritants can also cause changes in the mechanics of respiration such as increased pulmonary flow resistance or decreased compliance (a measure of elastic behavior of the lungs). One group of irritants, among which are sulfur dioxide, acetic acid, formaldehyde, formic acid, sulfuric acid, acrolein, and iodine, produce a pattern in which the flow resistance is increased, the compliance is decreased only slightly, and at higher concentrations the frequency of breathing is decreased. Another group, among which are ozone and oxides of nitrogen, have little effect on resistance, produce a decrease in compliance and an increase in respiratory rate. There is evidence that in the case of irritant aerosol, the irritant potency of a given material tends to increase with decreasing particle size, as assessed by the increase in flow resistance. Following respiratory mechanics measurements in cats exposed to irritant aerosols, the histologic sections prepared after rapid freezing of the lungs showed the anatomical sites of constriction. Long-term chronic lung impairment may be caused by irritants either as sequelae to a single, very severe exposure or as the result of chronic exposure to low concentrations of the irritant. There is evidence in experimental animals that long-term exposure to respiratory irritants can lead to increased mucus secretion and a condition resembling the pathology of human chronic bronchitis without the intermediary of infection. The epidemiological assessment of this factor in the health of residents of polluted urban atmospheres is currently a vital area of research.

Irritants are usually further subdivided into primary and secondary irritants. A primary irritant is a material which for all practical purposes exerts no systemic toxic action either because the products formed on the tissues of the respiratory tract are nontoxic or because the irritant action is far in excess of any systemic toxic action. Examples of the first type would be hydrochloric acid or sulfuric acid. Examples of the second type would be materials such as Lewisite or mustard gas, which would be quite toxic on absorption, but death from the irritation would result before sufficient amounts to produce systemic poisoning would be absorbed. Secondary irritants are materials that do produce irritant action on mucous membranes, but this effect is overshadowed by systemic effects resulting from absorption. Examples of materials in this category are hydrogen sulfide and many of the aromatic hydrocarbons and other organic compounds. Direct contact of liquid aromatic hydrocarbons with the lung can cause chemical pneumonitis with pulmonary edema, hemorrhage, and tissue necrosis. It is for this reason that in the case of accidental ingestion of these materials the induction of vomiting is contraindicated because of possible aspiration of the hydrocarbon into the lungs.

Asphyxiates

The basis of classifying these materials is their ability to deprive the tissue of oxygen. In the case of severe pulmonary edema caused by an irritant such as nitrogen dioxide or laryngeal spasm caused by a sudden severe exposure to sulfuric acid mist, the death is from asphyxia, but this results from the primary irritant action. The materials we classify here as asphyxiates do not damage the lung. Simple asphyxiates are physiologically inert gases that act when they are present in the atmosphere in sufficient quantity to preclude an adequate oxygen supply. Among these are such substances as nitrogen, nitrous oxide, carbon dioxide, hydrogen, helium, and aliphatic hydrocarbons such as methane and ethane. All of these gases are not chemically unreactive, and among them are many materials which pose a major hazard of fire and explosion. Chemical asphyxiates are materials that have as their specific toxic action the ability to render the body incapable of utilizing an adequate oxygen supply. They are thus active in concentrations far below the level needed for damage. Simple asphyxiates are carbon monoxide and cyanides. Carbon monoxide interferes with the transport of oxygen to the tissues by its affinity for hemoglobin. The carboxylhemoglobin thus formed is unavailable for transport of oxygen. Over the above familiar lethal effects, there is concern about how low level exposures will affect performance of such tasks as automobile driving, etc. In the case of cyanide, there is no interference with the transport of oxygen to the tissues. Cyanide transported to the tissues forms a stable complex with the ferric iron of ferric cytochrome oxidase, resulting in inhibition of enzyme action. Since aerobic metabolism is dependent upon this enzyme system, the tissues are unable to utilize the supply of oxygen, and tissue "hypoxia" results. Therapy is directed toward the formation of an inactive complex before the cyanide has a chance to react with the cytochrome. Cyanide will complex with methemoglobin. Thiosulfate is also given, as this provides the sulfate needed to promote the enzymatic conversion of cyanide to the less toxic thiocyanate.

Primary Anesthetics

The main toxic action of these materials is their depressant effect upon the central nervous system, particularly the brain. The degree of anesthetic effect depends upon the effective concentration in the brain as well as upon the specific pharmacologic action. Thus, the effectiveness is a balance between solubility (which decreases) and pharmacological potency (which increases) as one moves up a homologous series of compounds of increasing chain length. The anesthetic potency of the simple alcohols also rises with increasing number of carbon atoms through amyl alcohol, which is the most powerful of the series. The presence of multiple hydroxyl groups diminishes potency. The presence of carboxyl groups tends to prevent anesthetic activity, which is slightly restored in the case of an ester. Thus, acetic acid is not anesthetic, ethyl acetate is mildly so. The substitution of a halogen for a hydrogen of the fatty hydrocarbons greatly increases the anesthetic action, but confers toxicity to other organ systems, which outweighs the anesthetic action.

Hepatotoxic Agents

These are materials which have as their main toxic action the production of liver damage. Carbon tetrachloride produces severe diffuse central necrosis of the liver. Tetrachloroethane is probably the most toxic of the chlorinated hydrocarbons and produces acute yellow atrophy of the liver. Nitrosamines are capable of producing severe liver damage. There are many compounds of plant origin such as some of the toxic components of the mushroom *Amanita phalloides*, alkaloids from Senecio, and aflatoxins that are capable of producing severe liver damage and in some instances are powerful hepatocarcinogens.

Nephrotoxic Agents

These are materials which have as their main toxic action the production of kidney damage. Some of the halogenated hydrocarbons produce damage to the kidney as well as to the liver. Uranium produces kidney damage, mostly limited to the distal third of the proximal convoluted tubule.

Neurotoxic Agents

These are materials which in one way or another exert their main toxic symptoms on the nervous system. Among them are metals such as manganese, mercury, and thallium. The CNS seems particularly sensitive to organometallic compounds, and neurological damage results from such materials as methylmercury and tetraethyl lead. Trialkyl tin compounds may cause edema of the CNS. Carbon disulfide acts mainly on the nervous system. The organic phosphorus insecticides lead to an accumulation of acetylcholine because of the inhibition of the enzyme that would normally remove it and, hence, cause their main symptoms in the nervous system. NIOSH recently has cited over 93 organic solvents that cause these effects.

Agents that act on the blood of hematopoietic system include nitrites, aniline, and toluidine, which convert hemoglobin to methemoglobin. Nitrobenzene forms methemoglobin and also lowers the blood pressure. Arsine produces hemolysis of the red blood cells. Benzene damages the hematopoietic cells of the bone marrow. This damage can lead to diseases such as leukopenia and thrombocytopenia and eventually leukemia.

Agents that Damage the Lung

In this category are materials that produce damage to the pulmonary tissue, but not only immediate irritant action. Fibrotic changes are produced by materials such as free silica, which produces the typical silicotic nodules. Asbestos also produces a typical damage to lung tissue, and there is newly aroused interest in this subject from the point of view of possible effects of low-level exposure of individuals who are not asbestos workers. Asbestosis has been the subject of many conferences and governmental regulations. Other dusts, such as coal dust, can produce pneumoconiosis, which, with or without tuberculosis superimposed, has been of long concern in mining. Drinker and Hatch is a classic reference in this area, and Hunter discusses at length occupational exposures to dusts. Many dusts of organic origin, such as those arising in the processing of cotton or wood, can cause pathology of the lungs and/or alterations in lung function. The proteolytic enzymes added to laundry products are an occupational hazard of current interest. Toluenediisocyanate (TDI) is another material that can cause impaired lung function at very low concentrations, and there is evidence of chronic as well as acute effects.

OVERVIEW OF TERMS USED IN TOXICOLOGY

The toxicity of a material is not synonymous with its health hazard. Toxicity is the capacity of a material to produce injury or harm. Hazard is the possibility that a material will cause injury when a specific quantity is used under specific conditions. The key elements to be considered in evaluating a health hazard are these:

1. The toxicity of the materials used.
2. The physical properties of these materials.
3. The absorption probabilities of these materials by individuals.

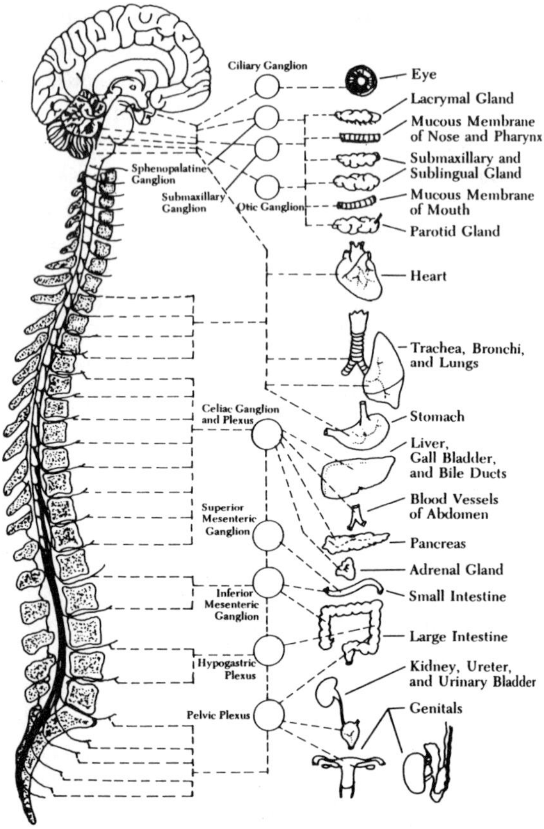

Figure 3.2. Body organs affected by exposure to common hazardous materials.

4. The extent and intensity of exposure to these materials.
5. The control measures used.

The degree to which a substance will affect living cells can be measured only after recognizable changes have occurred following absorption. Some changes — impaired judgment, delayed reaction time — may be produced at levels too low to cause actual cell damage. Toxicity is dependent upon the dose, rate, method, and site of absorption, plus many other factors including general state of health, individual differences, tolerance, diet, and temperature (see Figure 3.2).

ENTRY ROUTES FOR TOXIC MATERIALS

There are only three ways by which chemical agents can harm individuals:

1. By contact with or absorption through the respiratory tract.
2. By contact with or absorption through the skin.

3. By contact with or absorption through the digestive tract.

In general, systemic industrial poisonings usually result from

1. Inhalation of toxic agents. Local injury to the skin can be caused by direct contact with an irritant or corrosive material.
2. Ingestion of the toxic agent can occur to some extent; however, there would generally be considerable inhalation of the material where such conditions exist.
3. Absorption through the skin can occur upon exposure to some toxic agents. Some liquids and vapors are known to pass through the skin to such a degree that respiratory protection is not adequate against high concentrations. For example, hydrogen cyanide (HCN) is known to pass through the unbroken skin. Consideration should be given to the type of work clothes being worn; if they become saturated with solvents, they will act as a reservoir to bathe the body continually with the harmful material.
4. The absorption of harmful materials by the lungs is dependent upon the solubility in body fluids, the permeability of the lungs, the volume of inhalation, the volume of blood in the lungs, and the concentration gradient of vapors between the inhaled air and blood.

The respiratory system is composed of two main areas: (1) the upper respiratory tract airways — the nose, throat, trachea, and bronchial tubes, leading to the various lobes of the lungs; and (2) the alveoli, where the actual transfer of gases across the cell wall takes place. Chest X rays often detect silicosis before it becomes disabling. A reduction in the cholinesterase activity of the blood is a reliable indication of organic phosphate poisoning. Periodic medical examination in conjunction with analyses of atmospheric contaminants may be necessary to completely define the worker's exposure in relationship to this environment.

EXPERIMENTAL TOXICOLOGY

There are three axioms that form the basis for experimental toxicology. Although they appear self-evident, they are not susceptible to rigid proof.

1. Any substance contacting or entering the body will be injurious at some degree of exposure and will be tolerated without effect at some lower exposure.
2. The nature of the injuries which may develop in men can be determined by the study of the reactions of experimental animals.
3. From animal experiments, the degree of exposure that will not affect humans can be determined.

Experimental toxicology has done well for human safety by allowing regulation of exposure in accord with the results of animal experiments.

The first attempts in estimating the toxicity of a substance are usually made on the basis of animal experiments. Data from these experiments are expressed as lethal doses (LD) in milligrams of substance per kilogram of body weight of the test animal. The most commonly used expressions are these:

1. MLD — minimum lethal dose, the smallest dose which kills one of a group of test animals.
2. LD_{50} — lethal dose for 50%, the dose which kills one-half of a group of test animals (usually 10 or more).
3. LD_{100} — lethal dose for 100%, the dose which kills all of a group of test animals (usually ten or more).

These doses may also be expressed as lethal concentrations (LC) for airborne toxic substances.

ACCEPTABLE CONCENTRATIONS

The American National Standards Institute (ANSI) publishes the Z37 series of Acceptable Concentrations (ACs) for a number of substances.

The ANSI Z37 Committee has developed a number of innovations in concept. In the past, a standard was conceived as a single limit, threshold, or a boundary between what is and what is not healthful. However, there is no such single boundary; the pattern or exposure in space and time may be variable. Therefore, multiple boundaries are needed. As the boundaries are not sharp lines between what is and what is not healthful, an explanation of the basis and significance of each of these boundaries is needed. The different ACs for the same material follow:

1. An acceptable ceiling concentration for protection of health, assuming an 8-hr work day. (This is equivalent to the old "maximum allowable concentration.")
2. An acceptable time-weighted average for protection of health, assuming an 8-hr work day. (This is within the limits of the previous specified ceiling and defined peaks beyond this ceiling.)
3. An acceptable maximum for peaks above the acceptable ceiling for an 8-hr work day.

THE MEANING OF "ACCEPTABLE CONCENTRATIONS"

It is generally recognized that there are concentrations of air contaminants to which a person may be exposed without discomfort or ill effects. This standard defines such concentrations in light of the best present-day knowledge.

The immediate use of the ANSI Standard for ACs is for guidance in establishing engineering procedures to prevent objectionable concentrations of materials in the workplaces. These concentrations in themselves do not represent a scale of relative toxicity; they represent concentrations below which ill effects are unlikely under the specified conditions.

The term, acceptable concentrations, is not defined as a single concept, but is related to the duration and pattern of the exposure as set forth in the ANSI Standard for the individual substance.

In the application of an AC established as an ANSI Standard, it is important to understand the criteria upon which the standard was established. These criteria are avoidance of

1. Undesirable changes in body structures or biochemistry.
2. Undesirable functional reaction that may have no discernible effects on health.
3. Irritation or other adverse sensory effects.

One purpose for establishing AC is to provide a basis for interpretation of the results of air analysis as an indication of the severity of potential exposure, but not as a means of establishing the presence of occupational disease. Comparison of air analysis results with ACs indicates either acceptable conditions or otherwise the need, extent, and urgency of control measures. Sampling and analysis should be performed using a standard method, if available, and one that will provide a reliable indication of potential exposure.

It should be kept in mind that

1. The ACs are the highest concentration which can currently be justified consistent with the objective of maintaining unimpaired health or comfort of workers or both. It is good practice to maintain lower concentrations when this does not render industrial operations inefficient.
2. ACs are not precise values sharply dividing what is hazardous from what is safe under the particular circumstances and, therefore, are not appropriate for use as legal requirements.
3. These standards are considered to be guides for good industrial hygiene practices and should be applied and interpreted by persons with a full understanding of the basis and limitations of the information from which the standard has been developed.
4. The levels are applicable only to a single substance and to the exposure pattern as detailed in the standard.
5. Exposure to mixtures is the rule. When more than one of the components is of hygiene significance, ACs of the components should be based on the given mixture and situation. Look for an additive effect.
6. Routes of absorption other than inhalation, such as through the skin, should be considered in the evaluation of the exposure.
7. These standards do not apply to community atmospheric pollution as related to general populations.

Many individuals refer to a published TLV or AC as a “go” or “no go” value — something to be used in a specific situation — and overlook the important qualifications of these values set forth by the organizations establishing them. The

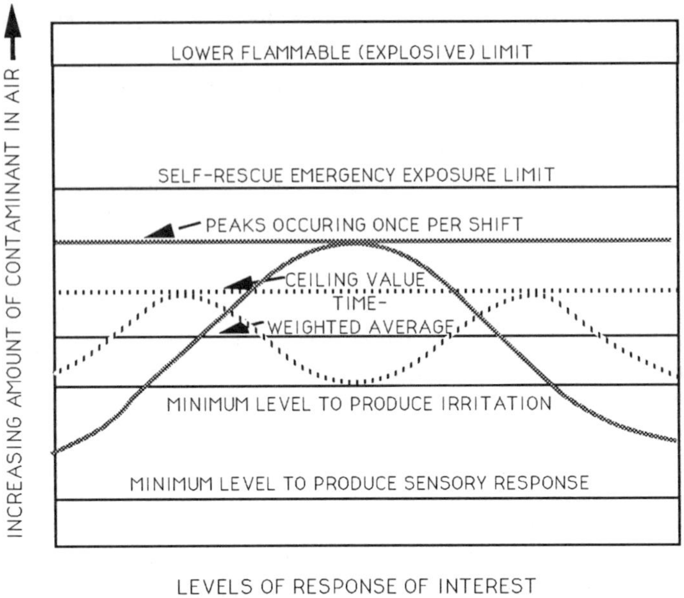

Figure 3.3. Knowlege fo the type of injury that would result from exposure to these various contaminant levels of interest to the safety or encironmental professional.

preface to the annual list of TLVs and the introduction to each of the ANSI Z37 series of ACs provide a basic explanation of the concepts on which these values are based. To use the values intelligently, their definitions should be understood before the values are applied to a specific situation. Knowledge of the type of injury that would result from dangerous overexposure is also needed (Figure 3.3).

Toxic effects may be broken down into four major classifications:

1. Acute toxicity
2. Chronic toxicity
3. Narcosis
4. Discomfort

The time-weighted concentration would be the appropriate guide where chronic toxic effects result from overexposure. When protection against narcosis is required, the concentration should never exceed some ceiling value. To minimize discomfort, the instantaneous excursions or peak concentrations are of interest (see Table 3.3).

Both the ANSI and American Conference of Governmental Industrial Hygienists (ACGIH), in defining their respective values, point out that an understanding of the criteria on which the values are established is necessary in deciding how to use the values.

Table 3.3. Odor Thresholds and Irritation Concentrations of Chemicals

Chemical Compound	Odor Low mg/m[1]	Odor High mg/m[1]	Description of Odor	Irritating Conc. mg/m[1]
Acenaphthene	0.5048	0.5048		
Acetaldehyde	0.0002	4.1400	Green, sweet, fruity	90.00
Acetic acid	2.5000	250.0000	Sour, vinegar-like	25.00
Acetic anhydride	0.5600	1.4400	Sharp odor, sour acid	20.00
Acetone	47.4666	1613.8600	Minty chemical, sweet	474.67
Acetonitrile	70.0000	70.0000	Ether-like	875.00
Acetophenone	0.8347	2.9460	Sweet, almond	
Acetyl acetone	0.0409	0.0409		
Acetylene	657.2000	657.2000		
Acrolein	0.0525	37.5000	Burnt, sweet	1.25
Acrylic acid	0.2820	3.1200	Rancid, sweet	
Acrylonitrile	8.1000	78.7500	Onion-garlic pungency	
Aldrin	0.2536	0.4027		
Allyl alcohol	1.9500	5.0000	Pungent, mustard	12.50
Allyl alcohol (N-)	150.0000	150.0000		
Allyl amine	14.5080	14.5080		187.20
Allyl chloride	1.4100	75.0000	Green, garlic, oniony	75.00
Allyl disulfide	0.0005	0.0005		38.06
Allyl glycidyl ether	44.0000	44.0000	Sweet	1144.00
Allyl isocyanide	0.0610	5.4240	Sweet, repulsive	17.02
Allyl isothiocyanate	0.0325	1.7052	Mustard oil	17.05
Allyl mercaptan	0.0002	0.0515	Garlic-like	454.50
Allyl sulfide	0.0007	0.0007		6500.64
Ammonia	0.0266	39.6000	Pungent, irritating	72.00
Amyl acetate (N-)	0.0265	37.1000	Fruity, banana, pear	530.00
Amyl acetate (sec-)	0.0107	0.0107		
Amyl alcohol (ios-)	25.2000	25.2000		
Amyl alcohol (N-)	0.4332	72.2000	Sweet	
Amyl alcohol (iert-)	0.8303	0.8303		
Amyl amine (N-)	56.6040	132.0760		

Table 3.3. (continued)

Chemical Compound	Odor Low mg/m^1	Odor High mg/m^1	Description of Odor	Irritating Conc. mg/m^1
Amyl mercaptan	0.0001	0.0018		
Amyl mercaptan (iso-)	0.0018	0.0018		
Aniline	0.0002	350.0000	Pungent, amine-like	
Anisole	0.2210	0.2210		
Apiole	0.0570	0.0570		
Arsine	0.8400	2.0000	Garlic-like	
Azetidine	51.9200	169.9200		
Azine	31.6480	66.2400		
Azole	134.2600	295.9200		
Benzaldehyde	0.0008	0.1823	Pleasant, bitter	20.01
Benzene	4.5000	270.0000	Sweet, solventy	9000.00
Benzene hexachloride	0.0015	142.8000		
Benzenethiol	0.0012	279.0000		
Benzothiazole	0.4424	2.2120		
Benzyl chloride	0.2350	1.5500	Solventy	41.00
Benzyl mercaptan	0.0132	0.2028	Unpleasant	22.81
Benzyl sulfide	0.0184	0.0184	Sulfidy	
Biphenyl	0.0062	0.3000		
Boron trifluoride	4.5000	4.5000	Pungent, irritating	
Bromine	0.3290	24.5000	Bleachy, penetrating	2.10
Bromoacetophenone	0.1221	1.3838	Unpleasant	0.33
Bromochloromethane	1680.0000	1680.0000		
Bromoform	5300.0000	5300.0000	Similar to chloroform	
Butadiene (1,3-)	0.3520	2.8600		
Butadiene dioxide	17.6000	17.6000		35.20
Butane	2.8500	14.6300		
Butenethiol (2-)	0.0001	0.0001		
Butyl acetate (iso-)	0.0090	90.0000	Pleasant, fruity	1350.00
Butyl acetate (N-)	33.1333	94.6666	Fruity	473.33

Butyl acrylate (iso-)	0.0110	0.0660	Sweet, musty	
Butyl alcohol (ios-)	0.3600	2225.0000	Mild, non-residual	300.00
Butyl alcohol (N-)	0.3600	150.0000	Sweet	75.00
Butyl alcohol (secondary-)	131.1500	131.1500	Strong, pleasant	
Butyl alcohol (tertiary-)	219.0000	219.0000	Camphor-like	
Butyl amine (N-)	0.2400	6.0000	Ammoniacal	30.00
Butyl cellosolve	0.4800	288.0000	Sweet, ester	
Butyl cellosolve acetate	0.7194	1.3080	Sweet, ester	
Butyl chloride (N-)	3.3352	6.3293	Pungent	
Butyl ether (N-)	0.3731	2.5051	Fruity, sweet	
Butyl formate	70.8900	83.4000		
Butyl furan (2-)	50.8000	50.8000		
Butyl lactate (N-)	35.0000	35.0000		
Butyl mercaptan	0.0016	0.0033	Stinks!	
Butyl sulfide	0.0897	0.0897		
Butyl toluene (P-, tertiary-)	30.0000	30.0000	Gasoline-like	48.00
Butylamine	3.0000	378.0000	Ammonia, fishy	30.00
Butylene	54.9600	54.9600	Gassy	
Butylene oxide	0.2058	2.0874	Sweet, alcohol	
Butylthiazole (2, iso-)	0.0202	0.0202		
Butyraidehyde	0.0136	26.5500	Sweet, rancid	
Butyric acid	0.0010	9.0000	Sour, perspiration	
Butyric acid (ios-)	29.1600	29.1600		
Camphor (synthetic)	7.8000	1200.0000		10.62
Caprolactam	28.0000	28.0000		
Capryl alcohol			Sweet, pungent	
Carbitol	1.1508	6.0280	Sweet, musty	
Carbitol acetate	0.1872	1.8936	Sweet	
Carbon disulfide	0.0243	23.1000	Disagreeable, sweet	
Carbon tetrachloride from CS	60.0000	128.4000	Sweet, pungent	
Carbon tetrachloride from CH_4	300.0000	1500.0000	Sweet, pungent	
Caryophyliene	0.5350	0.5350		
Cellosolve	2.0350	185.0000		
Cellosolve acetate	0.3024	270.0000	Sweet, musty	
Cellosolve solvent	1.1040	2.0240		

Table 3.3. (continued)

Chemical Compound	Odor Low mg/m[1]	Odor High mg/m[1]	Description of Odor	Irritating Conc. mg/m[1]
Chloral	0.2825	0.2825	Sweet	
Chlordane	0.0084	0.0419		
Chlorinated camphene toxaphene	2.3689	2.3689		
Chlorine	0.0300	15.0000	Bleachy, pungent	9.00
Chlorine dioxide	0.3000	0.3000	Sharp, pungent	15.00
Choloracetaldehyde	3.0000	3.0000	Sharp, irritating	3.00
Chloracetophenone (alpha-)	0.1020	0.1500	Sharp, irritating	0.05
Chlorobenzene	0.9800	280.0000	Sweet, almond-like	933.33
Chlorobenzylidene malonitrile (ortho-)			Peppery	1.52
Chlorobromomethane	2100.0000	2100.0000	Sweet	
Chloroform	250.0000	1000.0000		20480.00
Chlorophenol	0.0189	6.5224	Medicinal, empyrumatic	6801.18
Chloropicrin	5.4600	7.7000	Sharp, penetrating	2.10
Chlorotoluene	0.2350	0.2350	Pungent, irritating	
Citral	0.3738	0.3738		
Coumarin	0.0020	0.0120	Pleasant, vanilla	
Cresol	0.0012	22.0000	Sweet, creosote, tar	
Crotonaldehyde	0.1050	3.0000	Pungent, suffocating	23.01
Cumene	0.0392	6.3700	Sharp, aromatic	
Cyanogen	500.0000	500.0000	Pungent	32.00
Cyanogen chloride	2.0000	2.0000		
Cyclobutylamine	97.1750	340.8600		
Cycloheptylamine	309.2600	573.4000		
Cyclohexane	1.4350	1.4350	Sweet, aromatic	1050.00
Cyclohexanol	400.0000	400.0000	Camphor-like	200.00
Cyclohexanone	0.4800	400.0000	Sweet, pepperminty	100.00
Cyclohexylamine	106.0000	448.0000		
Cyclopentadiene	5.0667	5.0667		

Cyclopentyl acetate	0.1031	0.1031		
Cyclopentylamine	676.4000	2278.4000		
Cyclopropylamine	153.5170	153.5170		
DDT	5.0725	5.0725		
Decaborane	0.3600	0.3600		
Decalin	565.0000	565.0000		565.00
Decanoic acid	11.9510	112.4800		
Decanol	0.0006	43.2820		
Diacetone alcohol	1.3440	480.0000	Sweet	240.00
Diacetyl	0.0035	0.0880		
Diallyl sulphide	0.0005	0.1491	Garlicky	
Dibenzofuran	0.7752	1.6150		
Diborane	2.0000	4.0000	Repulsively sweet	
Dibromo-3-chloropropane (1,2-)	0.0965	0.2895		1.93
Dibutylamine	0.4224	1.4256		
Dibutylamine (N-)	0.4232	2.5392	Fishy, amine	
Dichloroacetic acid	1.2144	1.2144		
Dichloranisole (2,6-)	0.0003	0.0003		
Dichlorobenzene (ortho-)	12.0000	300.0000		150.00
Dichlorobenzene (para-)	90.0000	180.0000	Mothballs	240.00
Dichloroethane	445.5000	810.0000	Chloroform-like	
Dichloroethyl ether	90.0000	2160.0000		600.00
Dichloroethylene (1,2-)	0.3358	1975.0000	Acid, ethereal	
Dichlorophenol (2,4-)	1.4007	1.4007		
Dicyclo pentadiene	0.0297	0.0540	Sharp, sweet	2.70
Diethyl disulfide	0.0915	0.0195		
Diethyl ethanolamine	0.0536	0.1948	Amine	
Diethyl ketone	3.1725	49.3500		
Diethyl pyrazine (2,5-)	0.0336	0.1120		
Diethyl selenide	0.0617	0.0617	Putrid	
Diethyl sulphide	0.0177	0.0117	Foul, garlicky	
Diethyl trisulfide	0.0044	0.0044		
Diethylamine	0.0600	114.0000	Fishy, ammoniacal	150.00
Diglycidyl ether	25.0000	25.0000		50.00
Diisobutyl carbinol	0.1885	0.9424	Sweet, alcohol	

Table 3.3. (continued)

Chemical Compound	Odor Low mg/m^1	Odor High mg/m^1	Description of Odor	Irritating Conc. mg/m^1
Diisobutyl ketone	0.6600	1.8600	Sweet, ester	150.00
Diisopropyl amine	0.5200	3.4000	Fishy, amine	100.00
Dimethoxy dimethyl pyrazine	1.2366	1.2366		
Dimethyl acetamide	161.0000	163.8000		
Dimethyl amine	0.0846	0.0846		174.60
Dimethyl disulfide	0.0001	0.3465		
Dimethyl ethanolamine	0.0546	0.1638	Amine	
Dimethyl formamide (N,N-)	300.0000	300.0000		
Dimethyl naphthalene	0.0428	0.0428		
Dimethyl sulfide	0.0025	0.0508	Decayed cabbage	
Dimethyl trisulfide	0.0062	0.0062		
Dimethyl trithiocarbonate	0.0331	0.0331		
Dimethylacetamide (N,N-)	163.8000	163.8000	Amine, burnt, oily	
Dimethylamine	0.0378	55.8000	Fishy, ammoniacal	
Dimethylformamide (N,N-)	300.0000	300.0000	Fishy, pungent	
Dimethylhydrazine (1,1-)	12.0000	20.0000	Ammoniacal, amine-like	
Dioxane (1,4-)	0.0108	612.0000	Ether-like	792.00
Dioxane (para-)	20.1600	972.0000		720.00
Dioxaine (1,3-)	44.5400	335.3600	Sweet, musty	
Dipentene			Lemon-like	
Diphenyl ether (perfume)	0.6950	0.6950	Pleasant, geraniums	
Diphenyl sulfide	0.0026	0.0358	Burnt, rubbery	
Dipropylamine	0.0826	227.1500	Ammoniacal, amine	
Dipropylamine (N-)	0.4140	0.8280	Ammoniacal, amine	
Dipropylene glycol methyl ether	210.0000	6000.0000	Ether-like	450.00
Dodecanol (1-)	0.0152	0.0533		
Dodecyl Mercaptan (N-)	4222.8000	4222.8000		
Dowtherm A	0.7000	7.0000	Aromatic, disagreeable	21.00
Endrin	0.2808	6.3963		

Epichlorohydrin	50.0000	80.0000	Chloroform-like	325.00
Ethane	184.5000	1105.7700		
Ethanolamine	5.3333	10.6666	Ammonia	13.33
Ethoxy 3,4 dihydro 1,2 pyran (2-)	3.1440	3.1440	Sweet, fruity	
Ethoxy 3,4-dihydro 1,2 pyran	0.1048	3.1440	Sweet, fruity	
Ethyl acetate	0.0196	665.0000	Fruity, pleasant	350.00
Ethyl acrylate	0.0008	32.0000	Earthy, acrid, plastic	16.00
Ethyl alcohol (synthetic)	0.3420	9690.0000	Sweet, alcoholic	9500.00
Ethyl amine	0.4860	396.0000	Sharp, ammoniacal	180.00
Ethyl amyl ketone	31.2000	31.2000	Mild, fruity	260.00
Ethyl benzene	8.7000	870.0000	Aromatic	870.00
Ethyl benzoate	3.8068	3.8068		
Ethyl bromide	890.0000	890.0000	Ether-like	28925.00
Ethyl butanol (2-)	0.2919	3.2109	Musty, sweet	
Ethyl ether	0.9900	3.0000	Sweet-ether-like	300.00
Ethyl formate			Fruity	990.00
Ethyl hexanol (2-)	0.3990	0.7342	Musty	
Ethyl hexyl acetate	0.5132	1.4763	Sweet	
Ethyl hexyl acrylate (2-)	0.5497	1.3554	Sharp, musty	
Ethyl isothiocyanate	6.0520	6.0520	Mustard, unpleasant	65.15
Ethyl lactate	67.6200	67.6200		
Ethyl mercaptan	$3.2 = 10^3$	0.0920	Garlic	
Ethyl methyl disulfide	0.0487	0.0487		
Ethyl methylamine	21.6900	79.5300		
Ethyl morpholine (N-)	0.3680	1.1500	Ammonia	184.00
Ethyl phenylacetate	4.3615	4.3615		
Ethyl propyl amine	60.5200	181.5600		
Ethyl selenide	0.0003	0.0030		
Ethyl selenomercaptan	$8.0 = 10^6$	0.0054	Foul	
Ethyl silicate	722.5000	722.5000	Alcohol-like, sharp	5950.00
Ethyl sulfide	0.0009	0.0103		
Ethyl valerate	0.0266	0.0266		
Ethyl vinyl ketone	0.0004	0.0069		
Ethylene	299.0000	4600.0000	Olefinic	
Ethylene diamine	2.5000	28.0000	Ammoniacal, musty	250.00

Table 3.3. (continued)

Chemical Compound	Odor Low mg/m[1]	Odor High mg/m[1]	Description of Odor	Irritating Conc. mg/m[1]
Ethylene dibromide	76.8000	76.8000	Mild, sweet	
Ethylene dichloride	24.0000	440.0000	Sweet	
Ethylene glycol	62.5000	62.5000	Sweet	
Ethylene glycol dinitrate				0.18
Ethylene oxide	520.0000	1400.0000	Sweet, olefinic	
Ethylenediamine	25.0000	28.0000	Musty, ammoniacal	500.00
Ethyleneimine	4.0000	4.0000	Ammoniacal	200.00
Ethylidene norborene	0.0700	0.3650	Sweet, aromatic	30.00
Fluorine	6.0000	6.0000		50.00
Formaldehyde	1.4700	73.5000	Pungent hay	1.50
Formamide	150.0000	150.0000		
Formic acid	0.0450	37.8000	Pungent, penetrating	27.00
Furfural	0.0240	20.0000	Almonds	48.00
Furfuryl alcohol	32.0000	32.0000		
Glycol diacetate	0.5552	1.8626	Fruity, acid	
Heptachlor	0.3060	0.3060		
Heptanal	0.0140	0.0932		
Heptane	200.0000	1280.0000	Gasoline-like	
Heptyl alcohol (N-)	98.3250	98.3250		
Heptyl isobutyrate	0.0989	0.0989		
Heptyl propionate	0.0281	0.0281		
Hexachlorbutadiene	12.0000	12.0000		
Hexachlorocyclopentadiene	1.5000	3.3000		
Hexadiene	1.6750	127.3000		
Hexane (N-)				1800.00
Hexanol (1-)	0.0417	21.6840	Sweet, alcohol	
Hexylacetate (secondary-)	0.0120	600.0000	Unpleasant	600.00
Hexyl isobutyrate	0.0422	0.0422		
Hexyl propionate	0.0514	0.0514		

Hexylene glycol	250.0000	250.0000		250.00
Hydrazine	3.0000	4.0000	Ammoniacal, fishy	
Hydrochloric acid	7.0000	49.0000	Irritating, pungent	49.00
Hydrofluoric acid	0.0333	0.1333	Strong, irritating	4.17
Hydrogen bromide	6.6667	6.6667	Sharp, irritating	10.00
Hydrogen cyanide	0.9000	5.0000	Bitter almond	
Hydrogen peroxide			Slightly sharp	150.00
Hydrogen selenide	0.0016	12.0000	Decayed horseradish	6.00
Hydrogen, sulfide	0.0007	0.0140	Rotten eggs	14.00
Iodine	9.0000	9.0000		2.00
Iodoform	0.0062	0.0833		
Ionone	$4.63 = 10^{7}$	573.0500		
Isoamyl alcohol	36.0000	126.0000		360.00
Isobutyl 2-methoxypyrazine	13.5800	13.5800		
Isobutyl 3-methoxypyrazine	$1.3 = 10^{3}$			
Isobutyl 3-methyl pyrazine	0.2146	0.2146		
Isobutyl butyrate	12.3690	17.6700		
Isobutyl mercaptan	0.0020	0.0020		
Isobutyl pyrazine (2-)	2.2080	2.2080		
Isodecanol	0.1292	0.2713	Musty, alcohol	
Isopentanoic acid	0.0209	0.1084	Goaty	
Isophorone	1.0000	50.0000	Sharp, objectionable	50.00
Isopropyl acetate	0.1900	1520.0000	Fruity	380.00
Isopropyl alcohol	7.8400	490.0000	Pleasant	490.00
Isopropyl amine	0.5040	480.0000	Pungent, ammonia	24.00
Isopropyl ether	0.0714	1260.0000	Sweet, sharp, ether	1260.00
Isopropyl glycidyl ether	1440.0000	1440.0000		
Kerosene	0.5517	0.5517		122.60
Ketene			Sharp	41.40
Lauraldehyde	0.0151	0.0151		
Linalyl acetate	50.5260	50.5260		
Malathion	13.5000	13.5000		
Maleic anhydride	1.8400	1.9600	Acid	5.48
Mercaptobenzothiazole	12.0208	12.0208		
Mercaptoethanol	0.3828	2.0416		

Table 3.3. (continued)

Chemical Compound	Odor Low mg/m^1	Odor High mg/m^1	Description of Odor	Irritating Conc. mg/m^1
Mesityl oxide	0.0680	100.0000	Sweet	100.00
Methyl 2, cyanoacrylate	4.0000	12.0000		12.00
Methyl acetate	610.0000	915.0000	Fragrant, fruity	30496.90
Methyl acetylene-propadiene MAPP gas	180.0000	180.0000	Foul, objectionable	
Methyl acrylate	70.0000	70.0000	Sharp, sweet, fruity	262.50
Methyl acrylonitrile (alpha-)	6.0000	42.000		6.00
Methyl alcohol	13.1150	26840.0000	Sweet	22875.00
Methyl amine	0.0252	12.0000		24.00
Methyl amyl acetate	0.4123	2.3560	Sweet, ester	
Methyl amyl alcohol	1.3761	2.1684	Sweet, alcohol	
Methyl anthranilate	0.0581	0.0581		
Methyl benzyl alcohol	7235.5000	5235.5000		
Methyl bromide	80.0000	4000.0000	Sweetish	
Methyl butanol (2-)	0.0450	0.8280	Sour, sharp	
Methyl butanoic acid (2-)	0.0528	0.0528	Body odor	
Methyl butyl acetate	0.0266	0.0266		
Methyl cellosolve	0.2880	288.0000	Mild, nonresidual	368.00
Methyl cellosolve acetate	1.6320	240.0000	Sweet, ester	
Methyl chloride	21.0000	21.0000	Sweet, ethereal	1050.00
Methyl chloroform	542.8570	3800.0000	Chloroform-like	5428.57
Methyl cyclohexane	2000.0000	2000.0000	Faint, benzene-like	
Methyl cyclohexanol	2350.0000	2350.0000	Weak, coconut oil	2350.00
Methyl disulfide	0.0012	0.0039		
Methyl ethanolamine	3.0700	10.4380	Musty, ammoniacal	
Methyl ethyl ketone	0.7375	147.5000	Sweet, acetone-like	590.00
Methyl ethyl pyridine	0.0297	94.0500	Sour, pungent	
Methyl formate	500.0000	6875.0000	Pleasant	8750.00
Methyl furan (2-)	90.4500	90.4500		

Methyl glycol	186.6000	279.9000		
Methyl heptanoate	0.0236	0.0236		
Methyl hexyl ketone	1299.5200	1299.5200		
Methyl hydrazine	1.7500	5.2500		
Methyl iodide				21500.00
Methyl isoamyl alcohol	0.2919	0.8340	Pungent	
Methyl isoamyl ketone	0.0576	0.3360		
Methyl isobutyl carbinol	2.0800	200.0000	Sweat, mild odor	100.00
Methyl isobutyl ketone	0.4100	192.7000	Sweet, sharp	410.00
Methyl isocyanate				5.00
Methyl isopropenyl ketone	1.0222	1.0222		
Methyl mercaptan	$4.0 = 10^3$	0.0820	Sulfidy	
Methyl methacrylate	0.2050	1.3940	Arid, fruity, sulfidy	697.00
Methyl n-amyl carbinol	0.0989	0.3378	Sweet, alcohol	
Methyl n-amyl ketone	0.0940	0.0940		
Methyl n-propyl ketone	28.0000	45.5000		
Methyl naphthalene (2-)	0.0581	0.2905		
Methyl parathion	0.1328	0.1328		
Methyl pentaldehyde (2-)	0.3681	0.5562	Sweet, rancid	
Methyl pentane (2-)	0.2886	0.2886		
Methyl propanol (2-)	0.0027	0.1303		
Methyl propene (2-)	45.8000	45.8000	Gassy	
Methyl pyrazine (2-)	231.0000	231.0000		
Methyl salicylate	0.6220	0.8708		
Methyl styrene (alpha-)	0.2496	960.0000	Sweet, aromatic	960.00
Methyl thiocyanate	0.7475	0.7475	Sweet, unpleasant	480.19
Methyl vinyl ketone	0.5720	0.5720		
Methylamine	0.0252	12.0000	Fishy, pungent	30.00
Methylene chloride	540.0000	2160.0000	Sweet	8280.00
Methylene chlorobromide	2120.0000	2120.0000		
Methylene glycol	76.2000	76.2000		
Mineral spirits	157.5000	787.5000		
Monochlorobenzene	0.9800	0.9800	Chlorinated, mothball	
Morpholine	0.0350	0.4900	Fishy, amine	
Musk oil	$3.8 = 10^4$	0.0487		

Table 3.3. (continued)

Chemical Compound	Odor Low mg/m^1	Odor High mg/m^1	Description of Odor	Irritating Conc. mg/m^1
Mustard gas	0.0150	0.0150		
Myrcene	0.0723	0.0723		
Naphthalene	1.5000	125.0000	Mothball, tar-like	75.00
Nickel carbonyl	0.2100	21.0000	Musty	
Nitric acid	0.7500	2.5000		155.00
Nitric oxide	0.3600	1.2000		
Nitrobenzene	0.0235	9.5000	Shoe polish, pungent	230.00
Nitroethane	620.0000	620.0000	Mild, fruity	310.00
Nitrogen dioxide	2.0000	10.0000	Sweetish, acrid	20.00
Nitromethane	250.0000	250.0000	Mild, fruity	500.00
Nitropropane (1-)	1080.0000	1080.0000	Mild, fruity	360.00
Nitropropane (2-)	17.5000	1029.0000		
Nonane	3412.5000	3412.5000		
Nonanol (2-)	0.0005	20.6150		
Octadiene (1,3-)	54.0000	90.0000		
Octane	725.0000	1208.3300	Gasoline-like	1450.00
Octyl alcohol	0.6916	0.6916		
Oxygen difluoride	0.2000	1.0000	Foul	
Ozone	0.0010	1.0200	Pleasant, clover-like	2.00
Parathion	0.4760	0.4760	Garlic-like	
Pentaborane	2.5000	2.5000	Strong, pungent	
Pentachlorophenol			Pungent when hot	10.90
Pentane	6.6000	3000.0000	Gasoline-like	
Pentanedione (2,4-)	0.0409	0.0982	Sour, rancid	
Pentanol (n-)	0.7560	1.1160	Sweet, alcohol	
Perchloroethylene	31.3560	469.0000	Chlorinated solvent	1340.00
Perchloromethyl mercaptan	0.0075	0.0075		
Perchloryl fluoride	46.6666	46.6666	Sweet	
Phenol	0.1786	22.4200	Medicinal, sweet	182.40

Phenyl ether	0.0070	0.7000	Disagreeable	21.00
Phenyl ethyl alcohol (beta-)	35.0000	35.0000		
Phenyl sulfide	0.0026	0.0358		
Phenylacetaldehyde	0.0010	0.0196		
Phosgene	2.0000	4.0000	Musty hay, green corn	8.00
Phosphine	0.0280	3.6000	Oniony, mustard, fish	10.67
Phthalic Anhydride				30.00
Picoline (2-)	0.0532	0.1748	Sweet	
Picric acid	0.0005	0.0005		
Propane	1800.0000	36000.0000		
Propionaldehyde	0.0225	0.4029	Sweet, ester	
Propionic acid	0.0840	60.0000	Sour	
Propyl acetate (n-)	0.2100	105.0000	Sweet, ester	
Propyl alcohol	75.0000	500.0000		13750.00
Propyl alcohol (n-)	0.0750	150.0000	Sweet, alcohol	
Propyl mercaptan	0.0002	0.0746		
Propyl nitrate (n-)	210.0000	210.0000	Ether-like	
Propyl sulfide	0.0531	0.0531		2801.40
Propylene	39.5600	116.2720	Aromatic	
Propylene diamine	0.0424	0.2030	Sharp, amine	
Propylene dichloride	1.1667	606.6660	Sweet	
Propylene glycol dinitrate	1.2000	1.5600		
Propylene glycol isobutyl ether	60.5000	60.5000		121.00
Propylene glycol methylether	360.0000	360.0000		3600.00
Propylene oxide	24.7500	500.0000	Sweet, alcoholic	1125.00
Pyridine	0.0090	15.0000	Burnt, sickening	90.00
Pyrrolidine	58.0000	187.3400		
Quinone	0.4000	0.4000	Acrid	2.00
Rotenone	5.7960	5.7960		
Safrole	1.4586	1.4586		
Silicon tetrafluoride	4.2500	4.2500		42.50
Skatole	$4.0 = 10^2$	0.2680	Perfume	
Stoddard solvent	5.2500	157.5000	Kerosene-like	2100.00
Styrene (inhibited)	0.4300	860.0000	Solventy, rubbery	4300.00
Styrene (uninhibited)	0.2021	860.0000	Solventy, rubbery	430.00

Table 3.3. (continued)

Chemical Compound	Odor Low mg/m[1]	Odor High mg/m[1]	Description of Odor	Irritating Conc. mg/m[1]
Styrene oxide	0.3093	1.9640	Sweet	
Sulfur dichloride	0.0042	0.0042	Sulfidy	
Sulfur dioxide	1.1750	12.5000	*Yech!*	5.00
Sulfur monochloride			Nauseating	12.00
Sulfuric acid	1.0000	1.0000		1.10
Tetrachloroethane (1,1,2,2-)	21.0000	35.0000	Sickly sweet	1302.00
Tetrachloroethylene	31.3560	469.0000	Chlorinated solvent	710.20
Tetraethyl orthosilicate	30.6360	61.2720	Sweet, alcohol	
Tetrahydrofuran	7.3750	177.0000	Ether-like	
Tetrain	97.2000	97.2000		
Tetramethylenediamine	79.2000	79.2000		
Thiophene	0.0026	0.0026	Aromatic	
Thiophenol mercaptan	0.0012	382.5000		
Toluene petroleum	8.0250	150.0000	Rubbery, mothballs	750.00
Toluene from coke	17.5500	262.5000	Floral, pungent	750.00
Toluene 2,4 diisocyanate TDI	3.2000	17.1200	Sweet, fruity, acrid	4.00
Toxaphene	2.3660	2.3660		
Trichloro fluoromethane Freon 11	28.0000	1170.4000	Sweet	
Trichloro trifluoroethane Freon 113	342.0000	1026.0000	Sweet	
Trichlorobenzene (1,2,4-)	24.0000	24.0000		40.00
Trichloroethylene TCE	1.1340	2160.0000	Solventy	864.00
Trichloropropane (1,2,3-)			Strong, acid	300.00
Tricycloketone	1.8660	870.8000		
Triethyl amine	0.3600	1.1200	Fishy, amine	200.00
Trimethyl amine	0.0008	0.0008	Fishy, pungent	
Trimethyl phosphite	0.0005	0.0005	Pyridine-like	
Trimethylenediamine	3757.2000	11968.5000		
Trinitro tert-butylxylane	$3.8 = 10^6$	0.0487		

musk oil				
Turpentine	560.0000	1120.0000		560.00
Valeric acid	0.0026	0.0026		
Vanillin	$2.0 = 10^7$		Perfume	
Vinyl acetate	0.3600	1.6500	Sour, sharp	
Vinyl amyl ketone	0.5150	0.5150		
Vinyl butyl ketone	0.0321	0.0321		
Vinyl propyl ketone	0.0201	0.0201		
Vinyl pyridine	1.1670	1.9450	Nauseating	
Vinyl toluene	240.0000	240.0000	Disagreeable	240.00
Vinylidene chloride	2000.0000	4000.0000	Sweet, chloroformish	
VM&P naphtha	3.8700	3.8700		
Xylene	0.3480	174.0000	Sweet	435.00
Xylidene	0.0240	0.0240	Weak, amine-like	

It should be remembered that TLVs and ACs are, even though applied as intended, relative values. The concentration that just approaches the threshold limit is as hazardous or safe as the one that slightly exceeds it, and conversely. There is no fine line that can be drawn to provide complete safety for all individuals.

In general, it is desirable to limit air contamination to the lowest possible level — consistent with efficient plant operations.

Probably the greatest misuse of both the TLVs and the ACs is the practice of comparing toxic substances by ratio of their respective AC or TLV. It should also be remembered that neither the TLV nor AC should be considered the sole criterion of suspected occupational disease. ANSI and ACGIH specifically state that their values are intended to serve as guidelines for good engineering practice and are not intended to be legal requirements.

"HYGIENIC GUIDE" SERIES

The Hygienic Guides Committee of the American Industrial Hygiene Association (AIHA) has prepared guides on many commonly used industrial substances, with references for additional information. Each guide discusses briefly the physical properties, toxicity, hazards, and the engineering and medical controls necessary to control exposures. The Hygienic Guides contain recommended maximal atmospheric concentrations, short-exposure tolerance levels, and atmospheric concentrations immediately hazardous to life. The Hygienic Guides are published periodically in the *American Hygiene Association Journal* and are available from the AIHA.

EMERGENCY EXPOSURE LIMITS

The toxicology committee of the AIHA has published a series of emergency exposure limits (EEL), defining single brief accidental exposures to airborne contaminants that can be tolerated without permanent toxic effects. These limits are not intended to replace accepted safe practices and should be accompanied by appropriate medical surveillance.

EELs can be used in disaster planning to predict what type of emergency action to take in the event of an accidental spill, maximum credible accident, or other misoperations. The EELs can also be used to assist in the selection of the proper type of PPE. The AIHA is presently in the process of developing Emergency Response Preparedness Guidelines (ERPGs). These guidelines would state three levels of toxic effects. ERPG-1, which would stand for the maximum concentration to have no harmful effect. ERPG-2 would be the maximum concentration to have no effect when wearing PPE. ERPG-3 would be the maximum concentration that would cause severe harmful effects to a person. ERPGs are part of the chemical manufacturer's responsibility and are to be calculated by them under SARA Title III.

AIR CONTAMINANTS — PERMISSIBLE EXPOSURE LIMITS (TITLE 29 CFR PART 1910.1000)

Introduction

Soon after adoption of the OSH Act of 1970, OSHA promulgated PELs for many substances pursuant to the authority granted by Section (6a) of the Act, which allowed the agency to promulgate existing federal standards or consensus standards as enforceable OSHA standards. Accordingly, the PELs identified in the Z-Tables of Title 29 CFR 1910.1000 were adopted from regulations issued under the Walsh-Healey Public Contracts Act as existing federal standards. These, in turn, were adopted from the 1968 TLVs of the ACGIH. Some consensus standards from the American Standards Association (now the American National Standards Institute [ANSI]) also were adopted at that time, following the Section (6a) procedures.

Industrial experience, new developments in technology, and available scientific data clearly indicate that in many instances those adopted limits are now obsolete and inadequate. Also, many new toxic materials commonly used in the workplace are not covered. These inadequacies are evidenced by the lower allowable exposure limits recommended by many technical, professional, industrial, and government organizations in the U.S. and elsewhere. In addition, these organizations have identified many other substances for which allowable exposure limits are needed to supplement the existing Z-Tables. Many large industrial organizations feel obligated to supplement the existing OSHA PELs with their own internal corporate guidelines.

OSHA, in the past, has focused its priorities on the development of detailed regulations for some high-priority substances. This has resulted in substance-specific regulations for 24 substances. OSHA, however, has been unable to regulate the thousands of substances commonly found in the workplace or to review the scientific information to determine whether different limits are required for the more than 400 substances regulated prior to 1988 under the provisions of the Z-Tables.

OSHA, therefore, determined it was necessary to augment the substance-by-substance approach through the use of a generic rulemaking, which would simultaneously cover many substances. The Hazard Communication Standard is an example of a regulation using such an approach. At this time, OSHA is also in the process of developing generic standards to cover respiratory protection, medical surveillance, and exposure monitoring. A major part of this generic approach to rulemaking is to modify the existing PELs and to establish PELs for substances for which no exposure limits previously existed. The existing health literature and expert judgment indicate that such actions are required to protect against kidney and liver diseases, respiratory diseases, nerve disorders, carcinogenicity, irritation to various bodily organs, and many other deleterious health effects. Millions of employees are potentially exposed to substances of concern, and adoption of such a regulation represents a significant step in assuring the adequacy of health protection for workers.

Details of the Revised Standard

OSHA therefore amended its existing Air Contaminants standard, 29 CFR 1910.1000, including Tables Z-1, Z-2, and Z-3. This amendment is limited to making more protective 212 PELs listed in these three tables, setting new PELs for 164 substances not previously regulated by OSHA, and retaining certain PELs unchanged. These changes include the revision of the PELs, the inclusion of short term exposure limits (STELs) to complement the 8-hr time-weighted average (TWA) limits, the establishment of a designation for "skin" to protect against exposure by skin contact, and the ceiling limits as appropriate.

The final regulation permits the use of any compliance methodology, until December 31, 1992, to achieve each revised PEL. During this period, however, the established OSHA hierarchy of controls with preference for engineering controls will be applied to achieve the level of the existing PELs. A single new Table, Z-1-A, includes all of the revised PELs and the existing OSHA PELs. Tables Z-2 and Z-3 are temporarily maintained since they contain limits that cannot be conveniently included in the format used in Table Z-1-A. The original Table Z-1 has been deleted from the regulation because all of the substances in that table have been included in the new Table Z-1-A. The design of this new Table Z-1-A makes identification of all changes to PELs possible by simply comparing the Transitional Limits at the left side of the table with the Revised Limits shown at the right side of the table. Table Z-1-A is included in this booklet following the section on regulatory text. The final regulation and this booklet also include Tables Z-2 and Z-3 because these limits could not be included in the Transitional Limits columns of Table Z-1-A due to format differences.

Risk Reduction and Feasibility

OSHA reviewed health, risk, and feasibility evidence for all 428 substances for which changes to the PEL were considered. In each instance where a revised or new PEL was adopted, OSHA determined that the new limits substantially reduce significant risk of material health impairment among workers, and that new limits are technologically and economically feasible. This determination is based on a complete review of the material referenced in the proposed rule issued on June 7, 1988, and a detailed review of the entire record for the rulemaking, including all comments, evidence, testimony, and briefs submitted by the public during and after public hearings held from July 28 to August 16, 1988, in Washington, DC.

For a small number of substances used in specific operations, the full record indicates that it is presently technically infeasible to achieve, through engineering controls, the PELs necessary based on available health information. For these specific instances, engineering controls to achieve the new PELs are required only where the Assistant Secretary demonstrates that such controls are feasible. These substances and operations are carbon monoxide and sulfur dioxide in specific foundry operations, carbon disulfide in selected rayon fiber and sausage casing manufacturing processes, and styrene in selected open molding boat manufacturing processes. Any methods of control may be used (in these identified situations) to achieve the new PELs.

Protection

The revised PELs will protect workers against a wide variety of health effects that could cause material impairment of health or functional capacity. This includes protection against catastrophic effects such as cancer, cardiovascular, liver, and kidney damage, lung diseases, as well as more subtle effects resulting in CNS damage, narcosis, respiratory effects, and sensory irritation. For each substance, the health evidence in the record provides a substantial basis for establishing a new or revised PEL. OSHA utilized the already published and widely accepted Recommended Exposure Limits (RELs) developed by the NIOSH and the TLVs published by the ACGIH. OSHA, however, reviewed the entire record and made an independent judgment to determine that the revised PEL was feasible and that significant risk would be reduced.

The revised standards will provide additional occupational health protection to approximately 4.5 million workers at an annual cost of approximately $150 per employee protected. This cost is only a fraction of 1% of sales for all affected sectors.

Other Standards

OSHA will continue its practice of rulemaking for individual substances when substance-specific regulations are necessary and appropriate. OSHA also will continue its efforts to develop generic regulations to cover ancillary provisions required to supplement the PELs in 29 CFR 1910.1000.

Coverage

OSHA also has determined that it is appropriate to limit this rulemaking to the general industry sector. This segment involves the largest number of workers and will result in the greatest benefit. Application to the construction, maritime, and agriculture segments will require examination of feasibility for these sections and may require some modifications to this rule because of differences among these sectors. OSHA will pursue coverage for these sectors as part of second stage rulemaking and has informally notified the Advisory Committee on Construction Safety and Health of its plans.

The final regulation covers a total of 600 substances, including the following:

1. Addition of PELs for 164 new substances.
2. Adoption of more protective PELs for 212 substances.
3. No changes for 52 substances that were considered in this rulemaking.
4. No change to existing PELs for 9 substances that are currently undergoing Section (6b) rulemaking.

5. No change to existing PELs for 3 substances where some industry segments are not covered by the individual (6b) regulation.
6. No changes for PELs for 160 substances, that re unchanged, and were not evaluated during this rulemaking.

Compliance Dates

September 1, 1989 (6 months after the effective date of March 1), is the date for compliance with new or revised PELs, using any control methods. Approximately 4 years (December 31, 1992) following the regulation publication date, employers must comply with all PELs, with a preference for engineering controls. In the event that the planned OSHA regulation regarding version to methods of compliance is not completed by December 31, 1991, the December 31, 1992 date will be delayed 1 year until December 31, 1993.

RADIOACTIVE MATERIAL

Guides are published by the National Council on Radiation Protection and Measurements (NCRP), the International Commission on Radiological Protection (ICRP), and the Federal Radiation Council. These radiation protection standards for air, water, and body deposition of radionuclides and for exposure to external radiation are set at a level such that occupational exposure at the maximum permissible values recommended is not expected to entail appreciable risk to the individual.

WATER QUALITY

Standards are established for bacteriological quality, physical characteristics, chemical substances, and radioactivity by the U.S. Public Health Service pertaining to water supply systems used by carriers and others subject to federal quarantine regulations. These standards are accepted as good practice in evaluating the quality and safety of water supplies in industry.

It is highly advisable that the documented justification for the various limits or values be consulted for detailed information concerning mode of action and effect of exposure to a particular toxic material.

Thus far, the fundamentals of industrial toxicology, toxicity, routes of entry, modes of action, classification of harmful effects, intensity of action, contributions from experimental toxicology, and hygiene air quality standards have been reviewed.

The hazardous material operator may well ask how he or she can apply this knowledge to his/her work. A hazardous chemicals information sheet similar to the sample shown in Chapter 2, Table 2.1, could be prepared to elicit toxicological information from the supplier. The information in this questionnaire would be useful to the medical, purchasing, managerial, engineering, and safety departments in setting guidelines for safe use of these materials in industry. This information would be very helpful in the event of an emergency.

REFERENCES

1. Acceptable Concentrations, Z-37 Series (New York: American National Standards Institute).
2. Albert, A. *Xenobiosis: Food, Drugs & Poisons in the Human Body* (New York: Chapman & Hall Ltd., 1987).
3. API Toxicological Reviews (New York: American Petroleum Institute).
4. Broughton, A., and Thrasher, J. D. "Biological Monitoring for Immune Damage Due to Xenobiotics," Santa Ana, CA, 1989.
5. Brown, H. V. "Behavioral Implications of Exposure to Industrial Chemicals," *J. Occup. Med.* p. 561 (November 1966).
6. Browning, E. *Toxicity of Industrial Metals* (London: Butterworths, 1961).
7. *Toxicity and Metabolism of Industrial Solvents.* (New York: Elsevier).
8. Cook, W. A. "Maximum Allowable Concentration of Industrial Atmospheric Contaminants," *Ind. Med.* p. 933 (1945).
9. *Documentation of Threshold Limit Values,* (Cincinnati, OH: American Conference of Governmental Industrial Hygienists).
10. Doull, J., Klaasom, C. D., and Andur, M. O. Ed. *Casarett & Doull's Toxicology: The Basic Science of Poisons,* (New York: Macmillan Publishing Co., Inc., 1980).
11. Dubois, K. P., and Geiling, E. M. K. *Textbook of Toxicology* (New York: Oxford University Press, 1959).
12. Elkins, H. B. *The Chemistry of Industrial Toxicology* (New York: John Wiley & Sons,1959).
13. Hamilton, A., and Hardy, H. L. *Industrial Toxicology*, 4th ed. (Boston: John Wright-PSG Inc., 1983).
14. Hartman, D. E. *Neuropsychological Toxicology: Identification & Assessment of Human Neurotoxic Syndromes*, (New York: Pergamon Press, 1988).
15. Hatch & Gross. "Pulmonary Deposition and Retention of Inhaled Aerosols."
16. Olishifski, P.E., J.B., *Fundamentals of Industrial Hygiene*, (Chicago: National Safety Council, 1976), pp. 407–435..
17. Pendergrass, J. A. (Asst. Sec.) "Air Contaminants — Permissible Exposure Limits (Title 29 Code of Federal Regulations Part 1910.1000)," U.S. Department of Labor, OSHA 3112, U.S. Government Printing Office, Washington, DC (1989).
18. VonStille, T. Health Accounting and Assessment, Casselberry, FL (1989). (New software developed for environmental human epidemiology study.)
19. Williams, P. L., and Burson, J. L. Eds., *Industrial Toxicology: Safety and Health Applications in the Workplace*, (New York: VanNostrand Reinhold, 1985).

CHAPTER 4

Medical Program and Record Keeping

INTRODUCTION

The goal of this information will be to determine if individuals are unusually susceptible to illness or injury from possible exposures in the work environment and to establish the present medical status of the individual. It may be used for future epidemiological studies, to adjudicate claims and determine benefits, report medical conditions required by law to federal, state, and local agencies, and for other lawful purposes, including litigation.

MEDICAL PROGRAM

An effective medical control program will help prevent cases of occupational disease. Such a program can also serve as a check on the engineering controls because symptoms of exposure in a group of workers will indicate a failure that must be corrected. The extent of the medical program will depend upon the hazards and seriousness of the exposures. An industrial hygiene, toxicology, epidemiology, and environmental program should parallel the medical program. Both are essential to protect the health of employees.

The physical examination for new employees should include a thorough preemployment history with the occupational background given in detail. Chest X-rays should be made of all new employees who will be working in dust exposures that could produce disabling pneumoconiosis. The examining physician should decide on what physical capabilities are needed to perform the required work and to screen out those individuals who cannot meet these qualifications (Table 4.1).

Table 4.1. Occupational Medical Questionnaire

(for Medical Use Only)
Confidential

Name: ______________________________ Age: ___________ Date:

Last: ______________________________ First ______________ Middle:

Job Title or Category: ___________________________ Years at Present Job:

Sex: Male ( ) Female ( )

Have you ever had a MAJOR operation(s)? No ( ) Yes ( )

If yes, give details and dates:

____

____

Have you ever been a resident for more than 2 months outside the United States: No ( ) Yes ( )

If yes, please list location(s) and date(s):

____

List any occupational illnesses or injuries you have experienced on a job, past or present.

____

In you work are you now or have you ever been exposed to any significant levels of the following:

	Present	Past
Chlorinated solvents	___	___
Heavy metals (e.g., lead)	___	___
Aromatic hydrocarbons (e.g., benzene)	___	___
Oxides (e.g., nitrogen oxide)	___	___
Silica	___	___
Acids	___	___
Asbestos	___	___
Pesticides	___	___
Radiation	___	___
Noise	___	___
Other, Specify	___	___

Indicate any symptoms that you have experienced that might be due to exposure at work and indicate the suspected cause.

List any drugs you take regularly:

____

Are you allergic to any drugs? Which ones:

____

Table 4.1. (continued)

Have you been on any special diet(s) in the past year? No ( ) Yes ( )

If yes, describe type: ____

____

Do you clean your own home? No ( ) Yes ( )

What types of chemicals do you use to clean the house?______________________

____

Do you have your home fumigated: No ( ) Yes ( )

If yes, who sprays it and what do they spray for? (List the chemicals they use.) _________

____

Do you have city water or your own well for drinking? ______________________

Have you had any SEVERE or REOCCURRING conditions listed?

( ) Allergies
( ) Back pain
( ) Bowel problems
( ) Cancer
( ) Chest pain
( ) Chronic cough
( ) Cold or painful fingers
( ) Dental or gum problems
( ) Depression
( ) Diabetes
( ) Diarrhea
( ) Difficulty in sleeping
( ) Ear or hearing problem
( ) Edema (foot or leg swelling)
( ) Eye trouble (other than glasses)
( ) Fever
( ) Frequent or severe headaches
( ) Heart condition
( ) Breathing difficulty
( ) Abnormal bleeding
( ) Anemia
( ) Arthritis
( ) Asthma
( ) Cirrhosis of the liver
( ) Dermatitis
( ) Dizziness or fainting
( ) Epilepsy
( ) Erysipelas
( ) Gall bladder/stones
( ) Glaucoma
( ) Gout
( ) Hay fever
( ) Heart attack of heart trouble
( ) Hemorrhoids
( ) Hives
( ) Hot flashes
( ) Kidney problems
( ) Joint pains
( ) Leg cramps
( ) Liver problems
( ) Muscle aches or pains
( ) Nervousness
( ) Sickle cell disease or trait
( ) Skin disease
( ) Stomach disease
( ) Thyroid problem
( ) Tumors or cysts
( ) Unusual weakness
( ) Unexpected weight loss
( ) Hernia or rupture
( ) High blood pressure
( ) Jaundice
( ) Leukemia
( ) Mumps
( ) Nephritis
( ) Nervous breakdown
( ) Neuritis
( ) Paralysis of any type
( ) Polio
( ) Rheumatic fever
( ) Swollen joints
( ) Stroke
( ) Tuberculosis
( ) Ulcer (stomach or duodenal)
( ) Hepatitis

Family History

Table 4.1. (continued)

Indicate any blood relatives who ever has any of the following:

Disease	Mother	Father	Grand parent	Brother or Sister	My Children
Anemia	( )	( )	( )	( )	( )
Allergy, (asthma, eczema, hay fever)	( )	( )	( )	( )	( )
Bleeding disorders	( )	( )	( )	( )	( )
Cancer	( )	( )	( )	( )	( )
Diabetes	( )	( )	( )	( )	( )
Emphysema	( )	( )	( )	( )	( )
Epilepsy	( )	( )	( )	( )	( )
Heart attach	( )	( )	( )	( )	( )
High blood pressure	( )	( )	( )	( )	( )
Kidney disease	( )	( )	( )	( )	( )
Kidney stones	( )	( )	( )	( )	( )
Gall bladder disease	( )	( )	( )	( )	( )
Sickle cell disease	( )	( )	( )	( )	( )
Stomach ulcers	( )	( )	( )	( )	( )
Stroke	( )	( )	( )	( )	( )

If either of your parents are dead, list their age and cause of death, if known:

Mother died at age _____of_______________________________.
Father died at age _____of_______________________________.

Are you aware of any diseases or illnesses that run in your family? If yes, please list:

____

Do you drink alcoholic beverages? No ( ) Yes ( )

If yes, on the average of:____

Do you average drinking more than one bottle of beer per day? No ( ) Yes ( )

Do you drink more than a bottle of wine per week? No ( ) Yes ( )

Do you drink more than a fifth of liquor per week? No ( ) Yes ( )

Do you smoke? No ( ) Yes ( )

If no, are you a former smoker? No ( ) Yes ( )

If yes, how long ago did you quit? Years_______

How many years did you smoke before quitting? Years_______

How much were you smoking when you quit? (i.e., number of cigarettes, “pipes”, or cigars smoked per day. ___________

If yes, please answer the following:

Table 4.1. (continued)

How long have you smoked? Years______

How many of the following do you smoke per day?

Cigarettes	"Pipes" of tobacco	Cigars
______	______	______

Have you been vaccinated for:

( ) Tetanus Date: __________
( ) Poliomyelitis Date:__________
( ) Other Date:__________

What are your hobbies?

____

Do you own pets? No ( ) Yes ( )

If yes, what kind?____

Physician's Use Only:

Physician's summary and elaboration of all pertinent data. (Physician shall comment on all positive answers. Physician may develop by interview any additional medical history he/she deemed important, and record any significant findings here.)

Typed or Printed Name of Physician:

Signature: ______________________________

Date: ______________________________

Adapted from C. Tinkelpaugh, M.D., Medical Questionnaire (King of Prussia, PA: Workwell, Inc.) and M. E. Mitchell, Ph.D., Private Consultant, Miami, FL.

Routine periodic clinical examinations, such as porphyrin determinations and analysis of blood and urine samples, are practical methods for checking employees exposed to harmful materials. If unsafe exposures are found, further environmental control is mandatory. Affected employees should, of course, be given proper medical treatment.

Medical controls for employees who work with radioactive dusts must be more stringent than the medical controls for most industrial materials. An extensive bioassay sampling program is nearly always required.

PROGRAM COST

Medical monitoring is the responsibility of the employer. The employer will usually bear the entire cost. When a preexisting or non-job-related condition is detected in the course of an examination, the individual will be referred to his/her private physician for further evaluation, and/or treatment. The individual will be responsible for these additional costs.

EXAMINING PHYSICIAN

When possible, the medical monitoring shall be performed by or under the supervision of a physician "*certified*" in occupational medicine. The examining physician must be licensed by the state in which the physical is being performed. Any medical technicians involved should also be certified in the appropriate fields or licensed.

INITIAL OR BASELINE EXAMINATION

The medical examination will provide reference information for comparison with subsequent periodic examination findings and will be used to determine whether each employee is medically suitable for performing assigned tasks.

MEDICAL HISTORY

The medical history will be obtained by having each participant fill out a medical and occupational history questionnaire. The occupational exposure information should be reviewed by a qualified industrial hygienist and/or toxicologist to assist the examining physician by providing specific information on those chemicals. The health questionnaire will aid the examining physician by indicating potential disease by symptoms (see Table 4.1 for a sample questionnaire).

DIAGNOSTIC TESTS

Each individual will have the following tests performed:

1. *Blood tests.* Blood count and differentials — SMAC 25 or Blood Chemistries 160. (Comment: Since most blood work is done by machine, some tests the EPA considers unnecessary will be performed as part of the panel.)
2. *Urine tests.* Complete urinalysis including sedimentation. (Comment: This is a standard set of tests.)

3. *Chest X-rays.* Baseline chest X-rays consist of a minimum of one standard 14 x 17-in. film, posterior-anterior (P-A) views and one lateral view. The film should be read or reviewed by a board-certified radiologist (Class B reader). Subsequent periodic chest X-rays will be performed when clinically or subsequent exposures indicate.
4. *Electrocardiogram.* It will be a standard, 12-lead, resting type as a minimum and will be interpreted by and administered at the discretion of the examining physician.
5. *Pulmonary function tests.* The exam consists of simple tests of lung ventilation to measure forced expiratory volume in 1 sec (FEV1) and forced vital capacity (FVC); other factors such as FEF, MMEFR, MVV, FRC, RV, and TLC may be included. Special attention should be given to the fact that the individual may be required to wear a demand respirator.
6. *Audiometry.* Test will be performed 500, 1000, 2000, 3000, 4000, and 6000 Hz pure tone in an approved booth. Tests are to be administered by qualified technicians, and results are to be read by a certified audiologist (as indicated in 1910.95 Noise Std.).
7. *Visual acuity test.* Visual acuity, both near and far, and color vision are to be tested. Eye site must be correctable to 20/40.
8. *Additional tests.* Under some circumstances the examining physician may request the following tests. Whether the test will be performed or not should be determined by (1) previous exposures to certain chemicals, (2) existing or suspected medical condition, or (3) the potential for on-the-job exposure to certain chemicals.

   a. Sperm count (may include motility)
   b. Sputum cytology
   c. Proctosigmoidoscopy
   d. Laryngoscopy
   e. Colonoscopy
   f. Liver scan
   g. Tonometry
   h. Hemoccult
   i. Pesticide profile (if exposed to pesticides) and chemical profile
   j. Xenobiotic testing and/or myelin sheath damage

PHYSICAL EXAMINATION

Each individual will receive a "hands-on" physical examination by the attending physician. The physician will note any abnormalities observed during the examination. Temperature and blood pressure will be noted. The physician will include checking for perforated eardrums. A determination of whether the individual is capable of performing work while wearing a self-contained breathing apparatus as well as protective suits shall be included.

PERIODIC MONITORING

Individuals will receive periodic examinations because of

1. Conditions found in initial or baseline exams.
2. Real or possible exposure on a site such as spill, sudden release of chemicals, or failure of personal protective equipment.
3. Complaints from the individual that indicate a potential problem (physician's discretion).
4. Part of a routine program.

LABORATORY TESTS

1. Blood test — The following tests will be completed yearly or more often as indicated:
   a. Blood count and differential SMAC 23 or Blood Chemistries 160.
   b. Exposure to toxins in workplace or hazardous waste site.
2. Urine test — Complete urinalysis including sedimentation on yearly basis or as indicated. Include toxicant metabolites when exposed to chemicals.
3. Pulmonary function test — The test will be performed if individual has been exposed to toxic or irritant substances or when the individual has breathing difficulties such as when wearing a respirator.
4. Audiometry — Annual retests are required for personnel subject to high noise exposures (greater than 85 dB), those required to wear hearing protection, or as otherwise indicated.
5. If in a position that would expose one to blood-borne diseases such as hepatitis A or B, HIV, or AIDS, then the physician should test for those diseases.

PHYSICAL EXAMINATION

The examination is indicated if the test results are not within normal range or when the employer feels such is necessary for adequate evaluation.

LIMITATIONS OF TESTS USED FOR MONITORING

The examining physician will evaluate all test results that do not fall in the normal range to determine if the individual is suited for his/her work assignment. Special attention will be given to organs that perform detoxification.

BLOOD AND URINE TESTS

The results of these tests represent the fastest way of assessing the overall health of the individual. Deviations from normal may indicate an underlying problem.

CHEST X-RAYS

It is felt by many in the health profession that the use of X-rays should be limited to those times when they are clinically indicated. However, having a good quality P-A and lateral views can serve as a valuable reference in the future. Since some fibrotic lung diseases are hard to diagnose, both P-A and lateral views are indicated.

ELECTROCARDIOGRAM (EKG)

Some controversy exists about the use of EKG as a diagnostic tool. Many physicians believe the test only indicates gross abnormalities and do not identify the marginal patients. However, some potential problems can be identified so as to not make the test useless.

PULMONARY FUNCTION TEST

Spirometry is a valuable tool in assessing an individual's lung condition. By using an 80% of predicted for FVC as a cut off for referral for physician evaluation, 5% of the truly abnormal will be labeled "normal" and 5% of the normal will be labeled "abnormal" (80% is two standard deviations from statistical norm). This presents a minimal problem because the physician will take other factors into account when making his/her final assessment. Some race variations exist and should, therefore, be taken into consideration.

AUDIOMETRY

This test is useful in determining preexisting hearing loss. It may be used as an indicator of the effectiveness of a hearing conservation program.

VISUAL TESTING

Acceptable eyesight is required for an individual to perform his/her duties safely on a site. The use of corrective safety glasses may limit the individual's ability to perform by interfering with personal protective equipment. A minimum of 20/40 with correction is required.

ADDITIONAL TESTS

1. *Sperm counts.* Some sites contain chemicals that are known to reduce sperm production. Pre- and postexposure counts may be helpful in assessing site exposures.

2. *Sputum cytology.* It has been long debated if the use of sputum cytology would aid in the early diagnosis and treatment of lung disease. The test may be useful in individuals having clinical symptoms, but not for routine use.
3. *Tonometry.* This test is usually recommended for use on persons over the age of 40 years. Use at the discretion of the examining physician.
4. *Proctosigmoidoscopy, laryngoscopy, colonoscopy, liver scan, and hemoccult* are all useful diagnostic tests to be used only when clinically indicated and not as routine screening tests.

OSHA MEDICAL SURVEILLANCE REQUIREMENTS AND NIOSH RECOMMENDATIONS

The following surveillance information and questionnaire was produced by Bio Technology, Inc., Falls Church, Virginia. The document summarizes OSHA's Medical Surveillance Requirement in the General Industry Standard, 29 CFR 1910.120. It can be used by the examining physician to indicate what test should be performed if the possible chemicals are identified. This document can be extremely useful in an uncontrolled release situation where the individual may not be properly protected during exposure.

Program participation is for all individuals whose job normally requires them to

1. Enter a hazardous waste site.
2. Otherwise come in contact with hazardous materials, such as contaminated equipment, laboratory samples, etc. (30 or more days out of the year).
3. Perform physical requirement more strenuous than normal.

SPECIAL CONDITIONS

Use of Respiratory Protection Devices

During the physical examination the examining physician should pay special attention to the individual's ability to use a demand respirator and to carry an SCBA while performing tasks. Pulmonary function, cardiovascular status and weight carrying capacities must be evaluated. Ability to detect odors will be included.

Organ Impairment

Upon reviewing the results of the examination and laboratory tests, the examining physician must make a determination, should organ impairment be detected, whether the individual should be allowed to work on or near a hazardous waste site. For example, an individual may have an impaired liver. Since the primary site of intoxification of chlorinated hydrocarbons is the liver, the physician will have to weigh the risk of further challenging the organ. These decisions should be made by agreement of the individual, his/her employer and a physician.

Climatic Stress

Some waste site investigations are conducted under extreme climatic conditions. An individual who has limited impairment of, for example the kidneys, may be suited for most climatic conditions, but could not function in extreme heat. Certain prescription drugs may also affect an individual's ability to perform assigned duties. The examining physician should make note of an individual's limited capacities under extreme heat or cold.

Biological Monitoring

The purpose of a site health and safety plan is to prevent exposures. One way of assessing the program is by actually measuring the amount of chemicals absorbed. Biological monitoring measures the amount of chemical or its metabolite which is in or is excreted from the body. Typical examples are phenol in the urine from benzene exposure; lead in the blood; chlorinated hydrocarbon solvents in exhaled breath; and heavy metals in the feces and hair. Guides to the selection and proceed during can be found in the literature.

Education

It is most important when conducting medical monitoring programs that the individual be told that the program is being conducted to *protect their health.* Time should be spent explaining what the tests are, what the results will indicate and how the evaluations of the findings are made. Care should be taken to explain that test findings not in the normal range do not necessarily indicate an abnormality (that he/she is "sick"), nor will it automatically eliminate them from participating in hazardous waste site investigations. The employer and the examining physician should encourage the individual to seek consultation with their private physician when indicated.

Rules of Confidentiality

The employer or consulting physician will maintain the records in a confidential manner such that only authorized persons will have access to the records. These will include medical staff of the company or contracted medical personnel, the individual, the individual's personal physician, or the individual's designated representative. Upon written request, the individual may obtain a copy of the medical file. This should be provided within 15 days of the receipt of the written request. Information used for research, testing, statistical, or epidemiologic purposes will have all identifying data removed, including the identity of the individual. Any medical information or findings obtained which do not affect the individual's job performance will not be made available to the employer. This is to maintain the patient-physician confidentiality. Upon death, retirement, resignation, or other termination of services, the records will be retained by the employer or contracting physician. If the company goes out of business, employer is to send the medical records to NIOSH for maintaining and continuing record keeping.

ADDITIONAL HEALTH CONSIDERATIONS

Physical Stresses

Extremes of temperature, pressure, mechanical vibration, and repeated motion, as well as exposures to radiation and noise, produce stresses in the working environment. The industrial hygienist recognizes and evaluates these as having immediate or cumulative effects on the worker's health and safety. These physical stresses, therefore, are as important to him/her as are the chemical and biological hazards.

Noise

Noise may be defined as "unwanted sound." It is a form of vibration which may be conducted through solids, liquids, or gases. The effects of noise on man include the following:

1. Psychological effects, for example, noise can startle, annoy, and disrupt concentration, sleep, or relaxation.
2. Interference with communication by speech, and as a consequence, interference with job performance and safety.
3. Physiological effects, for example, noise-induced loss of hearing, or aural pain, nausea (when the exposure is severe).

Damage Risk Criteria

If the ear is subjected to high levels of noise for a sufficient period of time, some loss of hearing may occur. A number of factors can influence the effect of the noise exposure. Among these are

1. Variation in individual susceptibility.
2. The frequency distribution of the sound.
3. Other characteristics of the noise exposure, such as whether it is continuous, intermittent, or made up of a series of impacts.
4. The total daily time of exposure.
5. The length of employment in the noise environment.

Because of the complex relationship of noise and exposure time to threshold shift (reduction in hearing level) and the many possible contributory causes, establishment of criteria for protecting workers against hearing loss presents many difficulties. An added complication is whether such criteria should be designed to protect against any hearing loss or just loss in the speech frequency range, which is the only portion ordinarily considered for compensation purposes.

An effective hearing conservation program should be undertaken where exposure to industrial noise is capable of producing hearing loss. Such a program should include (1) noise exposure analysis; (2) control of noise exposure; and (3) measurement of hearing. (Note: Chemicals can cause neurological damage to the 8th peripheral nerve, which can cause hearing loss.)

Ionizing Radiation

A brief description of ionizing radiation hazards is given in this section.

Gamma radiation from radioactive materials and X-ray radiation are highly penetrating and may produce damage in any tissue of the body. What tissue will be damaged depends in part on the energy of the radiation and in part on the relative sensitivity of the tissue.

Alpha and beta radiation are generally also considered with gamma and X-ray radiation, although they are not as penetrating and are usually warded off by the skin. Their danger arises when the materials that produce them have been ingested and fixed in some tissue where the radiation can carry on destruction at close range.

In the handling of some types of radioactive materials and in the operation of the high-powered particle accelerators now being used, bombardment by high-speed protons and by both high-speed and thermal neutrons must also be considered as a possible hazard to health.

These types of radiation all cause injury by ionizing the tissue in which they are absorbed. Such injuries differ widely in location and extent, but the determining factors are the ability of the particular radiation to penetrate through the tissues and the amount of ionization produced by a given physical amount of the radiation. Alpha radiation, for instance, is classified as 20 times as biologically effective as X-ray because the alpha particles are heavy and possess a great deal of ability to disrupt and ionize the tissue molecules with which they come in contact.

X-ray radiation is produced by high-potential electrical discharge in a vacuum and should be anticipated in evacuated electrical apparatus operating at a potential of 10,000 V or more. Although X-ray radiation is certain to be produced in such apparatus, it is dangerous only if it penetrates the jacket of the apparatus and enters the inhabited part of the workroom.

If one is being overexposed to the radiation of a comparatively low-voltage-X-ray tube, an X-ray dermatitis of the hand is generally the first result to be seen. It is characterized by a rough, dry skin, a wartlike growth, and dry, brittle nails. With continuing exposure and somewhat more penetrating X-ray, bone destruction will develop.

		ADOPTED VALUES			
		TWA		STEL	
Substance	[CAS #]	ppm	mg/m³	ppm	mg/m³
Acetic acid	[64-19-7] (1976)	10	25	15	37
Biphenyl	[92-52-4] (1987)	0.2	1.3	-	-
Calcium hydroxide	[1305-62-0] (1978)	-	5	-	-
Diethylamine	[109-89-7] (1981)	10	30	25	75
Ethyl amyl ketone	[541-85-5] (1977)	25	131	-	-
Ferbam	[14484-64-1] (1986)	-	10	-	-
Methylamine	[74-89-5] (1977)	10	13	-	-
Nitric acid	[7697-37-2] (1976)	2	5.2	4	10
Oxalic acid	[144-62-7] (1976)	-	1	-	2
Perchloryl fluoride	[7616-94-6] (1976)	3	13	6	25
Quinone	[106-51-4] (1987)	0.1	0.44	-	-
Sodium hydroxide	[1310-73-2] (1977)	-	C 2	-	-
Tetrahydrofuran	[109-99-9] (1976)	200	590	250	737
Vinyl toluene	[25013-15-4] (1981)	50	242	100	483

Figure 4.1. Samples of TLVs listed by TLV levels by the American Conference of Governmental Industrial Hygienist. Reprinted from Threshold Limit Values (TLVs) and Biological Exposure Indices for 1990–91, 2nd Printing, Cincinnati, OH, pp. 10,11, by permission.

Although this discussion is concerned primarily with X-rays and gamma radiation because they are the best known, the same statements apply qualitatively to other types of penetrating radiation. The general method for preventing injuries from any penetrating radiations is separation from them by distance or by shielding with heavy substances in sufficient quantity to reduce the radiation received to below the maximum permissible dose and by limiting the time of exposure.

Nonionizing Radiation

Electromagnetic radiation has varying effects on the body, depending largely on the particular wavelength of the radiation involved. Following, in approximate order of decreasing wavelength and increasing frequency, are some hazards associated with different regions of the nonionizing electromagnetic radiation spectrum (Figure 4.1).

1. *Low frequency.* The longer wavelengths — including power frequencies, broadcast radio, short wave radio — can produce general heating of the body. The health hazard from these radiations is very small, however, since it is unlikely that they would be found in intensities great enough to cause significant effect.

2. Microwaves have wavelengths of 3 m to 3 mm (100 to 100,000 MHz). They are found in radar, communications, and diathermy applications. Microwave intensities may be sufficient to cause significant heating of tissues.

Effect is related to wavelength, power intensity, distance from the source, and time of exposure. Generally, the longer wavelengths will produce a greater temperature rise in deeper tissues than the shorter wavelengths. However, for a given power intensity there is less subjective awareness to the heat from longer wavelengths than there is to the heat from shorter wavelengths because of its absorption beneath the body's surface.

An intolerable rise in body temperature, as well as localized damage, can result from an exposure of sufficient intensity and time. In addition, flammable gases and vapors may ignite when they are inside metallic objects located in a microwave beam.

Power intensities for microwaves are given in units of watts per square centimeter. Areas having a power intensity of over 0.01 W/cm^2 should be avoided. In such areas, dummy loads should be used to absorb the energy output while equipment is being operated or tested. If a dummy load cannot be used, adjacent populated areas should be protected by adequate shielding.

- Infrared radiation does not penetrate below the superficial layer of the skin, so that its only effect is to heat the skin and the tissues immediately below it. Except for thermal burns, the health hazard upon exposure to conventional infrared radiation sources is negligible.

Excessive exposure of the eyes to radiation, mainly visible and infrared radiation, from furnaces and similar operation, upon many years of exposure can produce "glass blower's cataract" or "heat cataract." This condition is an opacity of the rear surface of the lens of the eye.

- Visible radiation and ultraviolet radiation also do not penetrate appreciably below the skin; their effects are essentially heating on the surface. The effects of the ultraviolet waves are much more violent than are those of the visible or infrared — a severe burn can be produced with no warning and significant damage to the lens of the eye can occur from excessive exposure.
- Ultraviolet radiation in industry may be found around electrical arcs or inspection lights (in silicon ship manufacturing). They should be shielded by materials opaque to the ultraviolet. Opacity in the ultraviolet has no relation to opacity in the visible part of the spectrum. Ordinary window glass, for instance, is almost completely opaque to the ultraviolet although transparent to the visible. A piece of plastic dyed a deep red-violet may be almost entirely opaque in the visible parts of the spectrum and transparent in the near ultraviolet.

Electric welding arcs and germicidal lamps are the most common strong producers of ultraviolet in industry. The ordinary fluorescent lamp generates a good deal of ultraviolet inside the bulb, but it is essentially absorbed in the glass bulb and its fluorescent coating.

The most common exposure to ultraviolet radiation is from direct sunshine. Men who continually work outdoors in the full light of the sun may develop tumors on exposed areas of the skin, and these tumors occasionally turn malignant.

Ultraviolet radiation from the sun also increases the skins effects of some industrial materials. After exposure to compounds such as cresols, the skin is exceptionally sensitive to the sun. Even a short exposure in the late afternoon when the sun is low is likely to produce a severe sunburn. There are other compounds that minimize the effect of UV-rays. Some of these are used in certain protective creams.

- Lasers emit beams of coherent light of a single color or wavelength and frequency, in contrast to conventional light sources, which produce random, disordered light wave mixtures of various frequencies. The laser (light amplification by stimulated emission of radiation) is made up of light waves that are nearly parallel to each other, all traveling in the same direction. Light-emitting atoms are "pumped" full of energy and stimulated to fall to a lower energy level, giving off light waves that are directed to produce the coherent laser beam.

The maser, the laser's predecessor, emits microwaves instead of light. Some companies call their lasers "optical masers." Proposed uses for the laser include machining and cutting of metals, welding of microscopic parts, and systems for high-capacity communications.

Since the laser is highly collimated (has a small divergence angle), it can have a large energy density in a narrow beam. Direct viewing of the laser should therefore be avoided. The work area should contain no reflective surfaces (such as mirrors or highly polished furniture), for even a reflected laser beam can be hazardous. Suitable shielding to contain the laser beam should be provided.

The eye is the organ most vulnerable to injury induced by laser energy. The reason for this is the ability of the cornea and lens to focus the parallel laser beam on a small spot on the retina.

The fact that infrared radiation of certain lasers may not be visible to the naked eye contributes to the potential hazard.

Lasers generating in the ultraviolet range of the electromagnetic spectrum produce corneal burns rather than retinal damage, because of the way the eye handles ultraviolet light.

Other factors that have a bearing on the degree of eye injury induced by laser light are

1. The pupil size — the smaller the pupil diameter, the less the amount of laser energy permitted to the retina.
2. The power of the cornea and lens to focus the incident light on the retina.
3. The distance from the source of energy to the retina.
4. The energy and wavelength of the laser.

5. The pigmentation of the subject.
6. The place on the retina where the light is focused.
7. Divergence of the laser light.
8. The presence of scattering media in the light path.

Effects of Temperature Extremes

Probably the most elementary factor of environmental control is control of the thermal environment in which people work. General experience shows that extremes of temperature affect the amount of work that a person can do and the manner in which he or she does it. The industrial problem is more often that of exposure of high temperatures rather than low temperatures.

The body is continuously producing heat through its metabolic processes. Since the body processes are so designed that they can operate only within a very narrow limit of temperature, they must dissipate this heat as rapidly as it is produced if the body is to function efficiently and well. A sensitive and rapidly acting set of thermostatic devices in the body must also control the rates of its temperature-regulating processes.

Sweating is probably the most important of the temperature-regulating and heat-dissipating processes. Almost all parts of the skin are provided with sweat glands that excrete a liquid (mostly water and a little salt) to the surface of the body. This goes on continuously even under conditions of rest.

In an individual who is resting and not under stress, the sweating rate is approximately 1 L/day, which is evaporated as rapidly as it is excreted. Under the stress of heavy work or high temperature, this sweating rate may increase to as much as 4 L (approximately 1 gal) per 4 hr. As much as 10 to 12 g (150 to 190 grains) of salt per day will be lost with the water. Both the water and salt loss must be made up rather promptly if good health is to continue. In terms of stress, an individual's ability to acclimate is variable, and such stress is dangerous for those with heart problems.

If the sweat can be evaporated as rapidly as it can be formed and if the heat stress is not such as to exceed the maximum sweating rate of which the body is capable, the body can maintain the necessary constant temperature.

The rate of evaporation depends upon the moisture content, the rate of air movement, and the temperature of the surrounding air. The industrial hygienist must simultaneously deal with all of these factors in any attempt to determine the effect of the thermal environment on workers.

Heat from radiation accounts for some of the body's thermal equilibrium with its surroundings. If an object in the surroundings is far below body temperature, such as a large glass window on a subzero day, a large amount of heat can be radiated and the person may feel chilled even if the air in his immediate environment is fairly warm. If an object in the surroundings, such as a furnace wall, is very much above body temperature, a person may receive a large amount of heat by radiation, and it may be extremely difficult to keep him cool enough by the other available means to maintain a constant body temperature.

Radiant heat is electromagnetic energy that does no heating whatsoever until it strikes some object, like a person, where it is absorbed. No relief from it whatsoever results from blowing air around. The only protection is to set up an invisible infrared "shadow." This can be done by placing any kind of opaque shield or screen between the man and the radiating surface. If the screen itself, however, does not reflect radiant heat, it will absorb the energy, heat up and reradiate. Fortunately, commercial aluminum foil, aluminum sheets, and corrugated aluminum siding are all highly reflective in the infrared — and are cheap. After locating the sources of radiant heat, shields placed between them and the employees will bring welcome relief.

Conduction through the clothing and dissipation into the air by conduction are sources of some cooling to the body. This usually is not an important means of cooling because the conductivity of clothing and the heat capacity of air are usually low.

Conduction and convection become an important means of heat loss when the body is in contact with a good cooling agent, such as water. For this reason, when people are exposed to cold water, they become chilled much more rapidly and effectively than when exposed to air at the same temperature.

Air movement cools the body by conduction and convection. More importantly, moving air removes from the body the film of saturated air that is formed very rapidly by evaporation of sweat and replaces it with a fresh layer capable of accepting more moisture.

Effects of high temperature are counteracted by the body's attempt to keep the internal temperature down by increasing the rate of heart beat. The capillaries in the skin then dilate to bring more blood to the surface so that both the rate of cooling and, gradually, the body temperature are increased.

If the thermal environment is tolerable, these measures will rather soon lead to an equilibrium where the heart rate and body temperature remain constant. If this equilibrium is not reached until the body temperature is about 102°F, corresponding to a sweating rate of about 2 L/hr, there is imminent danger of heatstroke.

Heatstroke (also known as sunstroke) is not necessarily the result of exposure to the sun. It is caused by exposure to an environment in which the body is unable to cool itself sufficiently. As a result, the body temperature rises and reaches a point where the heat-regulating mechanism breaks down completely. The body temperature then rises rapidly.

The symptoms are hot, dry skin, severe headache, visual disturbances, rapid temperature rise, and loss of consciousness. The condition is recognizable by the flushed face and high temperature. The victim should be removed from the heat immediately and cooled as rapidly as possible, most readily by being wrapped in cool wet sheets. Since the condition may be quickly fatal, medical help should be obtained as soon as possible.

Heatstroke is a much more serious condition than heat cramps or heat exhaustion. An important predisposing factor is excessive physical exertion.

Heat cramps may result from exposure to high temperature for a relatively long time, particularly if accompanied by heavy exertion, with excessive loss of salt and moisture from the body. Even if the moisture is replaced by drinking plenty of water, the excessive loss of salt may provoke heat cramps or heat exhaustion.

Heat cramps are characterized by the cramping of the muscles of either the skeletal system or the intestines. In either case the condition may be relieved in a few hours under proper treatment, although soreness may persist for several days.

Heat exhaustion may result from physical exertion in a hot environment when vasomotor control and cardiac output are inadequate to meet the increased demand placed upon them by peripheral vasodilation. Its signs are relatively low temperature, pallor, weak pulse, dizziness, profuse sweating, and cool, moist skin.

Prevention of heat exhaustion and heat cramps has become almost routine since the introduction of the use of salt tablets in the hot occupations. There is some resistance to the use of salt tablets on the part of a few individuals who say that they are nauseated by them. Usually, nausea results when the tablets are taken with an inadequate amount of water so that an excessive concentration salt is built up in the stomach. Generally, good diet can be relied on for adequate salt supply.

To prevent nausea, a full 8-oz glass of water should be taken with each salt tablet. It can also be prevented by the use of enteric-coated tablets, which do not dissolve until they have passed the stomach, or by the use of tablets impregnated with cellulose, which dissolve very slowly in the stomach contents.

Experience has shown that people do not stand a hot job very well at first, but develop tolerance rapidly and acquire full endurance within a week to a month.

Atmospheric Pressure

It has been recognized from the beginning of caisson work, dating back to about 1850, that men working under greater than normal atmospheric pressures are subject to various ills connected with the job. The main effect (decompression sickness commonly known as the "bends") results from the release of nitrogen bubbles into the circulation and tissues during decompression. The bubbles lodge at the joints and under muscles, causing severe cramps. To prevent this trouble, decompression is carried out slowly and by stages so that the nitrogen can be eliminated slowly and without the formation of bubbles.

Deep-sea divers are supplied with a mixture of helium and oxygen for breathing. Since helium is an inert diluent and is less soluble in blood and tissue than is nitrogen, it presents a less formidable decompression problem.

Under some conditions of work at high pressure, the concentration of carbon dioxide in the atmosphere may be considerably increased so that the carbon dioxide will act as a narcotic. Keeping the oxygen concentration high will minimize the condition although not prevent it. This procedure is useful where the carbon dioxide concentrations cannot be kept at a proper level.

One of the most common troubles encountered by workers under compressed air is pain and congestion in the ears from inability to ventilate the middle ear properly during compression and decompression. As a result, many workers under compressed air suffer from temporary and some from permanent loss of hearing.

The cause of this damage is considered to be an obstruction of the eustachian tubes, which prevents proper equalization of pressure on the middle ear.

The effects of reduced pressure on the worker are much the same as the effects of decompression from a high pressure. If pressure is reduced too rapidly, decompression sickness and ear disturbances similar to, if not identical with, the diver's conditions may result.

Men working at reduced pressure are also subject to oxygen starvation, which can have very serious and insidious effects upon the senses and judgment. There is a considerable amount of evidence that exposure for 3 to 4 hr to an altitude of 9000 ft above sea level without breathing an atmosphere enriched in oxygen can result in severely impaired judgment. Even if pure oxygen is provided, the altitude should be limited to that giving the same partial pressure of oxygen as air at 8000 feet.

Reduced pressure is not the only condition under which oxygen starvation may occur. Deficiency of oxygen in the atmosphere of confined spaces is commonly experienced in industry. For this reason, the oxygen content of any tank or other confined space should be checked before entry is made. Instruments such as the oxygen analyzer are commercially available for this purpose.

Normal air contains approximately 21% oxygen by volume. The first physiologic signs of a deficiency of oxygen (anoxia) are increased rate and depth of breathing. Oxygen concentrations of less than 16% by volume cause dizziness, rapid heat beat, and headache. One should never enter or remain in areas where tests have indicated such concentrations unless one is wearing some form of supplied-air or self-contained respiratory equipment.

Oxygen-deficient atmospheres may cause inability to move and a semiconscious lack of concern about the imminence of death. In cases of sudden entry into areas containing little or no oxygen, the individual usually has no warning symptoms, but immediately loses consciousness and has no recollection of the incident if he is rescued and revived.

Ergonomics

The term "ergonomics" literally means the customs, habits, and laws of work. According to the International Labor Office, it is

> "...The application of human biological science in conjunction with the engineering sciences to achieve the optimum mutual adjustment of man and his work, the benefits being measured in terms of human efficiency and well-being."

The ergonomics approach goes beyond productivity, health, and safety. It includes consideration of the total physiological and psychological demands of the job upon the worker.

In the broad sense, the benefits that can be expected from designing work systems to minimize physical stress on workers are

1. More efficient operation
2. Fewer accidents
3. Lower cost of operation
4. Reduced training time
5. More effective use of personnel

The human body can endure considerable discomfort and stress and can perform many awkward and unnatural movements — for a limited period of time. When, however, unnatural conditions or motions are continued for prolonged periods, the physiological limitations of the worker may be exceeded. To ensure a high level of performance on a continuing basis, work systems must be tailored to human capacities and limitations.

Biomechanics is that phase of engineering devoted to the improvement of the man-machine-task relationship in an effort to reduce operator discomfort and fatigue.

Biotechnology is a broader term. It encompasses biomechanics, human factors, engineering, and engineering psychology. To arrive at biotechnological solutions to work-stress problems, the sciences of anatomy, physiology, psychology, anthropometry, and kinesiology need to be brought into play.

Program areas of concern are

1. Strictly biomechanical aspects — the consideration of stress on muscles, bones, nerves, and joints.
2. Sensory aspects — the consideration of eye fatigue, color, audio signals, and the like.
3. External environment — such as lighting, glare, temperature, humidity, noise, atmospheric contaminants, and vibration.
4. The psychological and social aspects of the working environment.

Information obtained from a study of the above-mentioned factors can be translated into tangible changes in work environments. Among the changes that are proving to have positive effects insofar as fatigue and stress reduction are concerned are redesigned hand tools, adjustable chairs and workbenches, improvements in lighting, control of heat and humidity, and noise reduction.

Mechanical Vibration

A condition known to stonecutters as "dead fingers" or "white fingers," occurs mainly in the fingers of the hand used to guide the cutting tool. The circulation in this hand becomes impaired, and when exposed to cold the fingers became white and without sensation, as though mildly frost bitten. The condition usually disappears when the fingers are warmed for some time, but a few cases are sufficiently disabling that the men are forced to seek other types of work. In some instances both hands are affected.

The condition has been observed in a number of other occupations involving the use of fairly light vibrating tools, such as the air hammers used for scarfing and chipping in the metal trades. The condition is produced by vibration while the fingers are held in a strained position and is at least aggravated by chilling of the fingers at the same time. Prevention of this condition is much more satisfactory than treatment.

Preventive measures include directing the exhaust air from air-driven tools away from the hands so that they will not become unduly chilled, use of handles of a comfortable size for the fingers and, in some instances, substitution of mechanical cleaning methods for some of the hand methods which have produced most of the cases of "white fingers." In many instances, simply preventing the fingers from becoming chilled while at work has been sufficient to eliminate the condition.

Repeated Motion

Repetitive motions or repeated shocks like those in sorting and assembling jobs often cause irritation and inflammation of the tendon sheath of the hands and arms. The condition is generally known as tenosynovitis, and when once established, it is painful and disabling. The usual treatment is rest and diathermy.

The condition results from excessive strain many times repeated. It may occur to an employee who has been working at the same job for years if he is suddenly asked to put in considerable overtime. It is most likely to occur to a new employee or to an employee transferred to a new job. Prevention of the condition is, of course, much more satisfactory than treatment.

For example, in the printing industry cases of sore hands and wrists were prevented by lowering the surface of the jogging tables 2 inches so that the hands and wrists could fall into a relaxed position at the end of each motion.

The employees should be closely watched so that if they show signs of soreness in the backs of the hands, wrists, forearms, or shoulders they can be transferred temporarily to other work, or the job can be changed slightly to reduce the strain.

BIOLOGICAL STRESSES

Tuberculosis and AIDS

Tuberculosis and AIDS and other infections may be an occupational disease when they are contracted by nurses, doctors, or attendants caring for tuberculosis patients, or during autopsy or laboratory work where the bacilli may be present.

Fungus Infections

A number of occupational infections are common to workers in agriculture and in closely related industrial jobs. Grain handlers who inhale grain dust are likely to come in contact with some of the fungi which contaminate grain from time to time. Some of these fungi can and do flourish in the human lung, causing a condition called farmer's lung.

Fine, easily spread fungi, such as rust and smut, can be contacted once the covering of infected grain is broken and may produce sensitization. Infections from some of these fungi are endemic (of local nature) in the southwest and are not necessarily industrial. Also, an individual experiencing Sick Building Syndrome, which is due to improper ventilation in the building air conditioning, may become sensitized to Aspergillus (fungus) a bacteria found in the filters of the air conditioning unit.

Byssinosis

Byssinosis occurs in individuals who have experienced prolonged exposure to heavy air concentrations of cotton dust. Flax dust has also been incriminated.

The exact mode of action of the cotton dust is unknown, but one or more of these factors may be important:

1. Toxic action of microorganisms adherent to the inhaled fibers.
2. Mechanical irritation from the fibers.
3. Allergic stimulation by the inhaled cotton fibers or adherent materials.

It takes several years of exposure before manifestations are noticed.

Anthrax

Anthrax is a highly virulent bacterial infection. In spite of considerable effort in quarantining infected animals and in sterilizing imported animal products, this disease remains a problem. With prompt detection and modern methods of treatment, it is much less likely to be fatal than it was a few years ago.

"Q Fever"

There have been reports of infection — "Q fever" — with a rickettsial organism among meat and livestock handlers. It is similar to but apparently not identical with "tick fever," which has been known for many years. The exact mode of transmission is not known, but there is evidence that "Q fever" may come from contacting freshly killed carcasses or droppings of infected cattle. Probably in the latter case, transmission is by inhalation of the infectious dust. Prevention undoubtedly depends primarily upon recognition and elimination of the disease in the animal host.

Brucellosis

Brucellosis (undulant fever) has long been known as an infection produced by drinking unpasteurized milk from cows suffering from Bang's disease (infectious abortion). Since it can also be contracted by handling the animals or their flesh and is also transmitted by swine and goats, it is an occupational ailment of slaughterhouse workers as well as farmers, although more common among the latter. Here also, preventive measures are primarily proper testing and control of the animals to eradicate the disease.

Erysipelas

The other major bacterial infection to which slaughterhouse workers and fish handlers are subject is erysipelas, which seems to be especially virulent when contracted from the slime of fish.

Upper Respiratory Tract Infections

Workers exposed to dusts from vegetable fibers such as cotton, bagasse (sugar cane residues), hemp, flax, or grain may develop upper respiratory tract infections ranging from chronic irritation of the nose and throat to bronchitis, complicated by asthma, emphysema, or pneumonia, or a combination.

Dusts from many of these fibers may also be a source of allergens, histamine, and toxic metabolic products of microorganisms. However, most of the cases of illness arising from handling these fibers have been reported in the foreign literature.

Evaluation and Control

The degree of hazard arising out of exposure to harmful environmental factors or stresses would depend on

1. The nature of the material of energy involved.
2. The intensity of the exposure.
3. The duration of the exposure.

The properties of the specific agent and its possible physiological effect on employees must be ascertained by the magnitude of these three factors.

The toxicity of a material is not synonymous with its health hazard. Toxicity is the capacity of a material to produce injury or harm. Hazard is the possibility that a material will cause injury when a specific quantity is used under certain conditions.

The key elements to be considered when evaluating a health hazard are

1. How much of the material in contact with a body cell is required to produce injury?
2. The probability of the material being absorbed by the body to result in an injury.
3. Rate of generation of airborne contaminant.
4. Total time of contact.
5. Control measures in use.

Toxicity is dependent on dose, rate, method and site of absorption, general state of health, individual differences, diets, and temperature.

All materials can be handled safely if the proper precautionary measures are taken.

The particular exposure may be evaluated by chemical analyses of air samples, and a step-by-step analysis of the operations must be made to find the areas where employees are exposed to hazardous amounts of the material. The operational analysis should also determine how the material becomes airborne and is dispersed.

The importance of the sampling location, the proper time to sample, and the number of samples to be taken during the course of an investigation of the work environment cannot be overstressed. The duration of exposure (as well as concentration) is required to calculate the dosage a man receives. It is important that we actually test the air in the worker's breathing zone. An adequate number of tests should be taken to define the TWA exposure to be able to relate this to the TLV.

This might appear to be a routine, mechanical job. Actually, it is an art that requires intimate knowledge of the sampling equipment to be used and its shortcomings; where and when to sample; and how to weigh the many factors that can influence the sample — such as time of day, season, unusual operations, and masking effects of affiliated contaminants.

Air Quality Standards

The preface to the annual list of TLVs (see Figure 4.1) and the introduction of each of the American National Standard Z37 series of acceptable concentrations outline the basic concepts and purpose of establishing levels of air contaminants to which a person may be exposed without suffering discomfort or *known ill effects*.

The preface to the TLV list states that nearly all workers may be repeatedly exposed to the levels in this list without adverse effect. Certain values in this list preceded by a capital "C" should not be exceeded. The remaining values are to be considered as time-weighted averages. The basis for these TLVs may be *reasonable freedom from irritation, narcosis, nuisance,* or *impairment of health.* Some substances have the designation "skin" after their name. This refers to a potential contribution to the overall exposure by absorption through the skin or mucous membranes.

When extremes of temperatures and/or pressure are encountered, these TLV values should be adjusted to take into account the added stress.

The preface to the 1988 TLV list further states:

> "These limits are intended for use in the field of industrial hygiene and should be interpreted and applied only by persons trained in this field."

The Hygienic Guides (published by the American Industrial Hygiene Association) and National Safety Council Industrial Data Sheets also contain helpful information about air quality standards, in addition to the new updated OSHA Air Contaminants Standard 1910.1000.

Control

The type and extent of controls will depend upon the physical, chemical, and toxic properties of the air contaminant, the evaluation made of the exposure, and the operation that disperses the contaminant. The extensive controls needed for lead oxide dust, for example, would not be needed for limestone dust, since much greater quantities of limestone dust can be tolerated.

General methods of controlling harmful environmental factors or stresses include the following:

1. Substitution of a less harmful material for one that is dangerous to health.
2. Change or alteration of a process to minimize worker contact.
3. Isolation or enclosure of a process or work operation to reduce the number of persons exposed.
4. Wet methods to reduce generation of dust in operations such as mining and quarrying.
5. Local exhaust at the point of generation and dispersion of contaminants.
6. General or dilution ventilation with clean air to provide a safe atmosphere.
7. Personal protective devices, such as special clothing, eye, and respiratory protection.
8. Good housekeeping, including cleanliness of the workplace, waste disposal, adequate washing and eating facilities, healthful drinking water, and control of insects and rodents.
9. Special control methods for specific hazards, such as reduction of exposure time, film badges and similar monitoring devices, continuous sampling with preset alarms, and medical programs to detect intake of toxic materials.
10. Training and education to supplement engineering controls.

Sources of Help

The safety, industrial hygienist, and environmental professional who believes that he/she has an exposure problem should consult a competent specialist in the related field. Such help may be obtained from their own company, insurance carrier, the private consultants, professional organization, or state health or labor agencies.

Always remember when you are dealing with the human body and an exposure, the exposure may not just be one specific chemical, but a number of various chemicals or environmental factors. Always look for the "additive effect on the total picture."

REFERENCES

1. NIOSH, EPA, OSHA, and USCG. "Occupational Safety and Health Guidance Manual for Hazardous Waste Site Activities," U.S. Government Printing Office, Washington, DC (October 1985), Document No: 85-115, pp. 5-1–5-10.
2. ANSI 1984, American National Standard for Respiration Protection, ANSI Z88.6-184 (New York: American National Standards Institute, 1984).
3. Division of Technology Evaluation and Assistance, Laboratory Program Office, Center for Disease Control, Atlanta (Proficiency Testing).
4. College of American Pathologists, 7400 N. Skokie Blvd., Skokie, IL (Proficiency Testing).
5. American Association of Bioanalysts, 205 W. Lence Street, Brownsville, TX.
6. Tinkelpaugh, C. N., Workwell, Inc., The Merion Building, 700 S. Henderson, King of Prussia, PA (Medical/Work History Questionnaire). Mitchell, M. E., Private Consultant, Miami, FL.
7. OSHA. "How to Prepare for Workplace Emergencies," U.S. Department of Labor, U.S. Government Printing Office, Washington, DC OSHA-3088 (Revised) (1988).
8. OSHA. "Recordkeeping Guidelines for Occupational Injuries and Illnesses," U.S. Department of Labor, Bureau of Labor Statistics, O.M.B. No. 1220-0029, U.S. Government Printing Office, Washington, DC (April 1986).
9. NIOSH. "A Guide to the Work-Relatedness of Disease," Revised ed., U.S. Department of Health and Human Services, U.S. Government Printing Office, Washington, DC (August 1980).
   Kusnetz, M.S., Stanley
   Hutchinson, M.D., Marilyn K. (Editors)
10. Moutschen, J. *Introduction of Genetic Toxicology*, (New York: John Wiley & Sons, 1985).
11. Denman, S. D., and Bertram D. "The Industrial Environment — Its Evaluation and Control," NIOSH, U.S. Government Printing Office, Washington, DC (1973) pp. 197–206.

CHAPTER 5

Safety Planning and Principles at Hazardous Waste Sites

Most events in life represent a potential risk to the health, well-being, and safety of an individual. Therefore, we all constantly apply, often without consciously realizing it, the principles of risk assessment. This chapter will deal explicitly with the hazardous waste site activities involving safety planning.

ELEMENTS OF RISK ASSESSMENT

Recognition

Recognition of risks is defined as the identification of the potential hazard and the parameters that determine the degree of hazard. The recognition of a potential hazard is obviously the first and most important step in risk analysis.

Obvious dangers are readily avoided. Example: On the Chemdyne site in Hamilton, Ohio, field personnel walked through white powder that had spilled from a ruptured 55-gal drum. Next to the white powder were a dead rat and grasshopper. Shortly after coming in contact with the powder, the field team's outer protective rubber booties started to disintegrate. It was not difficult to assess the potential of risk and design protective measures.

Evaluation

Evaluation is defined as applying methods to reduce or control the risk associated with a material or an action so as to make it acceptable relative to the benefit achieved. The choice and use of respiratory protection devices is a prime example of evaluating respiratory protection.

Control

Control is defined as the application of methods to reduce or control the risk associated with a material or an action so as to make it acceptable relative to the benefit achieved. The choice and use of respiratory protection devices is a prime example of a risk control technique.

LONG-TERM RISKS

Many chemicals and hazardous materials do not produce immediate overt effects.

1. Asbestos was thought to be relatively harmless for many years. Early epidemiological studies reinforced that concept. Only after 20 years of study was a cause and effect relationship established between asbestos and cancer. Advancements in occupational medicine and human evaluations made by toxicologists specialized in this area have found a direct relationship of lessening the length of time due to other chemicals irritating the lungs and allowing the asbestos fibers to come in.
2. Other chemicals have been shown to be carcinogenic only after many years of study. Most chemicals found in hazardous waste sites have not been fully characterized as to their carcinogenicity. Therefore, since they are unknown, they must be treated as carcinogens now because in 20 years it will be unable to undo the damage caused by an exposure now.

Many chemicals that are inherently nontoxic or have a low to medium degree of toxicity can become highly toxic when there is a simultaneous exposure with another chemical. Instead of the toxicity of the two chemicals having an additive effect, the effect can become a multiple of the two individual effects, i.e., two, three, ten, or even twenty times more toxic.

SITE SAFETY PLAN EMERGENCY MEDICAL CARE AND TREATMENT

The site safety plan must address emergency medical care and treatment or response personnel, including possible exposures to toxic substances and injuries due to accidents of physical problems. The following items should be included in emergency care provisions:

1. Name, address, and telephone number of the nearest medical treatment facility should be conspicuously posted. A map and/or directions for locating the facility, plus the travel time, should be readily available.

2. The facility's ability to provide on-site care and treatment of personnel exposed to toxic or otherwise hazardous substances should be ascertained. If the facility lacks toxicological capability, arrangements may be needed for consultant services.

EMERGENCY RESPONSE PLANNING

The effectiveness of response during emergencies depends on the amount of planning and training performed. Management must show its support of *all* safety programs (internal, remote, or organizational) and the importance of emergency planning. If management is not interested in employee protection and minimization property loss, little can be done to promote a safe workplace. It is therefore management's responsibility to see that a program is instituted and that it is frequently reviewed and updated. The input and support of all employees must be obtained to ensure an effective program. The emergency response plan should be developed locally and should be comprehensive enough to deal with all types of emergencies. When emergency action plans are required by a particular OSHA standard, the plan must be in writing, except in firms with ten or fewer employees. The plan must include, as a minimum, the following elements:

1. Emergency escape procedures and emergency escape route assignments.
2. Procedures to be followed by employees who remain to perform critical plant operations or work site equipment shut-down, before evacuation.
3. Procedures to account for all employees of emergency evacuation.
4. Rescue and medical duties for those employees who are to perform them.
5. The preferred means of reporting fires and other emergencies.
6. Name or regular job titles of persons or departments (internal and external) to be contacted for further information or explanation of duties under the plan.

The emergency action plan should address all potential emergencies that can be expected in the workplace. Therefore, it will be necessary to perform a hazard audit to determine potentially toxic materials and unsafe conditions. For information on chemicals, the manufacturer or supplier can be contacted to obtain Material Safety Data Sheets (MSDSs). These forms describe the hazards that a chemical may present, list precautions to take when handling, storing, or using the substance, and outline emergency and first-aid procedures.

The employer, consultant, or project manager (if a hazardous waste site) should list in detail the procedures to be taken by those employees who must remain behind to care for essential plant operations until their evacuation becomes absolutely necessary. This may include monitoring area power supplies, water supplies, and other essential services that cannot be shut down for every emergency alarm.

For emergency evacuation, the use of floor plans or workplace maps that clearly show the emergency escape routes and safe or refuge areas should be included in the plan. All employees must be told what actions they are to take in the emergency situations that may occur in the workplace.

This plan should be reviewed with employees initially when the plan is developed, whenever the employee's responsibilities under the plan change, and whenever the plan is changed.

Chain of Command

A chain of command should be established to minimize confusion so that employees will have no doubt about who has authority for making decisions. Responsible individuals should be selected to coordinate the work of the emergency response team.

In larger organizations, there may be a site/plant coordinator in charge of plant/site-wide operations, public relations, and ensuring that outside aid is called in. Because of the importance of these functions, adequate backup must be arranged so that trained personnel is always available. The duties of the emergency response team coordinator should include the following:

1. Assessing the situation and determining whether an emergency exists that requires activating the emergency procedures.
2. Directing all efforts in the area including evacuating personnel and minimizing property loss.
3. Ensuring that outside emergency services such as medical aid and local fire departments are called in when necessary.
4. Directing the shutdown of site/plant operations when necessary.

Emergency Response Team

Emergency response teams are the first line of defense in emergencies. Before assigning personnel to these teams, the employer must assure that employees are physically capable of performing the duties that may be assigned to them. Depending on the size of the site/plant, there may be one or several teams trained in the following areas:

1. Use of various types of fire extinguishers
2. First aid, including cardiopulmonary recescitation (CPR)
3. Shutdown procedures
4. Evacuation procedures
5. Chemical spill control procedures
6. Use of self-contained breathing apparatus (SCBA)
7. Search and emergency rescue procedures

The type and extent of the emergency will depend on the site/plant operations and the response will vary according to the type of process, the material handled, the number of employees, and the availability of outside resources. Emergency response teams should be trained in the types of possible emergencies and the emergency actions to be performed. They should be informed about special hazards, such as storage and use of flammable materials, toxic chemicals, radioactive sources, and water-reactive substances, to which they may be exposed during fire and other emergencies. It is important to determine when *not* to intervene. For example, team members must be able to determine if the fire is too large for them to handle or whether search and emergency rescue procedures should be performed. If there is a possibility of members of the emergency response team receiving fatal or incapacitating injuries, they should wait for professional firefighters or emergency response groups.

Emergency Training

Training is important to the effectiveness of an emergency plan. Before implementing an emergency action plan, a sufficient number of persons must be trained to assist in the safe and orderly evacuation of employees. Training for each type of disaster response is necessary so that employees know what actions are required.

In addition to the specialized training for emergency response team members, all employees should be trained in the following:

1. Evacuation plans
2. Alarm systems
3. Reporting procedures for personnel
4. Shutdown procedures
5. Types of potential emergencies

These training programs should be provided as follows:

1. Initially when the plan is developed
2. For all new employees
3. When new equipment, materials, or processes are introduced
4. When procedures have been updated or revised
5. When exercises show that employee performance must be improved
6. At least annually

The emergency control procedures should be written in concise terms and made available to all personnel. A drill should be held for all personnel, at random intervals at least annually, and an evaluation of performance made immediately by management and employees. When possible, drills should include groups supplying outside services such as fire and police departments. In buildings with several places of employment, the emergency plans should be coordinated with other companies and employees in the building. Finally, the emergency plan should be reviewed periodically and updated to maintain adequate response personnel and program efficiency.

STRESS

Both physiological and psychological stress can affect emergency response personnel. Under certain conditions, stress can contribute significantly to accidents or harm workers in other ways. To reduce the potential for abnormal physical stress or mental anxiety, the procedures listed below should be followed.

1. Workers must be periodically examined by medical authorities to determine if they are physically and, if necessary, psychologically fit to perform their job.
2. Workers must be medically evaluated for prescription drugs or illegal drugs.
3. Workers must be in total frame of mind and physically fit to take on emergency response work.

GENERAL SAFETY PRACTICES

The following are *important personnel safety precautions* which would be enforced and posted on site as a minimum:

1. Eating, drinking, chewing gum or tobacco, smoking, or any practice that increases that probability of hand-to-mouth transfer and ingestion of material is prohibited in any area designated contaminated.
2. Hands and face must be thoroughly washed upon leaving the work area, and ingestion of material is prohibited in any other activities.
3. Whenever decontamination procedures for outer garments are in effect, the entire body should be thoroughly washed as soon as possible after the protective garment is removed.
4. No excess facial hair which interferes with a satisfactory fit of the mask-to-face seal is allowed on personnel required to wear respiratory protection equipment.
5. Contact with contaminated surfaces or with surfaces suspected of being contaminated should be avoided. Whenever possible, one should not walk through puddles, mud, or other discolored surfaces. Remember to bend your knees when leaning, lifting, or pulling equipment such as drums, containers, or vehicles over the ground.
6. Medicine and alcohol can potentiate the effects from exposure and toxic chemicals. Prescribed drugs should not be taken by personnel in response operations if there is a likelihood of such potentiation. If exposure has occurred after taking medication, see a physician.
7. Personnel and equipment in the contaminated area should be minimized, consistent with EPA-approved site operations.
8. Work areas for various operational activities must be specifically established.
9. Procedures for leaving a contaminated area must be specifically planned, marked out, and implemented prior to going to the site. Work areas and decontamination procedures must be established on the basis of prevailing site conditions.

CLIMATE/WEATHER CONSIDERATIONS

Adverse weather conditions are important considerations in planning and conducting site operations. Hot or cold weather can cause physical discomfort, loss of efficiency, and personal injury. Rain, sleet, thunder and lightning should initiate reevaluation of the sampling and health and safety plans of any operation.

HEAT STRESS PREVENTION

Of particular importance is heat stress resulting when protective clothing decreases natural body ventilation. One or more of the following will help reduce heat stress:

1. Provide plenty of liquids, made available in the support zone. To replace body fluids (water and electrolytes) lost because of sweating, use a 0.1% saltwater drink and increase salted foods intake. The commercial mixes may be preferable for those employees on a low-sodium diet.
2. Provide cooling devices to aid natural body ventilation. These devices, however, add weight, and their use should be balanced against worker efficiency. Long cotton underwear acts as a wick to help absorb moisture and protect the skin from direct contact with heat-absorbing protective clothing. It should be the minimum undergarment worn.
3. Install hose-down facilities to reduce body temperature and cool protective clothing.
4. In extremely hot weather, conduct nonemergency response operations in the early morning or evening.
5. Ensure that adequate shelter is available to protect personnel against heat, cold, rain, snow, etc., which can decrease physical efficiency and increase the probability of accidents.
6. In hot weather, rotate shifts of workers wearing impervious clothing.

HEAT STRESS MONITORING

For monitoring the body's recuperative ability to excess heat, one or more of the following techniques should be used as a screening mechanism. Monitoring of personnel wearing impervious clothing should commence when the ambient temperature is 70°F or above. Frequency of monitoring should increase as the ambient temperature increases or as slow recovery rates are indicated. When temperatures exceed 85°F, workers should be monitored for heat stress after every work period. The following are important considerations:

1. Heart rate (HR) should be measured by the radial pulse for 30 sec as early as possible in the resting period. The HR at the beginning of the rest period should

not exceed 100 beats/min. If the HR is higher, the next work period should be shortened by 10 min (or 33%), while the length of the rest period stays the same. If the pulse rate is 100 beats/min at the beginning of the next rest period, the following work cycle is shortened by 33%.

2. Body temperatures should be measured orally with a clinical thermometer as early as possible in the resting period. Oral temperature (OT) at the beginning of the rest period should not exceed 99°F. If it does, the next work period should be shortened by 10 min (or 33%), while the length of the rest period stays the same. However, if the OT exceeds 99.7°F at the beginning of the next rest period, the following work cycle should be further shortened by 33% or should be measured at the end of the rest period to make sure that it has dropped below 99°F.
3. Body water loss (BWL) due to sweating should be measured by weighing the worker in the morning and in the evening. The clothing worn should be similar at both weighings; preferably the worker should be nude. The scale should be accurate to plus or minus 1/4 lb. BWL should not exceed 1.5% of the total body weight. If it does, the worker should be instructed to increase his daily intake of fluids by the weight lost. Ideally, body fluids should be maintained at a constant level during the work day. This requires replacement of salt lost in sweat as well.
4. Good hygienic standards must be maintained by frequent change of clothing and daily showering. Clothing should be permitted to dry during rest periods. Persons who notice skin problems should immediately consult medical personnel.

EFFECTS OF HEAT STRESS

If the body's physiological processes fail to maintain a normal body temperature because of excessive heat, a number of physical reactions can occur ranging from mild (such as fatigue, irritability, anxiety, and decreased concentration, dexterity, or movement) to fatal. Standard reference books should be consulted for specific treatment.

Heat-Related Problems

1. Heat rash caused by continuous exposure to heat and humid air and aggravated by chafing clothes. This decreases the ability to tolerate heat and is also a nuisance.
2. Heat cramps caused by profuse perspiration with inadequate fluid intake and chemical replacement (especially salts). Signs: muscle spasm and pain in the extremities and abdomen.
3. Heat exhaustion caused by increased stress on various organs to meet increased demands to cool the body. Signs: shallow breathing; pale, cool, moist skin; profuse sweating; dizziness and lassitude.
4. Heat stroke, the most severe form of heat stress. Body must be cooled immediately to prevent severe injury and/or death. Signs and symptoms are red, hot, dry skin; no perspiration; nausea; dizziness and confusion; strong, rapid pulse; coma.

EFFECTS OF COLD EXPOSURE

Persons working outdoors in temperatures at or below freezing may experience frostbite. Extreme cold for a short time may cause severe injury to the surface of the body. Areas of the body that have a high surface area-to-volume ratio, such as fingers, toes, and ears, are the most susceptible.

Two factors influence the development of a cold injury: ambient temperature and the velocity of the wind. Wind chill is used to describe the chilling effect of moving air in combination with low temperature. For instance, 10°F with a wind of 15 mph is equivalent in chilling effect to still air at –18°F.

As a general rule, the greatest incremental increase in wind chill occurs when a wind of 5 mph increases to 10 mph. Additionally, water conducts heat 240 times faster than air. Thus, the body cools suddenly when chemical protective equipment is removed if the clothing underneath is soaked with perspiration.

Local injury resulting from cold is included in the generic term frostbite. There are several degrees of damage. Frostbite of the extremities can be categorized into

1. *Frost nip or incident frostbite.* The condition is characterized by sudden blanching or whitening of skin.
2. *Superficial frostbite.* Skin has a waxy or white appearance and is firm to the touch, but tissue beneath is resilient.
3. *Deep frostbite.* Tissues are cold, pale, and solid; extremely serious injury.

Systemic hypothermia is caused by exposure to freezing or rapidly dropping temperature. Its symptoms are usually exhibited in five stages:

1. Shivering
2. Apathy, listlessness, sleepiness, and (sometimes) rapid cooling of the body to less than 95°F
3. Unconsciousness, glassy stare, slow pulse, and slow respiratory rate
4. Freezing of the extremities
5. Death

Standard reference books should be consulted for specific treatments.

Table 5.1 reviews the ways to prevent hypothermia.

HOT WORK PERMITS

In an effort to establish some control over operations using open flames or producing sparks, many industrial firms have instituted hot work permit programs, which require that authorization be secured before equipment capable of igniting combustible materials is used outside areas normally specified. The first step in inaugurating a hot work permit program is the development of a policy statement by management, such as that given in Figure 5.1. The type and extent of the program will depend upon the size of the plant or facility, the complexity

Table 5.1. Hypothermia Prevention

COLD KILLS IN TWO DISTINCT STEPS

Step One:

The moment your body begins to lose heat faster than it produces it, you are undergoing exposure. Two things happen:

1. You voluntarily exercise to stay warm.
2. Your body makes involuntary adjustments to preserve normal temperature in the vital organs.

Either response drains your energy reserves. The only way to stop the drain is to reduce the degree of exposure.

THE TIME TO PREVENT HYPOTHERMIA IS DURING THE PERIOD OF EXPOSURE AND GRADUAL EXHAUSTION

Step Two: Hypothermia

If exposure continues until your energy reserves are exhausted,

1. Cold reaches the brain, depriving you of judgement and reasoning power. You will not realize this is happening.
2. You will lose control of your hands.

This is hypothermia. Your internal temperature is sliding downward. Without treatment, this slide leads to stupor, collapse, and death.

YOUR FIRST LINE OF DEFENSE: AVOID EXPOSURE

1. STAY DRY. When clothes get wet, they lose about 90% of their insulating value. Wool loses less; cotton, down, and synthetics lose more.
2. BEWARE THE WIND. A slight breeze carries heat away from bare skin much faster than still air. Wind drives cold air under and through clothing. Wind refrigerates wet clothes by evaporating moisture from the surface. WIND MULTIPLIES THE PROBLEMS OF STAYING DRY.
3. UNDERSTAND COLD. Most hypothermia cases develop in air temperatures between 30 and 50°F. Most outdoorsmen simply can't believe such temperatures can be dangerous. They fatally underestimate the danger of being wet at such temperatures.
   a. 50°F water is unbearably cold. The cold that kills is cold water running down neck and legs, cold water held against the body by sopping clothes, cold water flushing body heat from the surface of the clothes.
   b. DON'T ASK "HOW COLD IS THE AIR?" ASK INSTEAD, "HOW COLD IS THE WATER AGAINST MY BODY?"
4. USE YOUR CLOTHES. Put on rain gear before you get wet. Put on wool clothes before you start shivering.

YOUR SECOND LINE OF DEFENSE: TERMINATE EXPOSURE

If you cannot stay dry and warm under existing weather conditions, using the clothes you have with you, terminate exposure.

1. BE BRAVE ENOUGH TO GIVE UP REACHING THE PEAK OR GETTING THE FISH OR WHATEVER YOU HAD IN MIND.
2. Get out of the wind and rain. Build a fire. Concentrate on making your camp or bivouac as secure and comfortable as possible.

Table 5.1. (continued)

NEVER IGNORE SHIVERING

Persistent or violent shivering is a clear warning that you are on the verge of hypothermia. MAKE CAMP.

FORESTALL EXHAUSTION

Make camp while you still have a reserve of energy. Allow for the fact that exposure greatly reduces your normal endurance.

You may think you are doing fine when the fact that you are exercising is the only thing preventing your going into hypothermia. If exhaustion forces you to stop, however briefly,

1. Your rate of body heat production instantly drops by 50% or more.
2. Violent, incapacitating shivering may begin immediately.
3. You may slip into hypothermia in a matter of minutes.

Appoint a Four-Weather Leader

Make the best-protected member of your party responsible for calling a halt before the least-protected member becomes exhausted or goes into violent shivering.

YOUR THIRD LINE OF DEFENSE: DETECT HYPOTHERMIA

If your party is exposed to wind, cold, and wet, THINK HYPOTHERMIA. Watch yourself and others for symptoms.

1. Uncontrollable fits of shivering
2. Vague, slow, slurred speech
3. Memory lapses, incoherence
4. Immobile, fumbling hands
5. Frequent stumbling, lurching gait
6. Drowsiness (to sleep is to die)
7. Apparent exhaustion, inability to get up after a rest

YOUR FOURTH AND LAST LINE OF DEFENSE: TREATMENT

The victim may deny he's in trouble. Believe the symptoms, not the patient. Even mild symptoms demand immediate, drastic treatment.

1. Get the victim out of the wind and rain.
2. Strip off all wet clothes.
3. If the patient is only mildly impaired:
   a. Give him warm drinks.
   b. Get him into dry clothes and a warm sleeping bag. Well-wrapped, warm (not hot) rocks or canteens will hasten recovery.
4. If the patient is semi-conscious or worse:
   a. Try to keep him awake. Give warm drinks.
   b. Leave him stripped. Put him in a sleeping bag with another person (also stripped). If you have a double bag, put the victim between two warmth donors. Skin-to-skin contact is the most effective treatment.
5. Build a fire to warm the camp.

SUBJECT: HOT WORK PERMITS PROCEDURES

TO: ALL CONCERNED

Steps will be taken immediately to put into practice a permit tag system which should provide better protection against fire from welding and other hot work in hazardous areas. This program, as outlined below, is intended to be a practical one. It must be realized by everyone that for it to be effective, the wholehearted cooperation of all concerned must be secured.

1. After an inspection of the entire plant has been made by the fire chief and after a discussion with the various department heads, areas throughout the plant will be designated as hazardous for *any type of welding, burning, spark, open flame, or not work.* Such areas will be prominently marked, and before hot work is done within any such area permit tags must be secured in order to help ensure that the area will be as free as possible from fire hazards and that proper precautions will have been taken.
2. Tags have been prepared on which pertinent information must be filled out by the parties concerned. Each employee who may do hot work will be given a supply of tags which he will keep on his person or equipment. When he is sent to perform work in a hazardous area, he will first have his immediate foreman check the area with him to determine if necessary precautions to prevent fire have been taken.
3. It will be the maintenance foreman's responsibility to notify the foreman in the department in which the work is to be done, and together they will sign the permit for the employee to do welding or other hot work.
4. The fire chief will then check the area for fire-safe working conditions, see that standby fire extinguishers are present, and assign a fire watcher when necessary. When he is satisfied with the precautionary measures, he will sign the permit tag and return it to the employee who is to do hot work.
5. The signed permit card will be kept by the employee doing the work until the job is completed, at which time he himself will check the area for fire.
6. The area will subsequently be checked by the maintenance foreman who will see that any extinguisher that may have been used is designated for recharging.
7. The completed tag will be turned in at the end of the day to the maintenance department which will collect and forward all tags to the fire chief for filling and record purposes.

Figure 5.1. Example of a policy statement. This is an example of a policy statement that can be issued by a company inaugurating a hot work permit program in order to instruct employees.

of the operations, and the degree of hazard present at the work site and in surrounding areas.

Silent features of the program are to

1. Inspect area where work is to be done.
2. Establish fire watches.
3. Provide fire extinguishing equipment.
4. Communicate with and coordinate all departments concerned.
5. Isolate combustibles from sources of ignition.
6. Limit unauthorized use of flame- or spark-producing equipment.

A hot permit form of tag (Figure 5.2) is generally used to administer the program. Although standard forms are available through insurance companies, many plants have developed special forms that specifically relate to their individual operation. Detailed information on hot work permit program is contained in the MSDS, Hot Work Permits, available from the National Safety Council.

RECONNAISSANCE

Introduction

The team(s) initially entering the site must accomplish one or more of the following objectives:

1. Characterize the hazards that exist or potentially exist affecting the public health, the environment, and the response personnel.
2. Verify existing information and/or obtain data about the incident.
3. Evaluate the need for prompt mitigation actions.
4. Collect supplemental information to determine the safety requirements for personnel initially and subsequently entering the site.

Before the team enters the site, as much information as possible should be collected, depending on the time available, concerning the types of hazards, degree of hazards, and the risks that may exist. Using available information (shipping manifests, transportation placards, existing records, container labels, etc.) or offsite studies, the team assesses the hazards, determines the need to go to the site, and identifies safety requirements.

Preliminary On-Site Evaluation

The initial on-site survey is to determine, on a preliminary basis, hazardous or potentially hazardous conditions. The main effort is to rapidly identify the immediate hazards that may affect the public, response personnel, and the environment.

PERMIT NO. __________

For electric, acetylene and welding burning operations with portable equipment in all locations outside of shop.

Date ____________________

Time Started ________ Finishing ______

Building ____________________

Dept __________ Floor ________

Location on Floor ________________

Nature of Job __________________

____

Operator ____________________

Employee No./SS No. ______________

All precautions have been taken to avoid any possible fire hazard and permission is given for this work. Check for no flammable chemicals in area at all times.

Signed ________________________
(Supervisor)

Signed ________________________
(Safety/Plant Mgr)

PERMIT NO. __________

Date ____________________

Building __________ Floor ________

Nature of Job __________________

____

Operator ____________________

INSTRUCTIONS TO OPERATORS

This permit is good only for the location and time shown. Return the permit when work is completed.

1. Permits should be signed by the supervisor of the welder or cutter and by the safety engineer or plant manager.
2. Obtain a written permit before using portable cutting or welding equipment anywhere in the plant except in permanent safeguarded locations.
3. Make sure sprinkler systems are in service.
4. Before starting, sweep floor clean, wet down wooden floors, or cover them with sheet metal or equivalent in outside work. Don't let sparks enter doors or windows.
5. Move combustible material 25 feet away. Cover what can't be moved with a fire retardant material or sheet metal, carefully and completely.
6. Obtain standby fire extinguishers and locate at work site. Instruct helper or fire watcher to extinguish small fires.
7. After completion, watch scene of work a half hour for smoldering fires, and inspect adjoining rooms and floors above and below.
8. Don't use the equipment near flammable liquids or on closed tanks which have held flammable liquids or other combustibles. Remove inside deposits before working on ducts.
9. Keep cutting and welding equipment in good condition. Carefully follow manufacturer's instructions for its use and maintenance.

Figure 5.2. A typical Hot Work Permit (Tag should be tied or mounted at the work site.) (Reference: National Safety Council, Chicago.)

Of major concern are the real potential dangers — fire, explosion, oxygen-deficient atmospheres, radiation, airborne contaminants, containerized or pooled hazardous substances — that could affect workers during subsequent operations.

The preliminary survey of a site, incident, or other potentially hazardous work area provides the basis for the site safety plan, the organization of the work zone, evaluation of the hazards present, and the development of the work plan. A thorough review of existing information and assessment of present as well as past site conditions are necessary for the safety of the workers and the organization of the work to be done. An overall hazard assessment minimizes the risk to the workers by preparing them with available knowledge about the site.

Most of the preliminary site work is accomplished off site through background review. This establishes the known data base to date providing such information as hazardous substances on site, past activities, sample results, and any incidents of fire, explosion, or injury. It prevents duplication of work effort or finds a need to reinforce data or raises questions for further investigative work. Discussions with state officials who may have been on site, past site employees or present site employees, and neighbors often provide information not found in files. Coupled with the background data, a visual reconnaissance of the site completes the information to proceed with a safe investigation of the site.

Background Review and Visual Reconnaissance

The background review often requires a bit of detective work. The available information can range from file drawers of reports and sample data to a site name and vague location, depending of whether the work to be done is remedial action or initial site characterization. The collection and evaluation of the background data is vital to the safety of the workers and the efficiency of the work.

Information needed from the background review encompasses the elements of the site safety plan, the organization of the field team and work zones, the sampling strategy, the emergency procedures, and the remedial action plan. A chronology of events to data associated with the site helps establish present site conditions. Therefore, the historical data should be reviewed to try and determine what hazardous substances exist on site, how they were disposed — buried drums, lagoon, tanks; where they are concentrated, who was involved — do records or inventories exist; and what remedial activities have occurred.

To determine the risk to the worker and develop a safety plan, the important element is the nature of hazardous substances present. This is necessary to establish the size of the field team, the level of protection, and the work practices on site. Identification of the hazardous substances provides information such as fire or explosion hazard; respiratory protection requirements; protective clothing requirements; emergency procedures; and migration potential and/or pathways. When and how the substance was disposed and the containment used help to determine the likelihood of its persistence and presence on site.

Organic Vapors and Gases

If the type(s) of organic substance(s) involved in an incident is known and the material is volatile or can become airborne, air measurements for organics should be made with one or more appropriate, properly calibrated survey instruments.

When the presence or types of organic vapors/gases are known, instruments such as a photoionization detector (PID), a flame ionization detector (FID), and/or a portable gas chromatograph operated in the total readout mode, should be used to detect organic vapors. Until specific constituents can be identified, the readout indicates total airborne substances to which the instrument is responding. Identification of the individual vapor/gas constituents permits the instruments to be calibrated and used for more specific analysis. Remember these instruments do not indicate toxic metals, inorganics, biological agents, or radioactive sources.

Sufficient data should be obtained during the initial entry to map or screen the site for various levels of organic vapors. These gross measurements can be used on a preliminary basis to

1. Determine levels of personnel protection.
2. Establish site work zones.
3. Select candidate areas for more thorough qualitative and quantitative studies.

Higher than background readings on the PID or the FID may also indicate the displacement of oxygen or the presence of combustible vapors.

Inorganic Vapors and Gases

The ability to detect and quantify nonspecific inorganic vapors and gases is extremely limited. Presently, the PID has limited specific detection capability, while the FID has none. If specific inorganics are known or suspected to be present, measurements should be made with appropriate instruments, if available. Colorimetric tubes can be used if substances present are known (or can be narrowed to a few) and appropriate tubes are available. (For more detailed information on monitoring equipment, see Chapter 10.)

Oxygen Deficiency

At sea level, ambient air must contain at least 19.5% by volume of oxygen. At lower percentages, air-supplied respiratory protective equipment is needed. Oxygen measurements are of particular importance for work in enclosed spaces, low-lying areas, or in the vicinity of accidents that have produced heavier-than-air vapors, which could displace ambient air. These oxygen-deficient areas are also prime locations for taking further organic vapor and combustible gas measurements, since the air has been displaced by other substances. In addition, an oxygen-enriched atmosphere increases the potential for fires.

Combustible Gases

The presence or absence of combustible vapors or gases must be determined. If readings approach or exceed 10% of the lower explosive limit (LEL) or lower flammability level (LFL), extreme caution should be exercised in continuing the investigation. If readings approach or exceed 25% of the LEL, personnel should be withdrawn immediately (see Chapter 2).

Before resuming any on-site activities, project personnel in consultation with experts in fire or explosion prevention must develop procedures for continuing operations.

Levels of Protection

Personnel must wear proper protective equipment when response activities involve known or suspected atmospheric contamination, when vapors, gases, or particulates may be generated, or when direct contact with skin-affecting substances may occur. Respirators can protect the lungs, gastrointestinal tract, and eyes against air toxicants. Chemical-resistant clothing can protect the skin from contact with skin-destructive and absorbable chemicals. Good personal hygiene limits or prevents ingestion of material.

While personal protective equipment (PPE) reduces the potential for contact with harmful substances, ensuring the health and safety of response personnel requires, in addition, safe work practices, decontamination, site entry protocols, and other safety considerations. Together, these protocols establish a combined approach for reducing potential harm to workers. Additional specific information in PPE is found in Chapters 6 and 7.

Security

Personnel must take into account various types of security measures to assure the confidentiality of written reports, location, ground monitoring wells, drums of hazardous waste, etc. One must remember to make provisions for the safety plan to include security. Table 5.2 lists some emergency telephone numbers that security, safety, and environmental personnel may need to know and use.

Table 5.2. Emergency Telephone Numbers

Emergency (Chemtrec) (ONLY IN EMERGENCY) 800-424-9300

The Chemical Transportation Emergency Center (Chemtrec) is a centralized toll-free telephone service which has been set up to provide prompt information on the handling of transportation emergencies where hazardous chemicals are involved. This has been established by the Manufacturing Chemists Association (MCA) and is manned 24 hours a day, 7 days a week.

Hazardous Materials — Restricted Articles "HOT" Lines
Important 24-Hour Telephone Numbers
for Emergency Information

Bureau of Explosives, Association of
American Railroads (Explosives) 202-293-4048

Chemtrec (MCA) Chemical Transportation Emergency Center
(Chemicals & Pesticides) 800-424-9300
(Alaska and Hawaii) 202-483-7616
(District of Columbia) 202-483-7616

Communicative Disease Center
(Etiologic Agents — Biomedical Material) 404-633-5313

National Response Center, NRC (Oil/Hazardous Substances) 800-424-3802

Department of Transportation (DOT)
Office of Hazardous Materials Operations 202-426-0656

U.S. Coast Guard 800-424-8802
For reporting of major incidents only 202-426-1830

Other Telephone Numbers Concerned With Hazardous
Materials for Use During Normal Business Hours

United Parcel Service National Customer Service 203-622-6000
DOT Information Center (questions on regulatory matters) 202-426-9280
or
202-426-9281

National Agricultural Chemical Association
Pesticide Information 513-961-4300

National Response Team
Oil and Chemical Spills 800-424-8802

SARA Title III
Trade Secrets (EPA) 800-535-0202
or
202-479-2449

REFERENCES

1. NIOSH/OSHA/USCG/EPA. "Occupational Safety and Health Guidance Manual for Hazardous Waste Site Activities," Document No. 85-115, U.S. Government Printing Office, Washington, DC (1985), chap. 3.
2. National Response Team, NRT-1 Hazardous Materials Emergency Planning Guide, National Response Team, G-WER/12,2100 2nd St. SW, Washington, DC (March 1987).
3. EPA/FEMA/DOT. "Technical Guidance for Hazards Analysis: Emergency Planning for Extremely Hazardous Substances, Government Printing
4. OSHA. "How to Prepare for Workplace Emergencies", OSHA-3088 (Revised), U.S. Department of Labor, Government Printing Office, Washington, DC (1988).
5. Belding, H. S., NIOSH. "The Industrial Environment — Its Evaluation and Control," U.S. Government Printing Office, Washington, DC (1973) pp. 563–572.
6. Business and Legal Reports, Hazardous Waste Compliance Checklists for Supervisors, Bureau of Law and Business, Inc., Madison, CT (1985) pp. 35–51.
7. Bowman, V. A. *Release Prevention Control and Countermeasures: A Plan for SARA*, (Northbrook, IL: Pudvan Publishing Co., Inc.)
8. OSHA. 29 CFR General Industry Standards, "Hazardous Waste Site Operations and Emergency Response," U.S. Government Printing Office, Washington, DC (Revised) (July 1988).
9. Sacarello, H. L. A. "Decreased Latency Period for Asbestos When Exposed to Additional Chemical Irritants," *J. Occup. Med.* (submitted).

CHAPTER 6

Engineering Controls

INTRODUCTION

General methods of controlling environment factors or stresses which may cause sickness, impaired health, or significant discomfort among workers are given in this chapter. They include the following:

1. Substitution of a less harmful material for one which is dangerous to health.
2. Change or alteration of a process to minimize worker contact.
3. Isolation or enclosure of a process of work operation to reduce the number of persons exposed.
4. Wet methods to reduce generation of dust in operations such as mining, quarrying, and drilling.
5. Local exhaust at the point of generation or dispersion of contaminants.
6. General or dilution ventilation with clean air to provide a safe atmosphere.
7. Personal protective devices, such as special clothing or eye and respiratory protection.
8. Good housekeeping, including cleanliness of the workplace, waste disposal, adequate washing, toilet, and eating facilities, healthful drinking water, and control of insects and rodents.
9. Special control methods for specific hazards, such as reduction of exposure time, film badges and similar monitoring devices, continuous sampling with preset alarms, and medical programs to detect intake of toxic materials.
10. Medical controls.
11. Training and education to supplement engineering controls. A generalized diagram of these methods is shown in Figure 6.1.

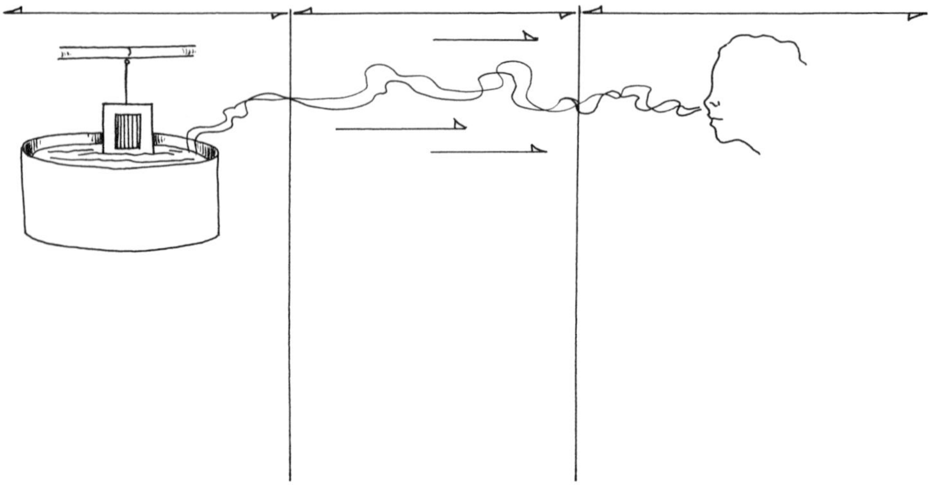

Figure 6.1. Generalized diagram of methods of hazard control.

SUBSTITUTION

Replacement of a toxic material with a less harmful one is a very practical method of eliminating an industrial health hazard. In many cases a solvent with a lower order of toxicity or flammability may be substituted for a more hazardous one.

In a solvent substitution, it is always advisable to experiment on a small scale before making the new solvent part of the operation or process. For example, carbon tetrachloride can be replaced by solvents such as methyl chloroform, dichlormethane, aliphatic petroleum hydrocarbons, or one of the fluorochlorohydrocarbons. (The precautions listed earlier should be reviewed.) Detergent-and-water cleaning solutions can be considered for use in place of organic solvents. Benzene can be replaced by toluene in most lacquers, synthetic-rubber solutions, and paint removers. Natural rubber cements with aliphatic hydrocarbon solvents can perform virtually the same function as benzene cements. The felt hat industry has controlled the hazard of mercurialism by substituting mercury-free carroting materials. Foundries using parting compounds that contain free silica can minimize the silicosis hazard by substituting relatively harmless powders. Sandstone grinding wheels have been largely replaced by artificial abrasive wheels usually made of aluminum oxide.

Figure 6.2. Solder grinders wear hoods to prevent foreign matter from entering their eyes, ears, noses, or mouths. Filtered air is supplied to the hoods at 12 to 15 psi; air is bled off around armpit openings.

CHANGING THE PROCESS

A change in process often offers an ideal chance to improve working conditions. Most such changes, of course, are made to improve quality or reduce cost of production, only occasionally to improve the in-plant environment. Yet, we must always keep this possible benefit in mind. This applies to the safety professional and industrial hygienist more than anyone else.

In some cases, a process can be modified to reduce the exposure to a dust or fume and thus markedly reduce the hazard. In the automobile industry, the amount of lead dust created by grinding solder seams with small, rotary, high-speed sanding disks was greatly reduced by changing to low-speed, oscillating-type sanders (Figure 6.2). Brush-painting or dipping instead of spray painting will minimize the concentration of airborne contaminants from toxic pigments (Figure 6.3). Other examples of process changes are arc-welding in place of riveting, vapor decreasing with adequate controls to replace hand-washing of parts in open containers, airless spraying techniques and electrostatic devices to minimize overspray as replacements for hand-spraying, and machine application of lead oxide to battery grinds which reduces lead exposure to operators in making storage batteries.

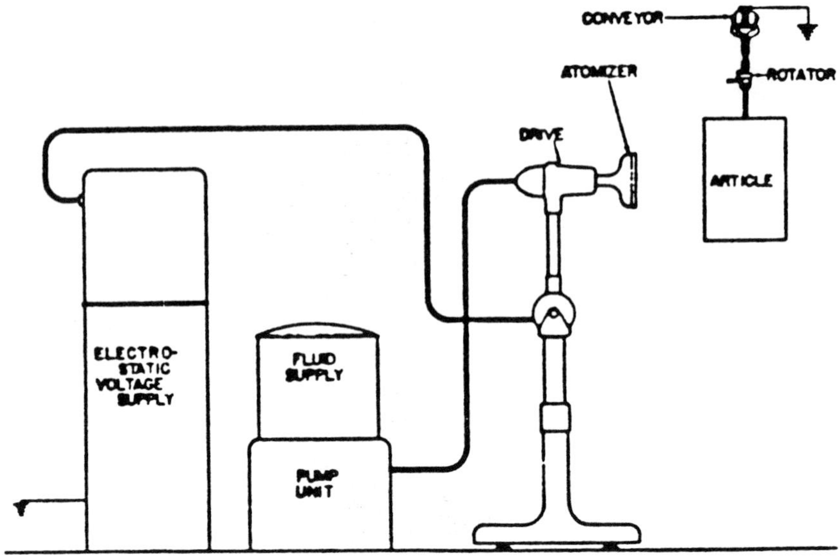

Figure 6.3. Electrostatic paint-spraying device minimizes overspray. Paint particles, passing through a negatively charged electrostatic field, assume negative polarity. They are then attracted to the work, that has been positively charged.

In buying individual machines, the need for accessory ventilation, noise suppression, and vibration and heat control should be considered before purchase. Often the best approach is to discuss these requirements with the supplier in advance.

ISOLATION OR ENCLOSURE

Some potentially dangerous operations can be isolated from the people nearby, which solves the exposure problem. The isolation can be by a physical barrier (such as an acoustic box to contain noise from a wining blower or a screaming rip saw), or by time (such as providing semiautomatic equipment so a person doesn't have to stay near the noisy machine constantly), or by distance (remote control).

Isolation is particularly useful for limited operations requiring relatively few workers or where control by any other method is too difficult or too expensive. The job is isolated from the rest of the operation and thus the majority of the workers are not exposed to the hazard. The workers actually concerned with the operation can then be protected by installing a ventilation system, which probably would not have been satisfactory if the operation had not been isolated and thus able to be enclosed (Figure 6.4).

It is not feasible to enclose the exhaust of certain abrasive blasting operations, such as those found in ship building. Especially where dry sand is used, it is desirable to isolate the operation. As much of the blasting as possible should be

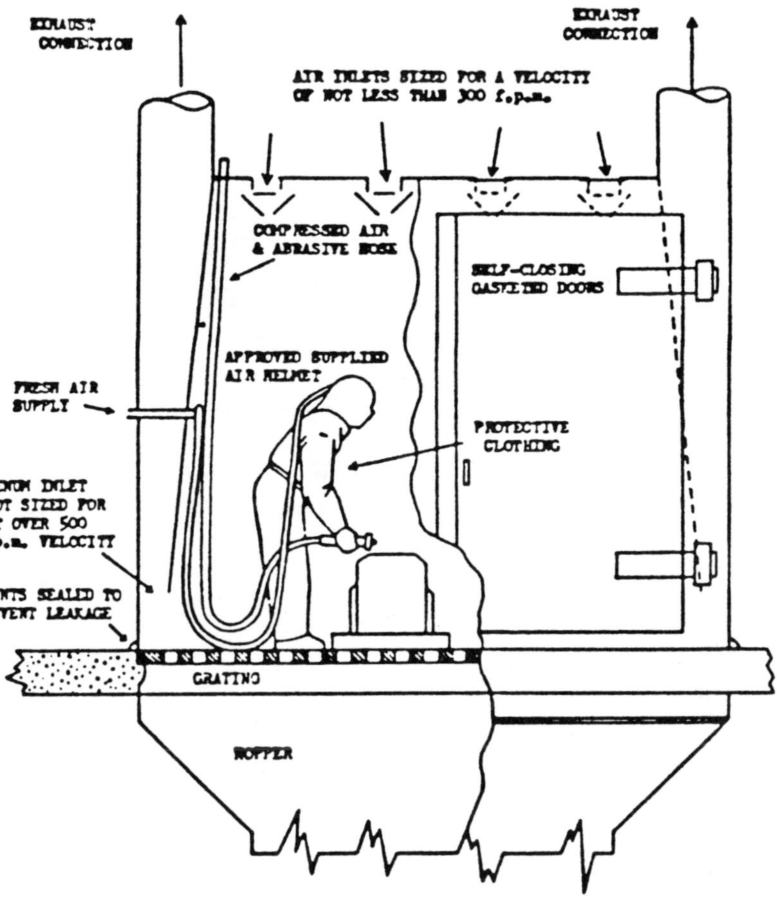

Figure 6.4. Air inlets and exhaust are arranged to sweep contaminated air away from worker's breathing zone in this enclosed sand blast area. Downdraft averages 80 fpm over the entire floor area. Air should exhaust downward, as shown here, or on two sides of the room at floor line.

done in a specified location which is as far as possible from employees other than the operator. Another way to isolate the blasting is to do it on an "off shift" when few other employees are working. Any employee working in the area should wear an air-supplied respirator.

Enclosing the process or equipment is a desirable method of control since the enclosure will prevent or minimize the escape of solvent vapor into the workroom atmosphere. Where some of the more highly toxic solvents are used, enclosure should be one of the first measures attempted, after considering substitution. In most instances, the enclosure should also be exhausted to avoid buildup of solvent vapors. Additional precautions must be taken when cleaning enclosed equipment or during startup or shutdown in order to avoid high concentrations of solvent vapors.

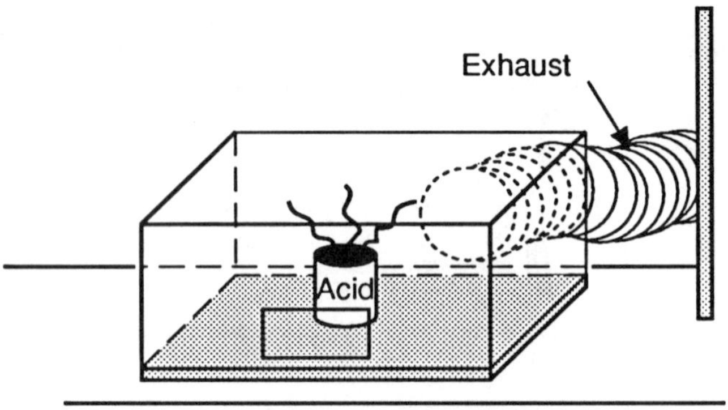

Figure 6.5. Booth vented outside for hazardous material, all self-contained.

Enclosure of an operation effectively prevents the dispersal of contaminants. Enclosed equipment is usually run tightly closed and is only opened during cleaning or filling operations. Further examples of where this type of control is effective are radium dial painting, glove booths (Figure 6.5), airless blast or shot blast machines for cleaning castings, and abrasive blasting cabinets.

In the chemical industry, the isolation of hazardous processes in closed systems is a widespread practice, which is one reason why the manufacturer of toxic substances is often less hazardous than their use under less well controlled conditions. In the mechanical industries, complete enclosure is frequently the best solution to severe dust or fume hazards, such as those from sand-blasting or metal-spraying operations.

WET METHODS

Dust hazards can frequently be minimized or greatly reduced by application of water or other suitable liquid at the source of dust, a method often used for silica and loose dusts (Figure 6.6). Wetting of floors before sweeping to keep down the dispersion of harmful dust is advisable when better methods, such as vacuum cleaning, cannot be applied.

"Wetting down" is one of the simplest methods for dust control. Its effectiveness, however, depends upon proper wetting of the dust. This may require the addition of a wetting agent to the water and proper disposal of the wetted dust before it dries out and is redispersed. Tremendous reduction in dust concentrations have been achieved by the use of water forced through the drill bits used in rock drilling operations. Many foundries successfully use water under high pressure for cleaning castings in place of sandblasting. Airborne dust concentrations can be kept down if molding sand is kept moist, if castings are wetted down before shakeout, and if the floors are wetted intermittently.

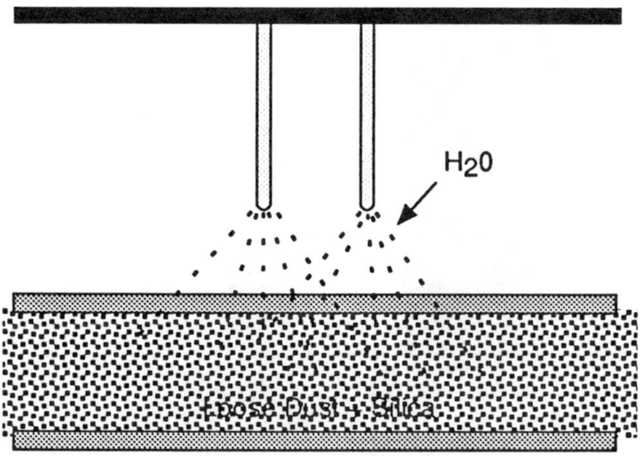

Figure 6.6. Water sprays reduce nuisance dust at transfer or discharge points. A fan or cone-shaped water spray is effective in beating down and wetting the dust.

LOCAL EXHAUST VENTILATION

A local exhaust system traps the air contaminant near its source so that a worker standing at the process is not exposed to harmful concentrations. This method is usually preferred to general ventilation, but should be used when the contaminant cannot be controlled by substitution, changing the process, or isolation or enclosure. Even though a process has been isolated, it may still require a local exhaust system.

After the system is installed and set in operation, its performance should be checked to see that it meets engineering specifications — correct rates of air flow, dust velocities, negative pressures, and the others. Its performance should be rechecked periodically as a maintenance measure (Figure 6.7).

GENERAL OR DILUTION VENTILATION

General or dilution ventilation — adding or removing air to keep the concentration of a contaminant below hazardous levels — uses natural convection through open doors or windows, roof ventilators, and chimneys, or artificial air currents produced by fans or blowers. Exhaust fans through roofs, walls, or windows constitute positive all-season dilution ventilation. Consideration must be given to providing makeup air, especially during winter months. Dilution ventilation is practicable only if the degree of air contamination is not excessive and particularly if the contaminant is released at a substantial distance from the worker's breathing zone. Under other more adverse conditions, the contaminated air will not be diluted sufficiently before inhalation.

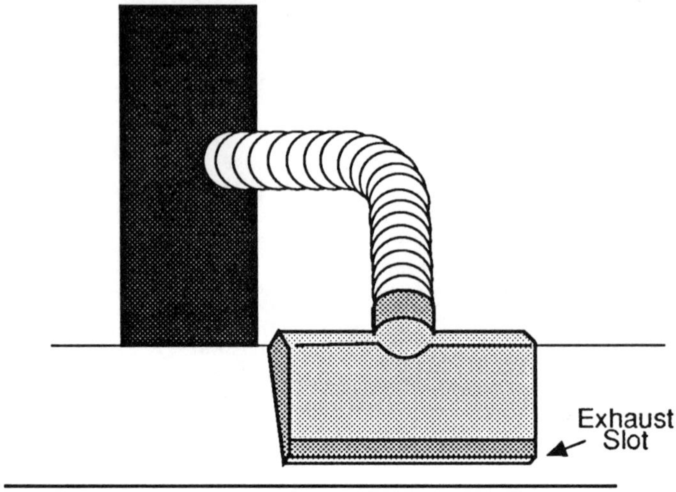

Figure 6.7. Local exhaust system traps contaminants near the source so that the worker is not exposed to harmful concentrations.

General ventilation should not be used where there are major, localized sources of contamination (especially highly toxic dusts and fumes); local exhaust is more effective and economical in such cases. Where comparatively small amounts of the less toxic solvents are vaporized, general or dilution ventilation can be a satisfactory method of control.

PERSONAL PROTECTIVE EQUIPMENT (PPE)

When it is not feasible to render the environment completely safe, it may be necessary to protect the worker from the environment. PPE is normally considered to be secondary to the controls mentioned previously: where it is not possible to enclose or isolate the process or equipment, provide ventilation, or other control measures; where there are short exposures to hazardous concentrations of contaminants. For example, where unavoidable spills occur, PPE should be provided and used.

Personal protective devices have one serious drawback — they do nothing to reduce or eliminate the hazard. Their failure means immediate exposure to the hazard, so the fact that a protective device may become ineffective without the knowledge of the wearer is particularly serious. Excellent equipment that meets accepted national standards and specifications is commercially available in great variety. Such equipment, however, is intended for emergency or temporary use only.

Eye and face protection includes safety goggles, face shields, and similar items used to protect against corrosive solids, liquids, vapors, and foreign bodies.

Figure 6.8. Properly attired supervisor (left) checks technicians job procedure. To protect against any slashed both men wear goggles, apron, and gloves.

Shaded lenses are used to screen out ultraviolet and infrared radiations. Of the many types of eye and face protection available, there is a correct type for each job, and it should be worn at all times on that job (Figures 6.8 and 6.9).

Ear Protection

Protection against noise-induced hearing impairment, such as ear plugs or muffs, may often present special difficulties. First, as with air contaminants, the real answer is to reduce the noise exposure. But in plants with old machines and processes, and even in new plants, this problem may loom as insoluble. It may be overwhelmingly costly to lower the noise. In some cases, it may be required to make wearing of ear protection mandatory (Figure 6.10).

Protective Clothing

Gloves, aprons, boots, coveralls, and other items made of impervious materials should be worn to control or eliminate prolonged or repeated contact with

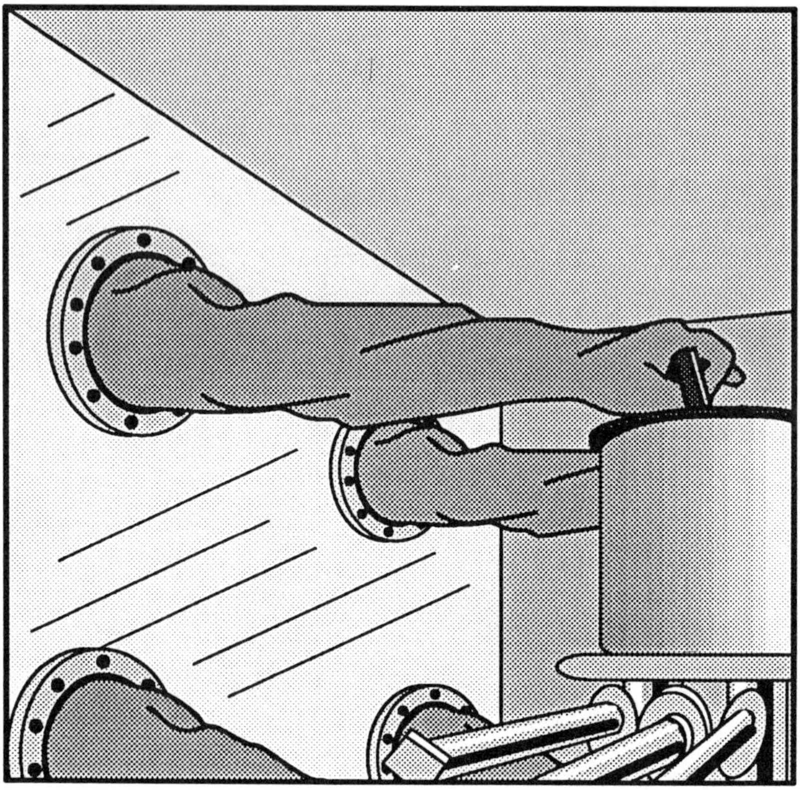

Figure 6.9. Where super hazards require "super protection" glove boxes can be used to protect the worker by providing a positive barrier.

dermatitis-producing solvents or chemicals that may cause systemic poisoning through skin absorption. Again, care must be taken in choosing the correct article for the specific application. For example, some types of rubber that will withstand trichloroethylene will become spongy and disintegrate in lacquer solvent.

For intermittent protection against radiant heat, reflective aluminized clothing can be used. Such garments need special care to preserve their essential shiny surface. Air-cooled jackets and suits are available to make endurable a high convective heat load. Lead-bearing materials are available to protect against ionizing radiation.

Respiratory protective devices are normally restricted to intermittent exposures or those that are impractical to control by other methods. Respiratory protection should not be substituted for engineering control methods. Exceptions are supplied-air devices for protection in sand blasting or for operations in confined spaces where an oxygen deficiency may exist.

There are two types of respiratory protective devices: (1) air purifiers, which remove the contaminant from the air by filtering or chemical absorption before

Figure 6.10. Personal ear protection must be used where engineering controls cannot reduce noise hazards to a safe minimum.

inhalation, and (2) air suppliers, which provide clean air from an outside source or oxygen from a tank. Full details of types approved by the U.S. Bureau of Mines and NIOSH should be obtained from the manufacturer.

Filter respirators cover the mouth and nose, but do not protect the eyes. For dust protection, there are a large number of respirators that have met the requirements of the U.S. Bureau of Mines which call for high filtering efficiency and low resistance to breathing. A small number of respirators have been approved for protection against metal fumes and mists.

Air-line respirators are usually preferable to chemical cartridge or mechanical filter respirators, because they are cooler, offer more positive protection, and offer no resistance to breathing. However, they require a proper compressor to supply fresh air (Figure 6.11).

Gas masks are preferable to respirators if the eyes must be protected or if higher concentration of gases or vapors are probable. Masks are available that protect against acid vapors, ammonia, organic vapors, carbon monoxide, and certain other vapors and gases.

Figure 6.11. An air-line respirator protects spray painter in enclosed area.

Self-contained oxygen apparatus — used for emergency and rescue work — have masks attached to oxygen cylinders or canisters which supply oxygen from chemical reaction. Such apparatus enables a worker to enter a contaminated or oxygen-deficient atmosphere and still retain some mobility for relatively short periods of time.

Selection of the proper type of respiratory protective equipment should be based on the following procedure:

1. Identify the substances against which protection is necessary.
2. Know the hazards of each substance and the significant properties of each.
3. Determine the levels of air contamination, oxygen deficiency, and the conditions of exposure.
4. Determine if any personal capabilities and characteristics are essential to the safe use of the devices or procedures required.
5. Determine what facilities are needed for maintenance.

Since a respirator often becomes uncomfortable after wearing for extended periods, the worker must fully realize the need for protection or he or she will not wear it. To get the worker's cooperation, these factors are important:

1. Prescribe respiratory protection equipment only after every effort has been made to eliminate the hazard.
2. Explain the situation fully to the worker.
3. Fit the respirator carefully.
4. Provide for maintenance and cleanliness, including sterilization before reissue.
5. Instruct the worker in the proper use of the respirator.

Protective creams and lotions help minimize skin contact with irritant chemicals. Their effectiveness varies, but if properly selected and correctly used, they can be very helpful. The cream or lotion must be selected on the basis of competent medical advice. The worker must then be instructed in the value of the protection and its proper application.

HOUSEKEEPING

Good housekeeping plays a key role in occupational health protection. Basically, it is another tool in addition to those already listed for preventing dispersion of dangerous contaminants. Dust on overhead ledges and on the floor can readily be dispersed to the in-plant atmosphere by traffic, vibration, and random air currents. Housekeeping is always important; where there are toxic materials, it is paramount.

Immediate cleanup of any spills of toxic material is a very important control measure. A regular cleanup schedule using vacuum cleaners or lines is the only true effective method of removing dust from the work area. An air hose for blowing away dust should not be used.

Good housekeeping is also essential where solvents are stored, handled, and used. Leaking containers or spigots should be remedied immediately by transferring the solvent to sound containers or by repairing the spigots. Spills should be cleaned up promptly by workers wearing protective equipment. All solvent-soaked rags or absorbents should be disposed of in airtight metal receptacles and removed daily from the plant.

It is impossible to have an effective health program unless maintenance housekeeping is good and the worker has been informed of the need for those measures. For example, if the thermostat on a vapor degreaser fails, or is accidentally broken, excessive concentrations of trichloroethylene might quickly build up in the work area unless the equipment is shut down and the necessary repairs are made immediately.

WASTE DISPOSAL

Orderly controlled disposition of all hazardous materials should be done by highly trained individuals under strict supervision. Most regulations prohibit the disposal of dangerous substances in any sewer system. Actually it is not desirable to dispose of any material that gives off objectionable odors in the sewer systems.

Procedures should be established for the disposal of unused dangerous chemicals, toxic residues and contaminated waste, containers of chemicals which are no longer needed, and containers from which labels have been lost or obliterated.

Final disposal of dangerous chemicals should be carried out at a disposal area remote from inhabited areas as required by RCRA.

The disposal of hazardous materials in the sewage system can create many problems; the commingling of materials such as acids and caustics with phosphorus, arsenic, and cyanide compounds, for example, can create health hazards. At any given time no one knows what a neighboring company may be dumping into a sewer, if it is permitted. Consequently, there can be disastrous results.

A competent chemist, by observing caution and using good chemical common sense, can neutralize a number of chemicals, dilute them highly with water, and safely dispose of them in the sewer system. Milliliter quantities of common flammable liquids which are miscible with water can also be disposed of in this manner, but the more complex dangerous chemicals do not lend themselves to safe disposal by this method. In each case, evaluation and approval from solid waste, RCRA, and local governmental agencies must approve each disposal to the sewer.

If a dangerous chemical is to be used in a plant or operation, it will be profitable to perform experimental investigations using microquantities to determine a means of neutralizing and rendering it harmless before full scale operations are begun. With this knowledge, the neutralizing agent can be prepared in advance and can be readily available for immediate use in an emergency. There are a number of methods published in the literature by which some dangerous chemicals can be rendered safe for disposal. Each case must be evaluated individually and must meet the requirement of RCRA as a treatment facility.

A number of systems and facilities have been suggested for disposal of dangerous chemicals — all of them are expensive and none of them provides a universal means of disposal for all hazardous materials. The deep well requires geological investigations and the approval of the authorities that have jurisdiction, plus a large monetary expenditure. The furnace type requires an adequate stack to carry off the products of combustion. This is also expensive, is subject to air pollution controls, and still generates off-gases and other products of combustion which may not be acceptable in a populated area. Some hazardous chemicals can be rendered safe by reaction with neutralizing reactants. This entails elaborate scrubber facilities, holding tanks, and final treatment facilities, all of which are expensive, require extensive investigation and study, and have limited application.

A new disposal method for organic solvents is the use of bacterial agents, supplemented with nutrients, which actually eat up the organic solvents. This process is known as bioreclamation. There are a few companies that specialize in this area of technology, such as Shell Oil and Bacardi Corporation.

Each disposal method should be considered individually before emergency disposal is imminent, and prior knowledge and the proper neutralizing materials should be available. If possible, have a remote disposal area, collect the materials in adequately designed containers, and have trained personnel to dispose of them safely without creating a nuisance or hazardous pollution problem. Every industry and government agency that generates any form of waste should be looking for waste reduction methods.

SPECIAL CONTROL METHODS

Many of the general methods mentioned previously (either along or in combination) can be used for the control of most occupational health hazards. A few additional situations, however, deserve special mention. Shielding is one of the best protective measures employed to reduce or eliminate exposures to ionizing radiation. Lead and concrete are two materials commonly employed to shield high-energy sources such as high-voltage X-ray, particle generators, and radioisotopes. Shielding is also used to protect against exposure to a radiant heat source.

Reduction of work periods is another method of control in limited areas where engineering methods are too costly or impracticable. In the forge industry, especially in hot weather, a shorter work day and frequent rest periods are used to minimize exposures to extremely high temperatures, thereby lessening the danger of heat exhaustion or heat stroke. For workmen who must labor in a compressed air environment, schedules of maximum length of shift and length of decompression have been published. The higher the pressure, the shorter the shift and the longer the decompression period.

Proper illumination of the work area is necessary to prevent eye strain leading to increased accidents or loss of visual acuity and to eliminate glare that may interfere with vision.

RESPONSIBILITY FOR CONTROL

Top management, the safety department, the purchasing department, the medical department or company physician, the maintenance department, the supervisors, and the workers themselves all have specific responsibilities for the prevention of industrial health diseases and the control of exposure to the body from these irritants.

In some processes, it is possible to reduce exposure by lowering the temperature of the process and decreasing the air motion around the operation. (Under some conditions, however, decreased air motion may promote rather than reduce exposure. It should be kept well in mind, therefore, that every exposure may require special study.)

An example of how this process should be applied would be for individual skin diseases. To control or eliminate dermatitis in a plant or unit, management should first recognize the scope of the problem and then delegate authority for action to the proper persons. Where it is necessary to have more than one department work on phases of dermatitis control or elimination, the activities of those departments should be coordinated at the management level. Periodic reports on the status of the dermatitis problem within the organization should be made to management by its delegated representatives.

The safety department should gather information on dermatitis hazards of materials used in the plant and should disseminate this information among foremen and other operating personnel. This department should make periodic surveys to check for exposures to skin irritants and should suggest means to correct any hazards found.

The purchasing department can sometimes help prevent unwilling exposure to skin irritants by procuring from the supplier information regarding the ingredients in a particular compound which have been known to give rise to skin irritation. Such information should then be passed on to the operating departments prior to use of the material.

Also, the medical and safety departments, should be constantly on the alert to gather and utilize published data pertaining to dermatoses-producing agents.

The medical department or company doctor should keep in mind the skin irritants to which the employee will be exposed and check especially for factors which may predispose the individual to dermatitis. The company physician should also make periodic physical examinations of workers exposed to skin irritants, watching particularly for skin affections. The medical department should work closely with the safety department and other departments to discover substances or conditions in the physical environment which may result in skin affections among employees.

NOISE CONTROL MEASURES

After the proper survey of existing noise levels for various environments and machines has been made, and after the measuring instruments have shown that an area is too noisy for comfort, what steps can the safety professional take to reduce noise levels for employees working in those areas? (See Figure 6.12.)

There are two approaches:

1. The environment and the machines can be redesigned so that the over-all noise level is reduced — noise reduction at the source.
2. Where it is impossible or not feasible to immediately reduce noise to a comfortable or safe level, individuals can be provided with protective devices designed to reduce possible aural injury.

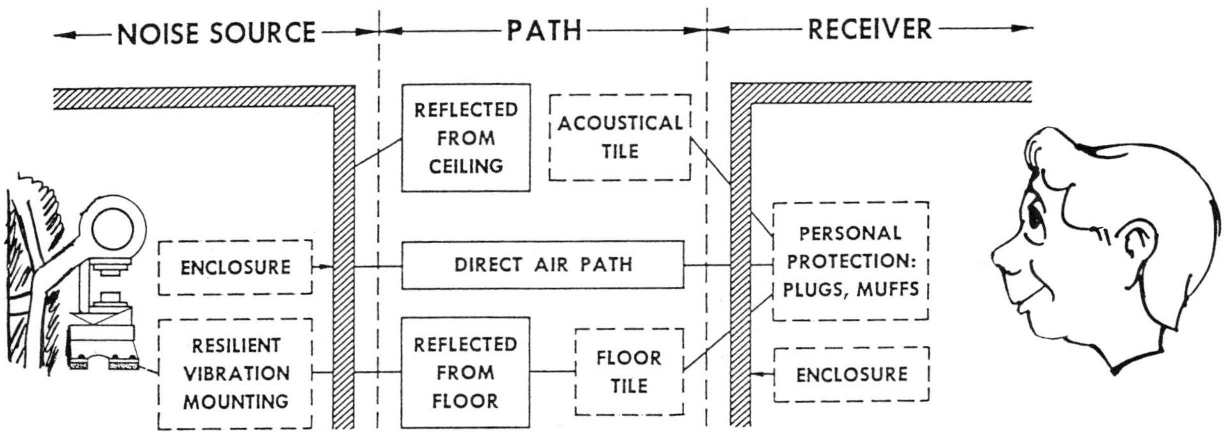

REDUCTION OF NOISE AT SOURCE BY:

1. Acoustical design
   a. Decrease energy for driving vibrating system.
   b. Change coupling between this energy and acoustical radiating system.
   c. Change structure so less sound is radiated.
2. Substitution with less noisy equipment.
3. Change in method of processing.

REDUCTION OF NOISE BY CHANGES IN PATH:

1. Increase distance between source and receiver.
2. Acoustical treatment of ceiling, walls and floor to absorb sound and reduce reverberation.
3. Enclosure of noise source.

REDUCTION OF NOISE AT RECEIVER BY:

1. Personal protection.
2. Enclosures — isolating the worker.
3. Rotation of personnel to reduce exposure time.
4. Changing job schedules.

Figure 6.12. Three places that noise can be reduced — at its source, along its path, and near the receiver. Diagram illustrates the use of systems analysis in the solution of noise control problems.

CONTROL OF STRUCTURE-BORNE SOUND

The general principles used to reduce structure-borne sound are

1. Isolate the source of vibration from a sound-radiating surface.
2. Place a sound barrier between the noise source and the equipment or machine exterior.
3. Damp the amplitude of vibration of the sound-radiating surface.

Vibration-isolation materials make use of a resilient mounting to separate the energy source from the vibrating surface. Products that reduce transmission of sound by conduction include: high-density glass wool materials, steel springs, elastomers, cork, and felt.

Sound barriers utilize a material capable of retarding air-borne sound transmission. They can be air-impervious materials and include glass wool blankets with impervious facings, asbestos, asphalt-treated gypsum board, hardboard or plywood, plastic sheeting, lead-loaded vinyl sheeting, and/or metal sheet or heavy foil.

Panel damping consists of the application of a material with high internal damping properties. The effect of such treatment is to reduce the ring or tinny sound of a metal surface. To decrease the amplitude of vibration of panels, the following may be used: asphalted felt (applied by adhesive), damping adhesives, damping tapes. Vibration isolators break the path of vibration because they are resilient and act as springs.

With sound barriers of the air-impervious sheet type, the incident sound waves set the sheet into vibration, and the sheet generates new sound waves of reduced intensity on the other side. The effectiveness of a barrier increases directly with its weight per square foot. For sound barriers of the faced glass wool type, in addition to the barrier action of the facing, the sound energy is absorbed in passing through the porous layer of the glass wool material and is converted into heat. Vibration dampeners are heavy, limp materials with high internal friction, which reduces the vibration of the panel.

Certain physical principles of noise control methods must be considered. For example, the property of resonance is basic to control by engineering methods. When a small object — such as a watch or a piano wire — is isolated, it produces relatively little sound. But, when it is in contact with a large acoustic conductor — such as the sounding board of a piano — there is a transfer of sound energy that results in considerable amplification. Machinery and equipment act the same way if insulators or isolators are not provided to prevent the noise from being transmitted and considerably amplified (Figure 6.13).

In some cases, it is possible to substitute quiet operations or devices for more noisy ones. Welding can be substituted for riveting in some operations without loss of strength or efficiency to the resulting product. Also, in many noisy environments, flasher lights instead of bells may be used as signals.

Too often, the rattles, squeaks, and thumps of machinery indicate that maintenance has been neglected. Frequently, considerable noise can be eradicated through a good housekeeping program, which includes regular inspection, tightening of

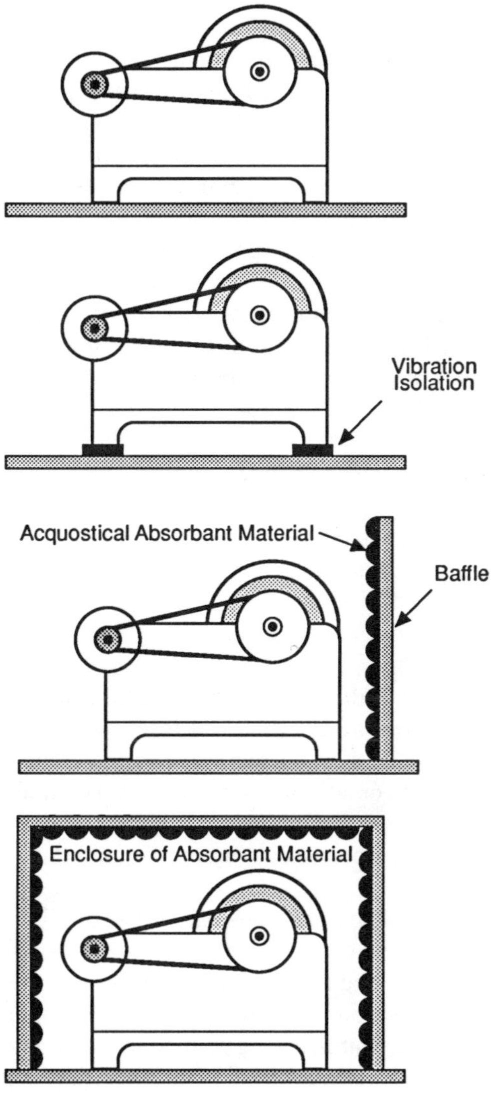

Figure 6.13. Noise reduction that is given by use of vibration isolators, a baffle of acoustical absorbent material, and enclosure of absorbent material. Original octave band analysis is shown as some dashed lines.

loose connections, and periodic lubrication of equipment, together with the installation of proper damping to reduce unnecessary vibration of parts.

Although proper maintenance is an extremely important phase of noise reduction, machinery eventually becomes too worn to accomplish properly the task for which it was designed. When that state of obsolescence is reached, substitution of more modern equipment, together with a complete survey of plant layout and operations, can often mean lower noise levels and increased safety and production.

When new construction is necessary, both sound and air conditioning of the new area should be considered. When both types of conditioning are utilized, the walls of the room should be constructed of materials of high density and mass and low elasticity, as compared to air. The result is low sound transmission. Combined with a layer of absorbent material, it provides good insulation for efficient air conditioning. The two types of conditioning then work together to isolate the individual from unnecessary noise.

Where the walls produce a high degree of sound reflection and there is no absorption, the echoes from a sound source do not have a chance to die away before another increment of sound is added. Under such conditions, the installation of acoustic materials on walls or ceilings can be of great benefit in bringing over-all sound levels down to the safe exposure point.

CONTROL OF AIR-BORNE SOUND

Engineering control of air-borne noise can be summed up in three words: absorption, isolation, substitution.

Absorption

Current practices include the installation of mechanical barriers that absorb noise within an area or prevent its movement from one area to another.

Absorbent materials reduce reflection or echoing, and may be acoustic wallboard, glass fiber, panels, or any other building material that combines high porosity and resistance to air flow. Numerous types of absorbent materials are available, and more are continually coming on the market.

Isolation

When isolation is the method used, noisy machines may be separated from one another so that their noise output is not so great within a single area; or they may be enclosed in highly absorbent housings with impervious outer walls. Usually a study of the overall operation may lead to changes that, together with better maintenance, will reduce the noise levels.

In order to have good sound isolation between two spaces, it is necessary to separate them by a heavy, impervious material — such as sheet metal, masonry, lead, or wood. The heavier the material, the more difficult it becomes from the incident sound wave to produce any vibration of the wall. It is the vibration of the wall that causes sound to be radiated into the adjoining space.

Studies within rooms enclosed by a sound isolating partition show that the reflected sound is almost as intense as the incident sound. The reflections on one side of the partition will cause a reverberant field to be built up. The increase in the intensity at the partition over that shown for a wall covered with absorbing material will cause unnecessary vibration of the partition.

Better sound isolation between two rooms can be obtained if the reverberant buildup in the first room is eliminated. A partition that combines the advantages of an absorbing material to reduce reverberation and a heavy, impervious panel to reduce the transmission of sound can obtain the optimum sound isolation possible with a single-wall construction. Double-wall construction may be necessary to control some industrial noise problems (Figure 6.14).

This combination of sound-absorbing and -isolating materials is the basic component of most designs for the control of air-borne noise. Often a protective facing is placed in front of the absorbing material to protect it from damage. Usually, this facing will be a perforated material such as sheet metal, press board, or mineral board; sometimes a thin plastic or glass wool bag may cover the absorbing material. A properly designed wall structure will find wide application to noise control problems.

In constructing an isolating partition, it is mandatory that no leaks — no matter how small — be permitted. In noise control, enclosure means that all cracks and openings are sealed so that the resulting structure is completely air tight.

To illustrate this point, an opening 1 in. in diameter in a concrete block wall 10 ft^2 would, in the mid-frequency range, transmit as much sound as the entire remainder of the wall. Similarly, a door placed in this wall that has a crack between it and the jamb of only a few thousandths of an inch will transmit as much sound through the crack as through the rest of the wall.

Therefore, all cracks should be sealed with an elastic caulking compound. Openings for ducts, conduits, piping, and control cables should have tight-fitting gaskets. Doors should be sealed with an extruded rubber gasket, made in the form of a hollow tube. The enclosure should be inspected and reinspected so that all leaks will be eliminated permanently.

It is often impossible to enclose a noise source and still maintain any production. In such a case, the smallest tolerable opening should be used. If an opening is much too large, and there is no absorbing material on the inside of the structure, there will be an increase in the noise at the operator's station.

A method of noise control that will permit operation may provide only a few decibels of quieting in the important speech frequency range. In many cases, this reduction may be sufficient to eliminate effectively the hazard or annoyance. If not, it may be necessary to utilize some method of remote automatic feeder control and to enclose the noise source. Surprisingly enough, remote control will often be less expensive than redesign of the machine to reduce the noise at the source. Some ideas are shown in Figures 6.15 through 6.19.

The approach to a noise control problem can best be summed up as follows. Consider the source:

1. Can a quieter machine be substituted?
2. Can the noise energy be reduced (low rpm)?
3. Can a useful change be made in the directivity pattern?
4. Can resilient mounts be used to reduce vibrations?
5. Can a muffler be used?

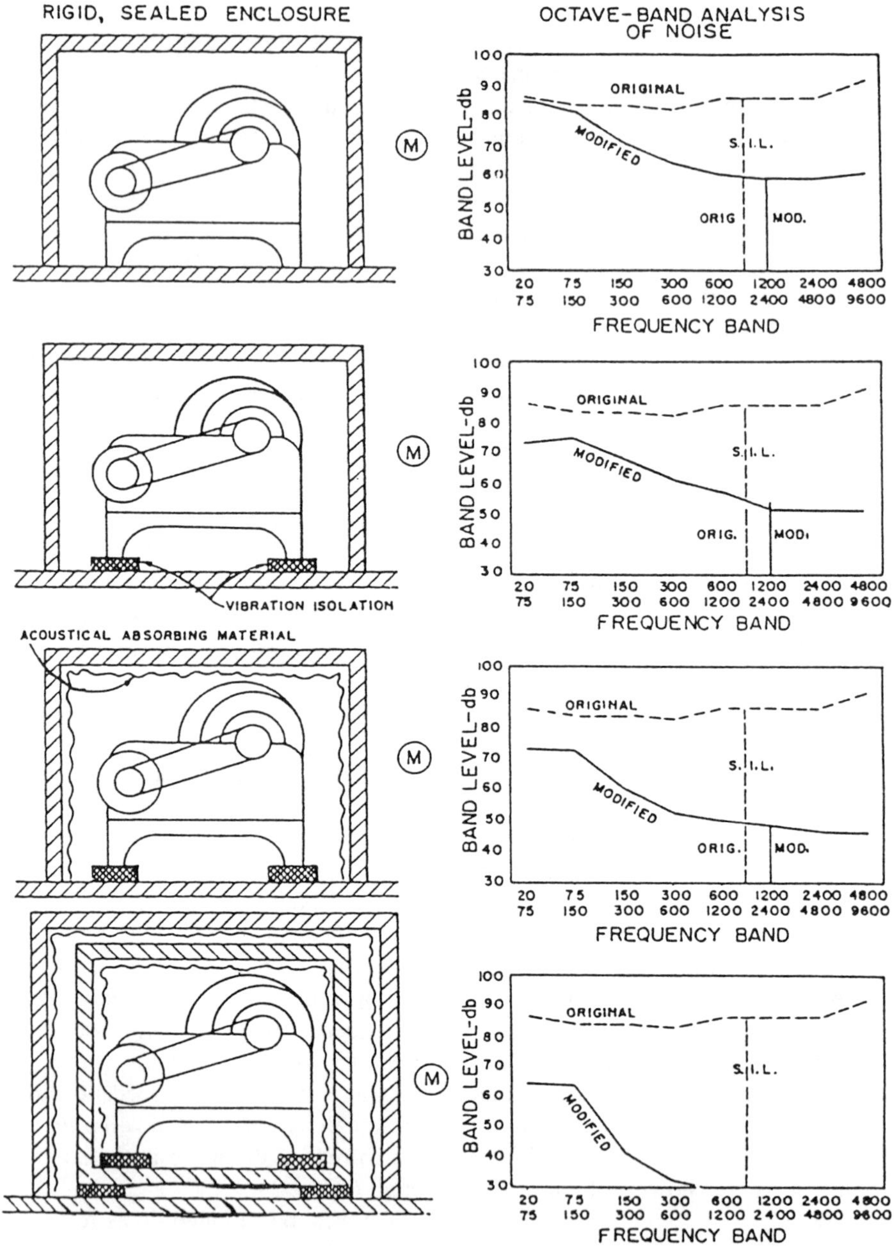

Figure 6.14. When a rigid, sealed enclosure is added, the noise is reduced to even a greater degree. "M" shows the location of measuring instrument.

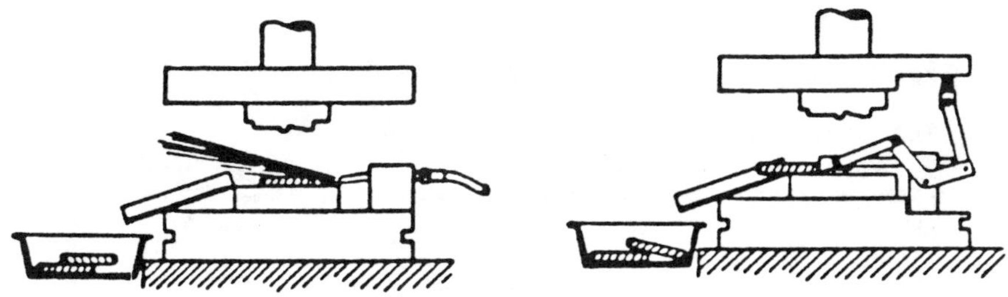

Figure 6.15. Quieter method cuts noise. At left is noisy air-ejection method; at right, quieter mechanical ejector.

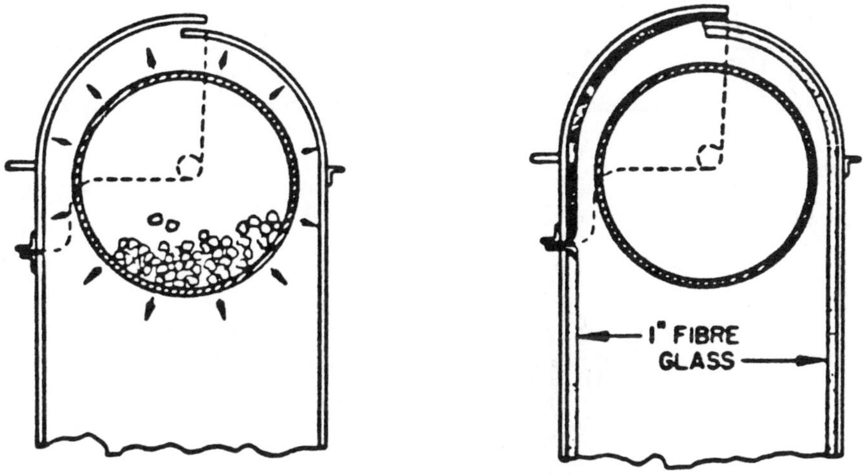

Figure 6.16. Enclosure house for a tumbler barrel (left) was lined inside with glass fiber; the inside of the door was coated with an under seal.

Consider the path from the source to the listener:

1. Can the source or the listener be readily moved to reduce the noise level?
2. Is acoustic treatment of the ceiling or walls a useful solution?
3. Should sound barriers be erected?
4. Is a total enclosure required?
5. Is personai protection required?

RADIANT HEAT CONTROL

Ventilation does not control radiant heat. No number of fans circulating air in a work space will reduce the radiant heat load on a person. Control comes only from

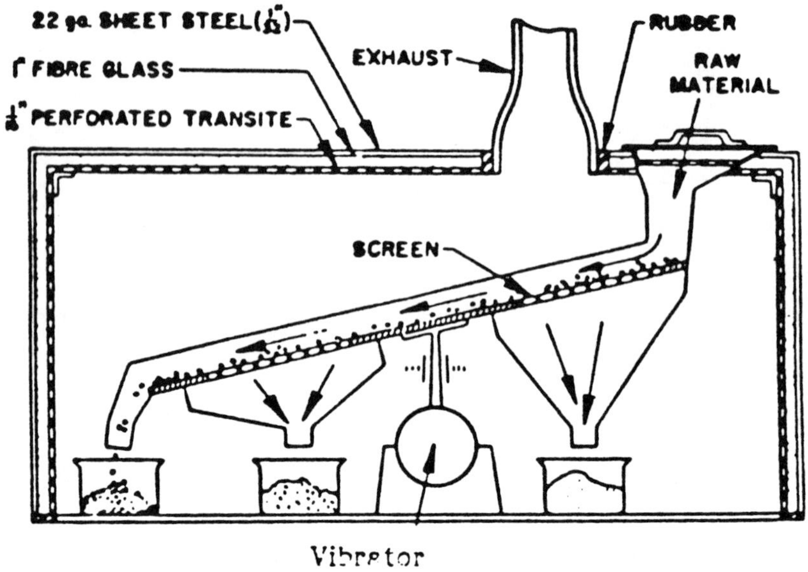

Figure 6.17. Total enclosure for a screen sifter reduced noise effectively.

Figure 6.18. Safety glass shield placed between the punch press and the operator serves as a sound barrier.

1. Reducing the temperature or thermal emissivity of the infrared source.
2. Installing shields which will set up infrared "shadows" for individuals to work in.
3. Providing personal protective clothing which is infrared reflecting.
4. Distance from source.

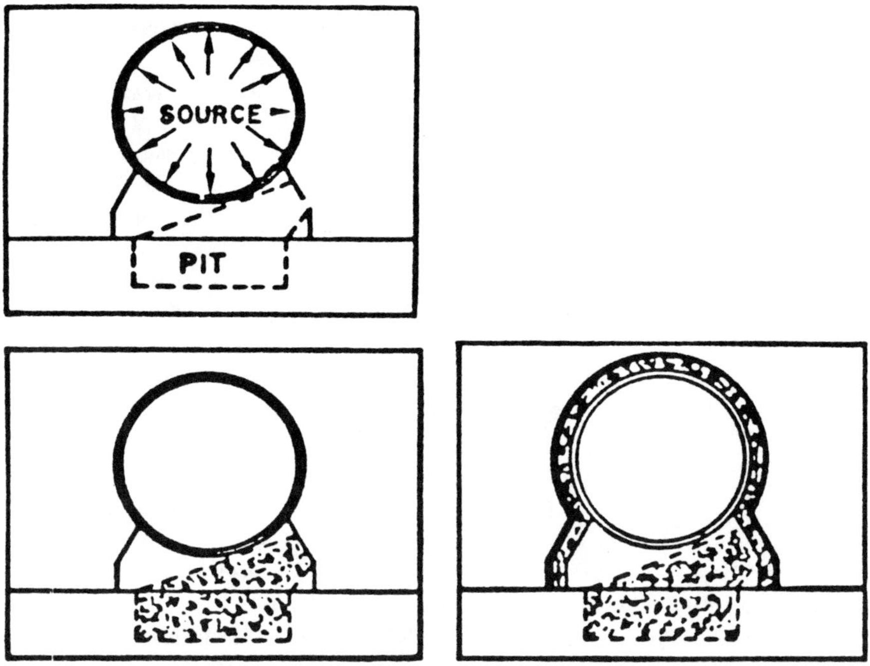

Figure 6.19. Noise source at a paper pulp mill (upper) drawing. Drawing at lower left shows pit filled with acoustical material to reduce reverberation build-up; below, the entire machine has been sprayed with acoustical material and covered with asphalt.

Wherever possible, conductive heat insulation should be applied to sources in which temperature cannot otherwise be lowered. This reduces the surface temperature, and so diminishes the radiant heat emitted.

SHIELDS

Heat is sometimes an inherent part of the operation or product, and the temperature cannot be lowered. For example, a red-hot ingot of steel, or molten glass, or a heat-treat bath must be at a certain high temperature. Then shields or clothing are the answer.

Shields have been commonly used for years, as illustrated by the variety of sheet iron, boiler plate, asbestos mat, and wooden shields which can be found in most hot plants. However, these materials are good absorbers of radiant heat. They get hot and reradiate to the individual working on the "cool" side. Such shield materials are unwisely chosen. The following will yield better results.

Reflective screens or shields made of corrugated or flat aluminum sheets (commercial siding and roofing). Polished copper, nickel, and chromium are also good, but of course, are impractical because of cost. Tinplate and new galvanized iron are satisfactory, but tend to tarnish fairly rapidly and then become ineffective. Excellent screens are made of aluminum foil, and foil backed by a supporting board such as gypsum board. If the foil becomes blackened or torn, it is easy to paste a new piece over the board. It is important to realize that the reflection is surface phenomenon. It does not matter how thick the aluminum is. To have reflecting surface is vital, and so the aluminum should never be covered or embedded in anything.

Heat exchanging screens can be made of iron or steel and can be water cooled. The heat absorbed by the screen is moved from the area by a film of water flowing over the metal.

Absorbing screens are a possibility only if they are covered with thermal insulation to keep their "cool" side from getting hot and reradiating to the individual.

Transparent screens can be made of glass. They may be (1) heat-absorbing glass, or (2) heat-reflecting glass. Ordinary flat glass absorbs long-wavelength infrared fairly well; there also are special glasses that accent this property. They all, however, get hot and reradiate copiously. Heat-reflecting glass has a thin permanent coating that does an admirable job of reflecting heat, while still being thoroughly transparent to light. Glass windows should be heat treated. Wire mesh can also be used as a semitransparent screen. Screens must be tailor-made for each operation. Sometimes they can be attached permanently to a hot surface with a gap between it and the shield. The width must be sufficient to allow air movement for removing heat by convection.

Such an arrangement is generally much cheaper than conductive insulation and does not upset the heat balance of the equipment — which is important for furnaces, annealing ovens, and the like. Sometimes the screens have to be movable to provide access to the equipment. Then they can be hung from above with hooks, and hinged. One plant uses overhead aluminum garage doors to open large control areas quickly. At some operations vertical standing screens on legs or wheels, or folding screens, fill a need.

Aluminized cloth and aluminum foil-cloth laminates are commercially available. They can be used as curtains, for example, hung by rings from a bar, like a shower curtain. These can be quickly pushed aside to work on controls or machinery. A variety of inexpensive paper-backed foil packaging and house insulating products are available. Some of these are wonderfully effective as heat reflectors, and can be made into expendable coverings for screens, frames, and the like.

A reflective booth, similar to a phone booth, is useful when there is a multitude of radiant heat sources and is of benefit if a worker can spend a fairly large portion of this time back from the machine. It must be provided with adequate ventilation, preferably using outside air. Small windows permit the operator to see out. The booth should be covered with aluminum foil, or made of sheet aluminum.

LASER HAZARD CONTROLS

An audible and/or visual signal should be employed to indicate that a laser is operating. For continuous operating lasers, the signal should be maintained during its activation.

If feasible, the laser beam should be fixed in one position to prevent the operator from directing the beam at an angle that could endanger himself and other employees.

Laser operations should be conducted in areas of high general illumination to keep the pupils constricted. To minimize specular reflections, all surfaces should be non-reflective in laser work areas. This includes the use of non-reflective paint.

When necessary to view the work piece while the laser is energized, it should be viewed by use of closed circuit television, optical comparator equipped with appropriate signs at each laser room. Capacitors should be discharged and the power turned off before making any adjustments that may require placing portions of the body in the beam path. Use all feasible means to isolate laser equipment away from heavily populated areas, employee travel routes, and then shield with curtains or a back drop. All electrical wiring should meet requirements of the National Electrical Code and all electrical equipment should be properly grounded. Provide periodic medical examinations for those working with and near lasers (Figure 6.20).

LASER CONTROLS

This section on laser controls is adapted from the Guide for Uniform Industrial Hygiene Codes on Regulations for Laser Installations, published by the American Conference of Governmental Industrial Hygienists and the Laser Institute of America.

The extent and type of control would vary depending on the type of laser being used and the manner of its use. Only properly indoctrinated and trained individuals should be placed in charge of laser installations and operations. Following are details relating to safe laser operation:

1. Personnel should not look into the primary beam or at specular reflections of the beam when power or energy densities exceed the permissible exposure levels.
2. Avoid aiming the laser with the eye to prevent looking along the axis of the beam, which increases the hazard from reflections.
3. Work with lasers should be done in areas of high general illumination to keep pupils constricted and thus limit the energy that might inadvertently enter the eyes.
4. Shatter-resistant safety eyewear, designed to filter out the specific frequencies characteristic of the system, affords partial protection. Laser safety glasses

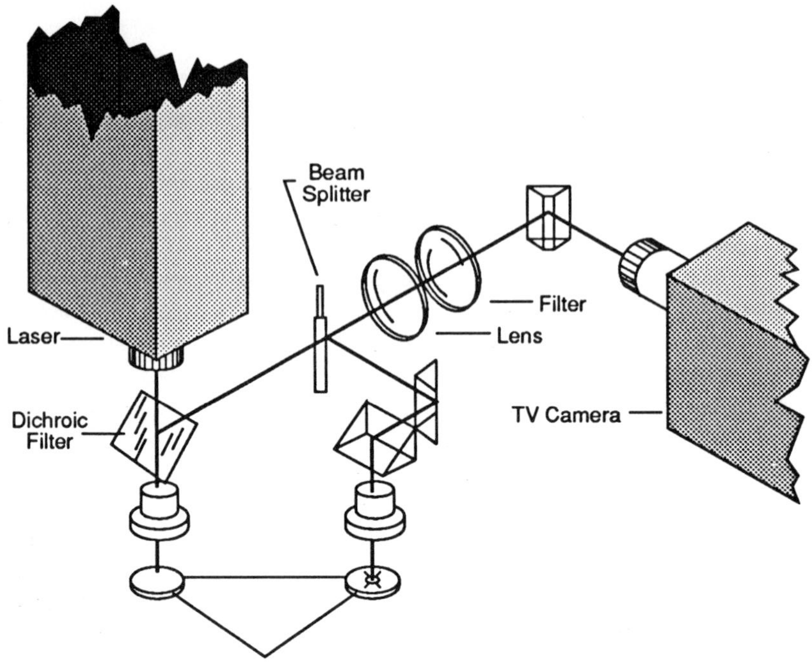

Figure 6.20. Left: low-energy (10 W) pulsed laser is used to "burn" holes half the thickness of a human hair in diamond chips. The bored chips are used wire-drawing dies. The diagram (above) shows the T.V. viewing system to monitor the operation. It is designed so that no intense light can be transmitted to the T.V. screen.

should be evaluated periodically to ensure maintenance of adequate optical density at the desired laser wavelength. There should be assurance that laser goggles, designed for protection from specific lasers, are not mistakenly used with different wavelengths of laser radiation. Distinctively colored frames are recommended, and the optical density should be shown on the filter. Laser safety glasses exposed to very intense energy or power density levels may lose effectiveness and should be discarded.

5. The laser beam should be terminated by a target material that is nonreflective and is fire resistant, and an area should be cleared of personnel for a reasonable distance on all sides of the anticipated path of the beam.
6. Avoid electrical shock from the potentially dangerous electrical sources of high and low voltage.
7. Special precautions should be taken if high-voltage tube rectifiers (over 15 kV) are used, because there is a possibility that X-rays will be generated.

Specific precautions (applicable to high-powered pulsed lasers):

1. Safety interlocks at the entrance of the laser facility should be so constructed that unauthorized or transient personnel be denied access to the facility while the laser power supply is charged and capable of firing.
2. Laser electronic firing systems should be so designed that accidental pulsing of a stored charge is avoided. The design should also incorporate a "fail-safe" system.
3. An alarm system including a muted sound, flashing lights (visible through laser safety eyewear), and a countdown procedure should be used once the capacitor banks begin to charge.
4. Installations using liquid gas coolant should be adequately ventilated. Water-cooled lasers are preferred.
5. Walls and ceilings should be painted with nongloss paint to produce a diffuse surface. Diffuse black is preferred in the target area and a light color surrounding will maximize the lighting distribution from general lighting fixtures.
6. Solid-state lasers should be operated by remote control firing with television monitoring, if feasible. This eliminates the need for personnel to be physically present in the same room. An alternative is to enclose the laser, the associated beam, and the target in a light-tight box.

Specific precautions (applicable to low-powered, continuous-wave gas and semiconductor laser systems):

1. General precautions with reference to aiming and the avoidance of specular reflection should be observed.
2. The laser beam should be terminated at the end of its useful beam path by a material that is a diffuse matte of such color or reflectivity to make positioning possible, but should minimize the reflection.
3. Reflective material should be eliminated from the beam area and good housekeeping maintained.

Carbon dioxide-nitrogen (CO_2-N_2) gas lasers:

1. The principal hazard associated with the CO_2-N_2 lasers is the fire hazard. A sufficient thickness of firebrick or other fire-retardent material should be provided as a backstop for the beam.
2. The laser assembly should be constructed of a material opaque to ultraviolet light generated by the gas discharge. Quartz tubing transmits ultraviolet light, whereas certain heat-resistant glass tubing is reasonably opaque at this part of the spectrum.
3. Reflections of the infrared laser beam should be attenuated by enclosure of the

beam and target area or by eyewear constructed of a glass-like material opaque to the CO_2 wavelength.

Specific precautions for outdoor laser installations:

1. Personnel should be excluded from the beam path at all points where the power or energy exceed the permissible limits. This should be accomplished by the use of physical barriers, administrative controls, the use of interlocks, and by limiting the beam traverse.
2. The tracking of nontarget vehicular traffic or aircraft, whether intentional or inadvertent, should be prohibited within the calculated hazardous areas.
3. Operation of laser systems while it is raining or snowing or when there is dust or fog in the air should be avoided without the use of laser protection eyewear.
4. The path covered by the beam should be cleared of all objects capable of producing reflections that are potentially hazardous. Concentrated light intensities generated by some lasers can be delivered over considerable distance (miles). Although these effects can be used to good advantage, they are potentially dangerous and should be given adequate consideration.
5. Warning signs and placarding of potentially hazardous areas should be accomplished in accordance with local standard operating procedures.

VENTILATION

Laser installations should be well ventilated to prevent accumulation of hazardous gases where continuous-flow gas lasers are used, or to control property the cryogenic liquids and gases used to purge and cool lasers.

1. Continuous-flow gas lasers may be used in sealed systems; however, cracking of the system may lead to the release of enclosed gases or vapors including chlorine, bromine, lead, mercury, carbon monoxide, and hydrogen cyanide. Rates of ventilation should be adequate to maintain exposure levels at or below the current TLV.
2. Both liquid nitrogen and liquid helium have been used to cool solid-state and semiconductor laser units. These cryogenic liquids may result in the formation of liquid oxygen, which can become an explosion hazard in the presence of organic matter. (Oxygen levels should be maintained to at least 18% equivalent to a PO_2 of 135 mmHg).

CONTROL OF MICROWAVE OVEN HAZARDS

An emerging problem for the safety professional and industrial hygienist is proper operation of food-warming microwave ovens to prevent radiation exposure to employees. The following material is adapted from "Microwave Oven Safety," a bulletin published by the National Automatic Merchandising Association.

Safe operations and performance of microwave ovens depend on these factors.

1. Structural integrity of the unit and freedom from shipping or delivery damage.
2. Reasonable care in use and freedom from vandalism, tampering, and abuse.
3. Proper state of repair and required maintenance.
4. Proper servicing practices by qualified personnel.

Here are some general safety suggestions for all microwave ovens. To plan your own safety program, check the manual of the model you use to obtain additional recommendations.

1. Read the instruction manual for each model oven before placing it in operation or performing repairs or adjustments (if an adequate manual is not available, a shop employee should be given training by the oven manufacturer or factory representative).
2. Examine each oven, upon delivery, for evidence of shipping damage. Pay particular attention to door alignment and tightness of closure.
3. Never permit the operation of ovens with safety interlocks to be bypassed or inactivated.
4. Require supervisors or other designated personnel to examine ovens routinely for signs of wear, damage, or tampering:
   a. Check door hinges (if present) for wear or malfunction.
   b. Examine oven cavity for breaks in seams or welds.
   c. Check door and hinge screws to make sure they are tight.
   d. Check door closure by closing door on a strip of paper inserted at several different points. If the paper pulls out without resistance, latch and hinge adjustment is usually indicated. (This test does not apply to doors which employ a choke-type door.)
   e. Examine door safety interlocks for wear, tampering, or malfunction.
   f. Examine door seals for arcing, grease accumulation, missing screws, or other damage.
5. With power supply off, clean oven cavity, door, and seals with water and mild detergent frequently (do not use scouring pads, steel wool, other abrasives, or commercial cleaning agents).
6. Periodically determine the level of microwave radiation being emitted when the oven is in operation. Radiation and radiation leakage is measured with monitoring instruments called densitometers or power density meters. Radiation leakage monitoring is usually done as close to the equipment surface as possible, checking all four sides of the door opening (with a nonmetallic container of water or other load in the oven).

TRAINING AND EDUCATION

Proper training and education are supplements to the engineering controls. Here, the occupational nurse and safety professional sometimes can be key figures. For example, in an effective hearing conservation program the responsibility of seeing that ear protectors are properly fitted lies with the resident nurse. Also, by observing the workers on their visits to the dispensary, the nurse has first-hand knowledge of where health problems exist.

The safety professional, perhaps by persuading a worker to change his or her manner of weighing a toxic material, or of handling a scoop or shovel, can do much to minimize dust exposure.

The worker must also know the proper operating procedures that make engineering controls effective. If the worker performs an operation away from an exhaust hood, he or she will not only defeat the purpose of the control, but will also contaminate the work area. Workers can be alerted to safe procedures through booklets, instruction signs, labels, safety meetings, and other educational devices.

Since new materials are constantly being marketed and new processes developed, reeducation and follow-up instruction must also be part of an effective industrial hygiene program.

SUMMARY

The most effective method of control is at the source — prevent exposure to harmful physical energy stresses and contamination of workroom air and thus prevent inhalation of harmful chemical agents. The importance of personal hygiene should not be overlooked. Periodic medical examinations provide a good opportunity for instruction of employees in various personal hygiene measures.

Good washing facilities, clean lunchrooms, and clean work clothes can help prevent additional, even though minor, exposure to toxic materials. Also, contaminated work clothes should not be taken home where a toxic dust could contaminate the home or expose other members of the family.

REFERENCES

1. NIOSH Criteria Document 80-116. "Compendium of Materials for Noise Control," U.S. Government Printing Office, Washington, DC, 1980.
2. NIOSH Criteria Document 76-179. "Abrasive Blasting Operations: Engineering Control and Work Practices Manual," U.S. Government Printing Office, Washington, DC, 1979.
3. NIOSH Criteria Document 76-180. "Engineering Control Research Recommendations," U.S. Government Printing Office, Washington, DC, 1976.
4. NIOSH Criteria Document 76-186. "Recirculation of Exhaust Air," U.S. Government Printing Office, Washington, DC, 1976.
5. NIOSH Criteria Document 78-141. "The Recirculation of Exhaust Air... Symposium Proceedings," U.S. Government Printing Office, Washington, DC, 1978.

6. NIOSH Criteria Document 78-165. "Control of Exposure to Metal Working Fluids," U.S. Government Printing Office, Washington, DC, 1978.
7. NIOSH Criteria Document 79-143A Vol. I and 79-143B Vol. II. "Validation of a Recommended Approach to Recirculation of Industrial Exhaust Air, Vol. I (Spring Grinding, Chrome Platting, Dry Cleaning, Welding, and Vapor Degreasing Operations); Vol. II (Lead Battery, Woodworking, Metal Grinding and Enamel Blending Operations)," U.S. Government Printing Office, Washington DC, 1979.
8. NIOSH Criteria Document 81-118. "Control of Emission from Seals and Fittings in Chemical Process Industries," U.S. Government Printing Office, Washington, DC, 1981.
9. NIOSH Criteria Document 81-121. "An Evaluation of Engineering Control Technology for Spray Painting," U.S. Government Printing Office, Washington DC, 1981.
10. NIOSH Criteria Document 84-102. "Engineering Control of Occupational Safety and Health Hazards: Recommendations for Improving Engineering Practice, Education and Research, Summary Report," U.S. Government Printing Office, Washington, DC, 1984.
11. NIOSH Criteria Document 76-183. "Laser Hazard Classification Guide," U.S. Government Printing Office, Washington, DC, 1976.
12. Contract Report 099-72-0054. Basic Course Microwave and Lasers. U.S. Government Printing Office, Washington, DC, 1972.
13. NIOSH Document 79-117. "Industrial Noise Control Manual," U.S. Government Printing Office, Washington, DC, 1979.
14. NIOSH Document 76-120. "List of Personal Hearing Protectors and Attenuation Data," U.S. Government Printing Office, Washington, DC, 1976.

CHAPTER 7

Respiratory Protection and Fit Testing

INTRODUCTION

No chapter can cover every situation in the workplace or site where respirators should be worn. Nonetheless, this chapter presents considerable information on respiratory equipment and complements the relevant safety and health regulations and manufacturer's requirements. OSHA regulation governing the use of respirators in general industry are found in the Code of Federal Regulations (CFR) Title 29 Title 1910.134; those governing use in the construction industry are cited in CFR 29 Part 1926.103; and those for the maritime industry are CFR 29 Parts 1915.152 and 1918.102. Copies of these standards are in an appendix of this book. All are available in many libraries and from the Government Printing Office.

When it is not feasible to render the environment completely safe, it may be necessary to protect the worker from contact with airborne contaminants. Personal protective equipment (PPE) should be provided and used where it is not possible to enclose or isolate the process or equipment, provide ventilation, or use other control measures or where there are short exposures to hazardous airborne concentrations of contaminants. Respiratory protection must not be a substitute for engineering control procedures to eliminate the hazard.

Designs of respiratory protective devices vary in application and protective capability. The user must, therefore, assess the inhalation hazard and understand the specific use and limitations of available equipment to assure proper selection.

Respiratory protective devices are tested and approved by the U.S. Bureau of Mines (USBM) and the Department of the Interior for protection against a wide range of inhalation hazards, from oxygen deficiency and/or highly toxic atmospheres to those containing "nuisance" dusts.

HISTORICAL OVERVIEW

A respirator is defined as a device covering the mouth or nose to prevent the inhalation of toxic substances in the atmosphere. The oldest record we have of the use of such a device is that given in the first century A.D. by Pliny the Elder, who wrote of mineral refiners wearing bladders over their faces to avoid inhaling mercury-bearing dust. Gas masks, oxygen-supplying apparatus, and hose masks (diving helmets) were introduced into Europe during the 19th century, but they did not come into general industrial use until after World War I. The oxygen-breathing apparatus was introduced into the U.S. about 1907. Its first general usage came in 1910 with its adoption for mine rescue work. The canister gas mask was introduced into this country when the USBM was assigned the problem of developing a military mask for the American soldiers in 1917. This work later grew into the Research Division of the Chemical Warfare Service. A good serviceable mask for war uses was developed, and at the end of the war, work was continued by the USBM with a view toward the development of an industrial mask.

After World War I many of the military masks of the acid-gas and organic-vapor type were thrown on the consumer market by salvage dealers. Thus, believing that they were to be used for protection against all gases. This error was encouraged by some of the dealers who made extravagant claims concerning the masks. As a result there were many poisonings, mainly nonfatal, around ammonia refrigerating plants and among fire fighters who were exposed to carbon monoxide. The alarm caused by these poisonings produced an extensive educational campaign through newspapers, the technical press, and personal instruction. Research gradually developed canisters for various gases and the universal canister that protects against all common gases or combinations of gases. The next step was the development of the hose mask, which is simple and foolproof and offers respiratory protection in any situation that can be reached by 150 ft of hose.

In each instance, however, personal respiratory protection apparatus that really offered dependable protection was not made available in the U.S. until the USBM set up approval schedules and procedures for permissibility tests. These schedules are revised from time to time to conform with advances in research and development. It was not until 1935 that effective dust respirators became available to American industry. When the first respirator passed the official schedule of tests. Scores of inferior, and in many instances useless, devices were discarded after these tests demonstrated their unsatisfactory performance.

These tests now cover oxygen-breathing apparatus, self-contained breathing apparatus, canister gas masks, hose masks, and abrasive-blasting and other air-line respirators. The tests specify the minimum permissible performance and the maximum permissible resistance to inhalation and exhalation. Certificates of approval issued by the USBM for the accepted devices are the basis for all lists of approved industrial respiratory protection equipment published by any agency in the U.S.

The data are published, enabling the public to choose the respirator giving the best performance, and lists all devices meeting the minimum requirements. However, these requirements are raised occasionally to keep pace with the more progressive manufacturers, and holders of early approvals are encouraged to improve the performance of their devices to meet these higher standards in order to retain the approval label. Nominal fees charged for approval testing vary with the time required for individuals to perform the scheduled tests. All approved devices are stamped with an approval label or number, such as BM-1400, in which the first two digits of the number refer to the schedule of tests under which the device was approved, and the last two are specific for the particular device.

RESPIRATOR CERTIFICATION

Approved Respirators

The term "approved respirator" has become a part of the language of industrial hygiene. For a period of 50 years "approved respirator" designated approval by the USBM. This agency has had an interest in the use of respirators in mines since the early 1900s. As mentioned earlier, this experience was used during World War I to develop a gas masks for the army. After the war, interest remained in the development of adequate respiratory protection of civilian use. Finally, the bureau was assigned the task of setting standards for respiratory protection.

In 1920 the bureau began to publish "Approval Schedules" for specific types of respirators. Manufacturers could voluntarily submit their devices for testing by the bureau, and if the equipment was satisfactory, it could be sold as USBM approved. Since this system proved to be of value to the consumer, industrial health regulatory agencies began to make it mandatory that only USBM-approved respiratory equipment be used.

While the federal OSHA was being set up, most of the respiratory approval testing was assumed by NIOSH. The present approval program is carried out in the NIOSH TCB in Morgantown, West Virginia. Presently, respiratory prodevices must be approved jointly by the Mine Safety & Health Administration of the Department of Labor and NIOSH of the Department of Health and Human Services for the contaminant or situation to which the employee is exposed. (The minimal respirator program requirements under OSHA regulations are listed in Table 7.1.)

NIOSH APPROVAL REQUIREMENTS

The specific regulations concerning respirator approval are to be found in the 30 CFR 11. These regulations have been gradually changed through the years and have had the effect of upgrading respirator performance. They should be studied

Table 7.1. OSHA's Requirements for a Minimal Respirator Program

1. "Written standard operating procedures (S.O.P) governing the selection and use of respirators shall be established."
2. "Respirators shall be selected on the basis of hazards to which the worker is exposed."
3. "The user shall be instructed and trained in the proper use of respirators and their limitations."
4. [Reserved] "where practicable, the respirators should be assigned to individual workers for their exclusive use." (OSHA recommended only.)
5. "Respirators shall be regularly cleaned and disinfected. Those used by more than one worker shall be thoroughly cleaned and disinfected after each use."
6. "Respirators shall be stored in a convenient, clean, and sanitary location."
7. "Respirators used routinely shall be inspected during cleaning. Worn or deteriorated parts shall be replaced. Respirators for emergency use, such as self-contained breathing devices, shall be thoroughly inspected at least once a month and after each use."
8. "Appropriate surveillance of work area conditions and degree of employee exposure or stress shall be maintained."
9. "There shall be regular inspections and evaluations to determine the continued effectiveness of the program."
10. "Persons should not be assigned to tasks requiring use of respirators unless it has been determined that they are physically able to perform the work and use the equipment. The local physician shall determine what health and physical conditions are pertinent. The respirator user's medical status should be reviewed periodically (for instance, annually)."
11. "Approved or accepted respirators shall be used when they are available."

carefully because the methods of approval have limitations that affect the use of an approved respirator.

All respirators must pass a fit test to determine that the face-to-face piece seal is effective, and in the case of hoods and helmets, to ensure that they are not likely to develop a negative pressure during use. Note that except from sand blasting hoods, the test atmosphere is isoamyl acetate. The means of detection used is the wearer's sense of smell. Isoamyl acetate fit tests require careful use if the results are to be meaningful. Subjects should be highly motivated and trained before using this equipment.

Significant points of interest for each approval category are listed below.

CLASSES OF RESPIRATORY PROTECTIVE DEVICES

Respiratory protective devices fall into two classes: air purifying and atmosphere supplying.

Air-Purifying Devices

Air-purifying devices remove contaminants from the atmosphere and can be used only in atmospheres containing sufficient oxygen to sustain life (at least 16% by volume at sea level) and within specified concentration limitations of the specific device (Figure 7.1). Various chemicals remove specific gases and vapors, and mechanical filters remove particulate matter. The useful life of an air-purifying device is dependent upon the concentration of the contaminants, the breathing volume of the wearer, and the capacity of the air purifying medium.

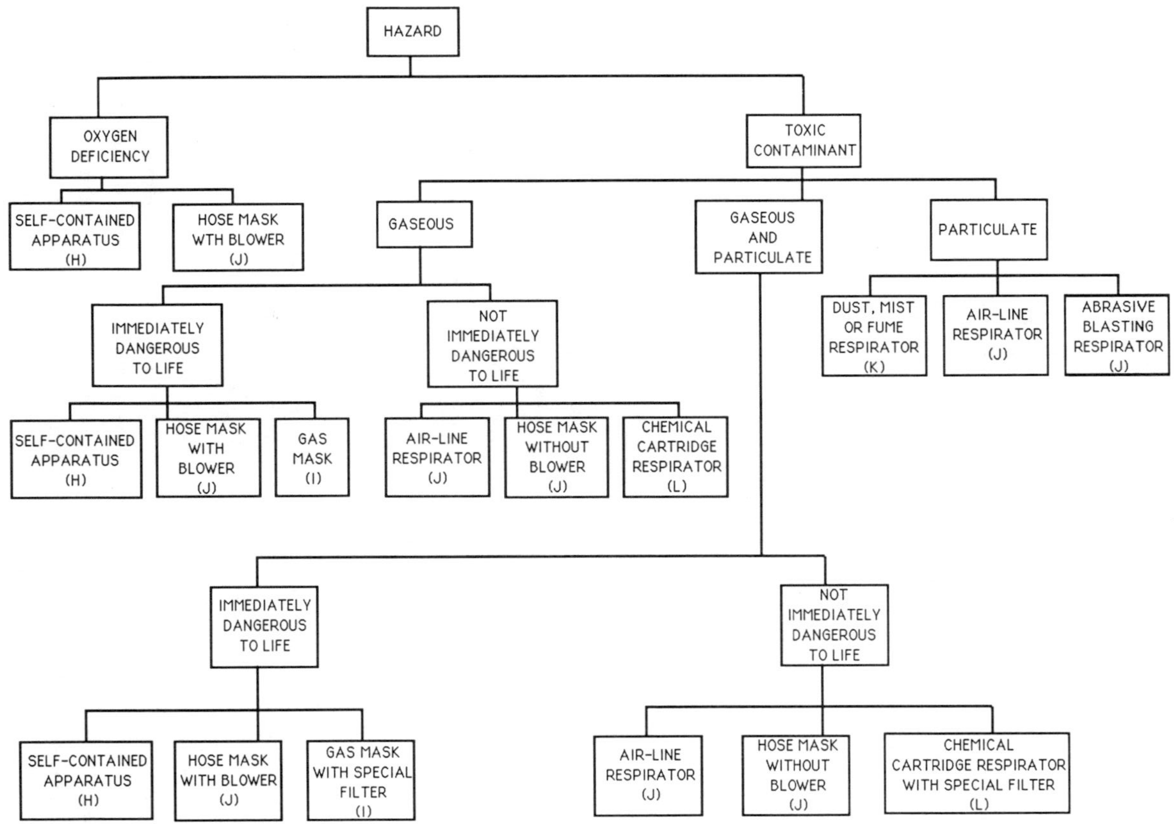

Figure 7.1. Suggested outline for selecting respiratory protective devices. Letters in parentheses refer to Subparts of Title 30, Code of Federal Regulations, Part II, which discusses these items. The text refers to the former Bureau of Mines schedules.

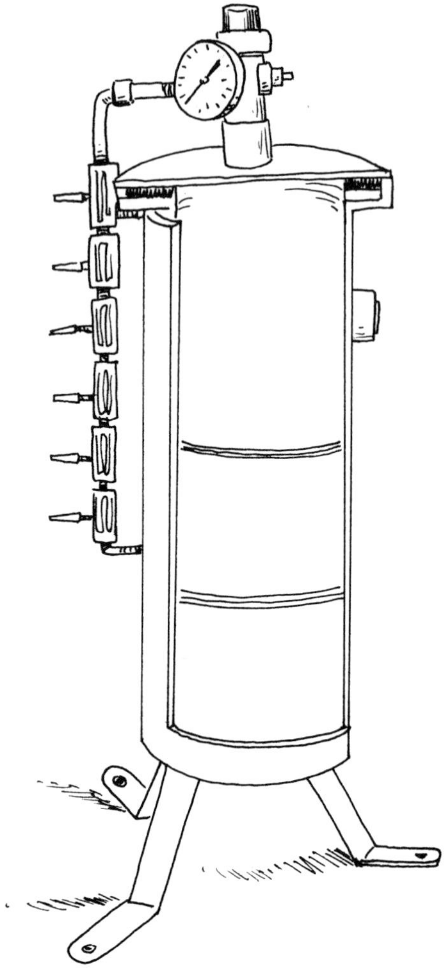

Figure 7.2. Cutaway view shows air path through purifier and the filtering materials for air-line respiratory equipment. Note manifold (at left) that permits six air lines to be connected.

The basic types of air-purifying devices are three.

Mechanical filter respirators provide respiratory protection against particulate matter such as nonvolatile dusts, mists, or metal fumes. Selection of the appropriate respirator is based on the type, toxicity, and particle size of the particulate matter (Figure 7.2). Specific types of mechanical filter respirators are approved under USBM Approval Schedule 21 and its revisions.

Chemical cartridge respirators provide respiratory protection against certain gases and vapors in concentrations not in excess of 0.1% (by volume). Specific types of chemical cartridge respirators are approved under USBM Approval Schedule 23 and its revisions.

Combinations of chemical cartridge and mechanical filter respirators provide respiratory protection where exposure is both gaseous and particulate.

Gas masks provide respiratory protection against certain specific gases and vapors in concentrations up to 2% (by volume), or as specified on the canister label, and against particulate matter. Specific gas masks are approved under USBM Approval Schedule 14 and its revisions.

Air-Supplied Devices

Air-supplied devices deliver breathing air through a supply hose connected to the wearer's face piece. The air delivered must be free of contaminants; the air source must be located in clean air and monitored frequently. With the exception of hose masks with blowers, these devices should be used only in atmospheres not immediately dangerous to life.

The basic types of supplied air devices are air-line respirators — with constant flow or demand flow — and hose masks — with or without blower.

Air-supplied devices are approved under USBM Approval Schedule 19 and its revisions.

Self-Contained Breathing Apparatus

Self-contained breathing apparatus provide complete breathing protection for various periods of time based on the amount of breathing air or oxygen supplied and the breathing demand of the wearer.

The basic types of self-contained breathing apparatus are oxygen cylinder rebreathing type and chemical oxygen rebreathing (self-generating type air or oxygen demand type). Self-contained breathing apparatus are approved under USBM Approval Schedule 13 and its revisions (Figures 7.3 and 7.4). Except for some special application equipment not covered by the approval schedule, all self-contained breathing apparatus should bear USBM approval.

The proper selection depends upon the application and service time required, but complete respiratory protection is provided and the problem of proper protection for a specific hazard is eliminated.

How and When to Use Each Type

Selection of the proper type of respiratory protective equipment should be based on the following procedures:

1. Identify the substance or substances against which protection is necessary.
2. Know the hazards and significant properties of each substance.
3. Determine the conditions of exposure and level of air contaminant present.
4. Ascertain whether any human capabilities and limitations are essential to the safe use of the devices.

Figure 7.3. Mechanical filter respirator.

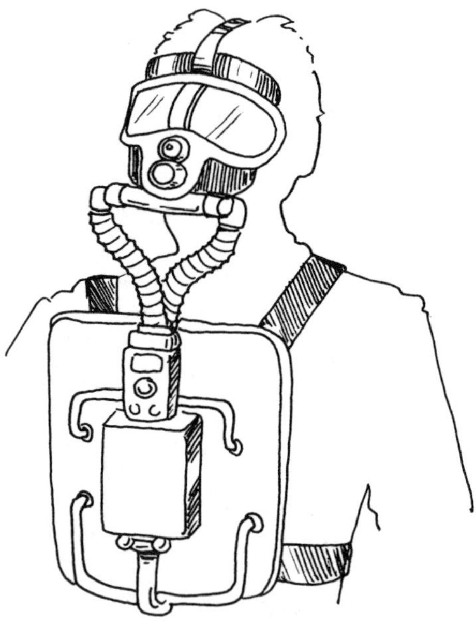

Figure 7.4. Self-generating oxygen breathing apparatus.

5. Fit the respirator carefully and train the worker in its use.

There is one limitation applicable to all respiratory protective equipment. Certain gases can harm the body by means other than through the respiratory tract. For example, ammonia in concentrations of approximately 3% or higher can cause skin burns (particularly on most skin). To avoid that possibility, rubberized clothing should be worn in addition to the proper respiratory protection.

Similarly, protective clothing should be worn when appreciable amounts of gases such as hydrocyanic acid are present. Hydrocyanic acid — a vapor at slightly above room temperature — is capable of penetrating the skin and causing systemic poisoning; although to do so, concentrations have to be considerably higher than those required for poisoning through the respiratory tract.

The proper selection of respirators must be made according to the guidance of American National Standards Institute (ANSI) publication, "Practices for Respiratory Protection," ANSI Z88.2-1969. (A later edition of this standard, Z88.2-1980, has been issued by ANSI.)

Another factor to consider when using respirators is the air supply rates. The wearer's work rate determines the volume of air breathed per minute. The volume of air supplied to meet the breathing requirements is of great significance when using atmosphere-supplying respirators such as self-contained and air-line respirators that use cylinders because this volume determines their operating life. The useful service life of these respirators under even moderate working conditions may be significantly less than under conditions of rest.

MECHANICAL FILTER RESPIRATORS

Mechanical filter respirators offer respiratory protection against air-borne particulate matter including dusts, mists, metal fumes, and smokes. They consist essentially of a soft resilient face piece, of either half-mask or full-face design. Directly attached is one of several types of mechanical filters made up of some fibrous material which removes the harmful particles by physically trapping them as air is inhaled through the material.

The filter must be highly efficient, however, to stop small yet harmful particles. Mechanical filter respirators do not provide protection against gases, vapors, or oxygen deficiency. There are many classes of mechanical filter respirators consistent with the various classes of air-borne particulate matter. Although one respirator can be made to provide effective protection against all true particulates, in most cases it will be too expensive and perhaps too cumbersome for the great majority of exposure users. Therefore, many special-purpose respirators are available to provide the most economical and efficient protection against specific particulate hazards.

The USBM — the official approval agency for respiratory protection equipment — tests and certifies mechanical filter respirators under Approval Schedule 21. It approves respirators for one or any combination of particulate hazards-nuisance, fibrosis-producing, and/or toxic dusts, mists, and fumes (Table 7.2).

The most desirable compromise must be worked out for each classification with respect to breathing area, resistance to breathing, efficiency in filtering particulates of specific size ranges, and the time required to clog the filter. For example, in choosing a respirator for protection against particulates more toxic than lead and which will have to filter effectively particles as small as tobacco smoke, the user must accept an increase in bulk, weight, and cost more than those for a respirator needed to protect only against so-called nuisance dusts.

Table 7.2. Respirator Approval Classifications Under Bureau of Mines Schedules

Schedule 21A	Schedule 21B
1. Respirators for mechanically generated dusts, including pneumoconiosis-producing dusts such as silica dust, and toxic dusts not significantly more toxic than lead	1. Respirators for dusts having a TLV of not less than 0.1 mg/m^3 or 2.5 mppcf
2. Respirators for fumes of various metals not significantly more toxic than lead	2. Respirators for metal fumes having a TLV not less than 0.1mg/m^3
3. Respirators for mists of materials which do not produce harmful gases or vapors, such as silica mist and chromic acid mist	3. Respirators for mists having a TLV not less than 0.1 mg/m^3 or 2.4 mppcf
4. Respirators for various combinations of above	4. Respirators for dusts, mists, and fumes having a TLV less than 0.1 mg/m^3 where the exposure concentration is no more than 10 times the TLV (protection factor of 10)
	5. Same as 4, except limited to exposure concentrations up to 100 times the TLV (protection factor of 100)
	6. Same as 4, except limited to exposure concentrations up to 1000 times the TLV (protection factor of 1000)
	7. Respirators for various combinations as above
	8. Respirators that are combinations of mechanical filter and other types

CHEMICAL CARTRIDGE RESPIRATORS

Chemical cartridge respirators afford protection against light concentrations (0.05 to 0.1% by volume, depending upon the contaminant) of certain acid gases and organic vapors by utilizing various chemical filters to purify the inhaled air. They differ from mechanical filter respirators only in that they use small cartridges containing chemicals to remove harmful gases and vapors. It is necessary that sufficient oxygen be present at all times to support life.

Many gases and vapors are dangerous or irritating to breathe. Some are hazardous, discomforting, or both in extremely low concentrations. Although not usually immediately harmful to life, light concentrations of some gaseous materials cause nausea and headache, reduce worker output, and sometimes produce chronic disorders which may eventually result in occupational disease.

The simplest and most convenient type of respiratory protective equipment to use in those cases are the chemical cartridge respirators, equipped with the proper chemical filtering elements. The USBM Approval Schedule 23 covers organic vapor respirators, and the bureau will issue certificates of approval to manufacturers whose products meet their performance requirements. Approval numbers are shown on the respirator and/or packing carton. It is well to note that chemical cartridge respirators are *"non-emergency" respiratory protective devices and should never be used in dangerous atmospheres.* Do not use them for protection against gaseous material that is extremely toxic in very small concentrations.

There are three other important rules applying to chemical cartridge respirators:

1. They should not be used for exposures to harmful gaseous matter which cannot clearly be detected by odor.
2. They should not be used against any gaseous material in concentrations which are highly irritating to the eyes.
3. They cannot be used for protection against gaseous material which is not effectively stopped by chemical cartridges utilized, regardless of concentrations.

Combination respirators use dust, mist, or fume-mechanical filters, plus a chemical cartridge for dual or multiple exposure. Responsibility for proper respiratory selection in application of agricultural chemicals has been recently shifted from the USDA to the USBM. Test schedules are presently pending, and users of agriculture chemicals may check with vendor companies for guidance in proper respiratory selection.

Respirators with independently replaceable mechanical filters are preferred when using this type, because the dust filter normally plugs before the chemical cartridge is exhausted.

One combination mechanical-chemical filter respirator employs a back-mounted filter element and is especially well suited for spray painting and welding operations where the air contaminant is concentrated in front of the worker (example: Racal & Wilson Units).

GAS MASKS

Application and limitations. Gas masks have been used effectively for many years for respiratory protection against certain gases, vapors, and particulate matter which otherwise might be harmful to life or health (Figure 7.5). They provide simplicity of operation, compactness, ease of maintenance, and economy. However, because gas masks are air-purifying devices, designed solely to remove specific contaminants from the air, it is essential that their use be restricted to atmospheres which contain sufficient oxygen to support life (at least 16%, by volume, at sea level) and which contain generally no more than 2% concentrations of toxic gases and vapors by volume. It is imperative that the user assess the exposure conditions carefully before selecting a specific mask for respiratory protection. To do so properly requires the use of instruments to measure the concentration of oxygen and harmful gases and vapors present in the atmosphere under the various exposure and operating conditions. If the specific exposure concentrations are suspected of exceeding the specific limitations, only a self-contained breathing apparatus or a hose mask with blower should be used.

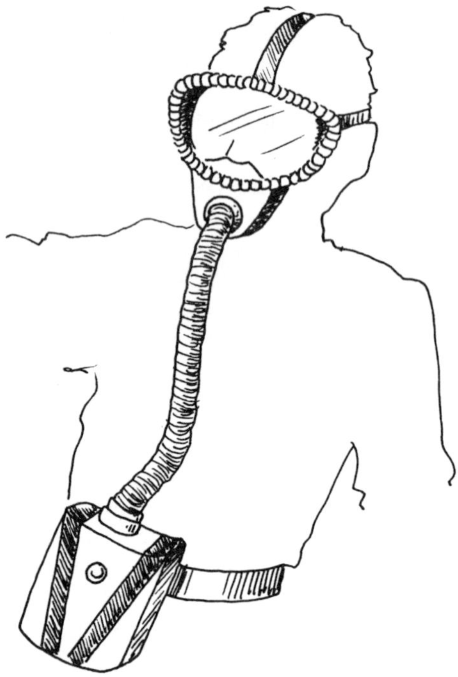

Figure 7.5. Gas mask.

From a practical standpoint, gas masks are generally suitable for ventilated areas not subject to rapid change, but should never be used in confined spaces below or above ground where oxygen deficiency and high gas concentrations may occur.

The normal content of oxygen in the atmosphere is 20.9%. Oxygen concentrations below 16% by volume are considered unsafe for human exposure because of harmful effects on bodily functions, mental processes, and coordination. At very low oxygen concentrations, collapse can be immediate, without warning. While 16% oxygen (at sea level) is generally considered the lower limit for safe human exposure, the pressure of oxygen within the lung is the important factor. Therefore, under conditions of significantly reduced atmospheric pressure, which occur at high altitudes, the lower safe limit is higher than 16% (by volume).

In assessing exposure conditions, it is important to remember that oxygen deficiency can occur in confined spaces by displacement of air by other gases or vapors, or by means of oxidation processes (such as fire, rusting, and aerobic bacteria) where the oxygen is consumed.

Various gas masks with conventional-sized canisters are tested and approved by USBM under Approval Schedule 14 for respiratory protection against specific gases and vapors in concentrations up to 2% by volume (3% for ammonia), or as specified on the canister label. Each canister is specifically labeled and color coded to indicate the type of protection afforded (Table 7.3). Chin-type canisters,

Table 7.3. Color Code for Respirator Canisters

Atmospheric Contaminants to be Protected Against	Colors Assigned
Acid	White
Hydrocyanic acid gas	White with 1/2-in. green stripe completely around the canister near the bottom
Chlorine gas	White with 1/2-in. yellow stripe completely around the canister near the bottom
Organic vapors	Black
Ammonia gas	Green
Acid gases and ammonia gas	Green with 1/2-in. white stripe completely around the canister near the bottom
Carbon monoxide	Blue
Acid gases and organic vapors	Yellow
Hydrocyanic acid gas and chloropicrin vapor	Yellow with 1/2-in. blue stripe completely around the canister near the bottom
Acid gases, organic vapors, and ammonia gases	Brown
Radioactive materials, except tritium and noble gases	Purple (magenta)
Particulates (dusts, fumes, mists, fogs, or smokes) in combination with any of the gases or vapors	Canister color for contaminant, as designated above, with 1/2-in. gray stripe completely around the canister near the top
All of the above atmospheric contaminants	Red with 1/2-in. gray stripe completely around the canister near the top

Reference: ANSI K13.1 — 1967 Gas Mask Canister.

because of their smaller size, should be limited to concentrations not in excess of 0.5% by volume (Figure 7.6). Most canisters contain a filter for the removal of dust and other particulate matter, indicated by a gray stripe around the top of the canister.

Service life of an air-purifying device depends on the following factors:

1. The design, including the quality and amount of chemical fill, packing uniformity, and density.
2. Variable exposure conditions, including concentration of contaminants in the air, breathing rate of the wearer, temperature, and humidity. Generally, higher concentrations, breathing rates, and humidity conditions adversely affect service life.

Figure 7.6. Mask with chin-type canister.

Since the exposure conditions are subject to wide variation, it is most difficult to estimate the service life of a gas mask canister (Figure 7.7). However, for guidance purposes, USBM approval Schedule 14F stipulates the following minimum service requirements at an average breathing rate of 25 L/min in concentrations of 2% for most gases and vapors or 3% for ammonia:

- Industrial-size canisters — 30 min
- Type N canister — ?????
- Acid gases — 15 min
- Organic vapors — 25 min
- Ammonia — 15 min
- Carbon monoxide — 30 min

Super-sized canisters, because of their greater volume of chemical fill, will last approximately twice as long as the equivalent industrial-size canister. Chin-style canisters, because of their small size, should be used in concentrations not in excess of 0.5 %.

It is generally recommended that canister gas masks used for emergency purposes should be replaced after each use. Specific indications for canister replacement and/or return to fresh air are

1. If canisters with window indicators show the specified color changes.
2. If any leakage is detected by smell, taste, eyes, nose, or throat irritation.
3. If high breathing resistance develops.
4. If the canister shelf life is exceeded.

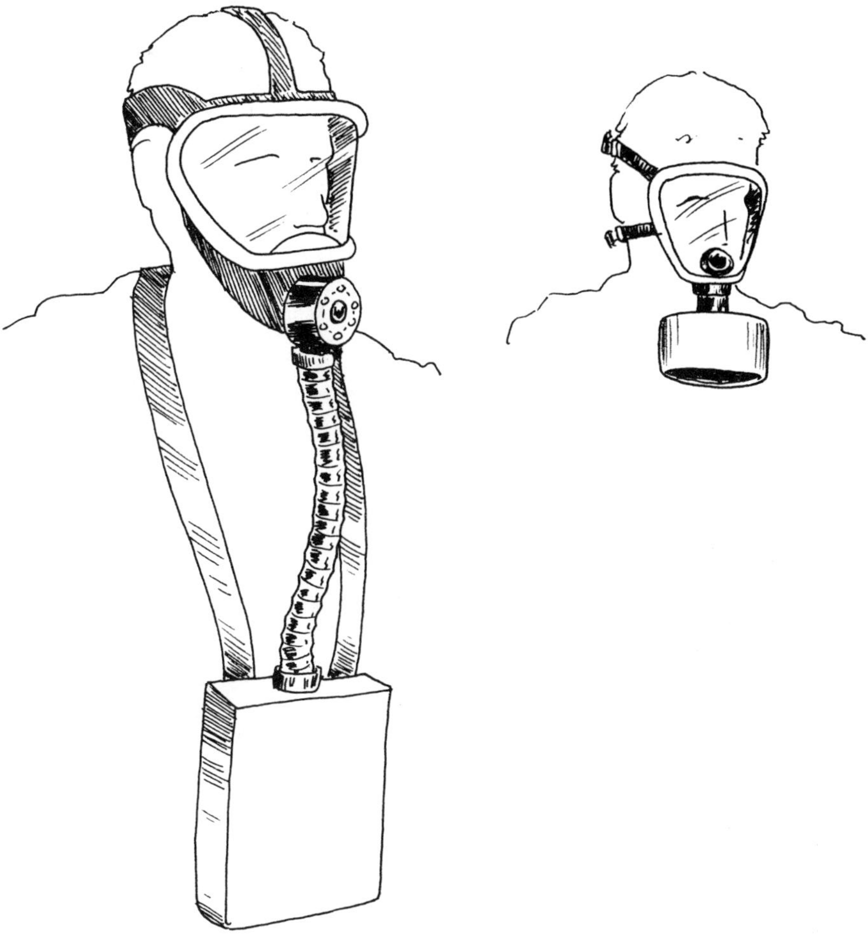

Figure 7.7. Two different designs of canister-type gas masks. Note the speaking, diaphragms, and the face pieces that allow wide, panoramic vision. (Left) mask has rigid, adjustable head harness; (Right) mask has conventional strap-type head harness.

Specific warning signals which require immediate return to fresh air are

1. Uncomfortable heat in the inhaled air. (A properly operating canister will become warm on exposure to certain gases or vapors, but a canister which becomes extremely hot indicates that concentrations above the canister limit have been encountered.)
2. A feeling of nausea, dizziness, or ill-being.

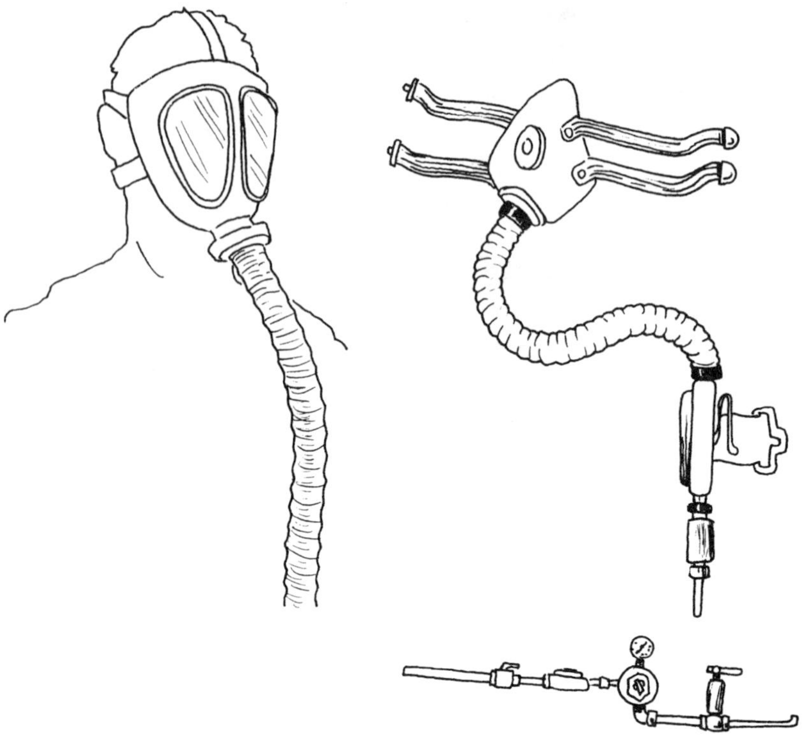

Figure 7.8. Air-line respirator. Picture at left shows the full-face mask. Diagram (right) shows parts and connections.

AIR-LINE RESPIRATORS

Air-line respirators should be used only in atmospheres not immediately harmful to life or from which the wearer can escape without the use of the respirator. This limitation is necessary because the air-line respirator is entirely dependent upon an air supply which is not carried by the wearer of the respirator. If this air supply fails, the wearer is without respiratory protection and might not escape from the immediately hazardous atmosphere. Another limitation of air-line respirators is that the air supply hose limits the wearer to a fixed distance from the air supply source.

The air-line respirator is connected to a suitable compressed air source by a hose of small inner diameter and air is delivered to the user continuously or intermittently in sufficient volume to meet the wearer's breathing requirements (Figure 7.8).

Accessory equipment such as pressure regulators, pressure relief valves, and air filters may be necessary to make sure that the air is at the proper pressure and quality for breathing. Air-line respirators are furnished in many types, but there are three basic classes — constant flow, demand flow, and pressure demand flow.

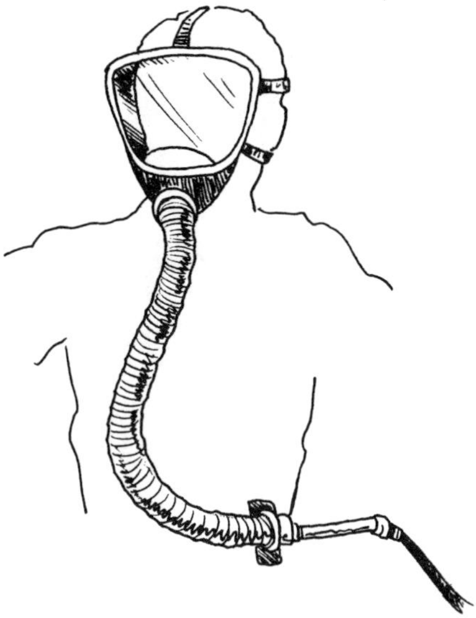

Figure 7.9. Constant-flow air-line respirator.

The respirators are equipped with half masks or full face pieces. If eye protection is required, a full face piece must be used.

Constant flow units are normally used where there is an ample air supply, such as provided by an air compressor. Damand types are generally used where only compressed air cylinders are available.

The USBM approves air-line respirators under Approval Schedule 19, which has these significant requirements — the maximum hose length for which approval is granted is 250 ft, and the maximum permissible inlet pressure is 125 psi. Approval is sought for specific hose lengths and inlet pressures. With the longest hose length for which approval is sought assembled to the respirator, and the lowest inlet pressure introduced to the air supply hose, constant flow units must deliver at least 4 cfm (cubic feet per minute) measured at the face piece. The exhalation resistance at 85 L/min should not exceed 1 in. (ce) — one inch of water — column height.

When helmets or hoods are used, the same requirements must be met, except that the flow rate must be at least 6 cfm. For both types of assemblies with the highest inlet pressure and shortest hose length, the maximum flow should not exceed 15 cfm.

Constant-flow air-line respirators with face pieces alone are used where respiratory protection only is needed (Figure 7.9). A hood can be added to the face piece for protection against sand or shot blast, and frequently a helmet is used for this application with a hood or cape fitted to it. Such units are used also for lead grinding.

Demand-type air-line respirators with half masks or full face pieces deliver air flow only during inhalation, with exhalation to the atmosphere (Figure 7.10). Such

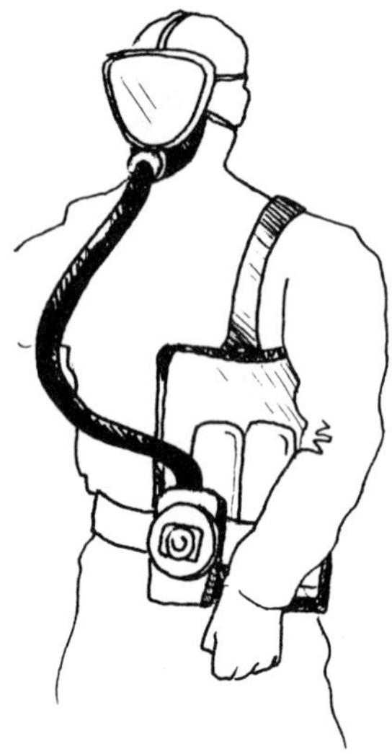

Figure 7.10. Demand-type breathing apparatus.

respirators are normally used where the air supply is restricted to high-pressure compressed air cylinders. A suitable pressure regulator is required to make sure that the air is at the proper pressure for breathing. The same requirements for approval apply to the demand type as to the constant flow units, except the minimum flow at a maximum resistance of 2 in. (wc) measured at the face piece should be at least 4 cfm and not more than 15 cfm with all hose lengths within the inlet pressure range for which approval is sought. The exhalation resistance at 85 L/min (2.8 cfm) should not exceed 1 in. (wc).

For those conditions where the possible inward leakage (caused by the negative pressure during inhalation always present in demand system) is unacceptable and where there cannot be the relatively high air consumption of the constant-flow units, a pressure demand air-line respirator may be the best choice. It provides a positive pressure during both inhalation and exhalation and must meet the same requirements for demand systems, except that the static pressure in the face piece should not exceed 1.5 in. (wc) and the exhalation resistance at 85 L/min should not exceed the static pressure in the face piece by more than 2.0 in. (wc).

The air supply is the responsibility of the user, and the air-line respirator is approved for use only when it supplies respirable air at the correct pressure and flow. The compressed air should meet the most recent requirements of Compressed Gas Association Specification G7.1 for Type I, Class D gaseous air. This currently requires that the carbon monoxide level not exceed 20 ppm, and condensed hydrocarbons not exceed 50 mg/m^3.

With internally lubricated piston-type compressors, overheating may produce carbon monoxide; therefore, routine testing for carbon monoxide of the installation of a constant monitoring analyzer is desirable.

Some air compressors are manufactured specifically to provide respirator air. They use compression seal liquids such as water or diaphragms for delivering the air.

HOSE MASKS

Hose masks, which supply outside air to the wearer through a length of hose, are available either with or without blowers. Those with blowers may be approved by the USBM for respiratory protection in any atmosphere regardless of the degree of contamination or oxygen deficiency, providing clean breathable air can be reached within the distance of the permissible hose length. (Up to 300 ft of hose can be used on this type of equipment.)

The USBM approves either hand-driven or motor-hand-driven blowers. Motor driven units are also available, but are not presently approved.

The air hose must have a large inside diameter, approximately one in. In case of blower failure, the wearer can breathe through the hose while escaping from the contaminated air. It must be highly resistant to petroleum vapors as well as being able to withstand crushing weight. This accounts for the heavy wire reinforce construction of hose mask hoses. The entire hose mask assembly including face pieces, harness, air hose, and blower is usually packaged in a portable trunk. The blower operator serves also as a standby or guard, ready to aid the workman in case of mishap.

Hose masks without blowers are used when uncontaminated air can be reached within a distance of 75 ft. Those units, however, carry only limited approval and cannot be used in atmospheres immediately harmful to life. USBM Approval Schedule 19 covers this class of equipment.

SELF-CONTAINED BREATHING APPARATUS

Self-contained breathing apparatus provides complete respiratory protection in toxic gases and where there is oxygen deficiency. The wearer is independent of the surrounding atmosphere because he/she is breathing with a system admitting no outside air. The oxygen or air supply of the apparatus itself takes care of respiratory requirements. Self-contained breathing apparatus is of three basic types — oxygen cylinder rebreathing type, self-generating type, and the demand type.

The oxygen cylinder rebreathing type (in use today) is the "lung-governed" type, which automatically compensates for the varying breathing demand of the user. It consists of a relatively small cylinder of compressed oxygen, reducing and regulating valves, a breathing bag, face or mouth piece plus nose clip, and a chemical container to remove carbon dioxide from the exhaled breath. The self-generating type uses the principle of rebreathing, but it has no mechanical operating components. The two types of cylinder rebreathing units now manufactured are approved by the USBM for 2- or 3-hr duration, and both function in the same manner and will be discussed together.

The high-pressure oxygen from the cylinder is reduced in pressure to a breathing level by means of a reducing and regulating valve. Some units have a small constant-flow valve plus a lung-controlled valve which adds any required additional flow. Other types have only an admission valve which delivers the oxygen from the breathing bag to the wearer's face.

Exhaled breath passes down another tube into the container holding the carbon dioxide-removing chemical and then through a cooler. Finally, the purified exhalation flows into the breathing bag where it mixes with the incoming oxygen from the cylinder.

The rebreathing principle permits the most efficient utilization of the oxygen supply. The exhaled breath contains both oxygen and carbon dioxide as the human body extracts only a small part of the oxygen inhaled. As the user exhales into the container, the carbon dioxide is removed by the chemical and the oxygen that is left is reused. That method of operation applies to all oxygen cylinder rebreathing type apparatus as well as to the self-generating type.

The oxygen cylinder must be refilled and the carbon dioxide-removing chemical replaced after each use. As is true of all respiratory protective equipment, training in proper use and maintenance is essential for the most efficient operation.

Demand-type apparatus are available in different models for specific applications. All consist of a high-pressure oxygen or air cylinder, a demand regulator connected either directly or by a high-pressure tube to the cylinder, a face piece and tube assembly with an exhalation valve or valves, and a method of mounting the complete apparatus on the body. In use, the wearer opens the cylinder valve after putting on the face piece, inhales oxygen (or air) at breathing pressure through the demand regulator to the face piece, and then exhales through a valve in the face piece to the surrounding atmosphere.

The term "demand regulator" means that the air flow is on inhalation demand, automatically regulating itself to the desired level to compensate for variations in breathing needs. All demand apparatus are relatively inefficient when compared with the rebreathing types, because the exhaled oxygen is released to the atmosphere instead of being reused.

Of particular application to some industries is a short-duration unit known as an "escape mask." It is normally worn by a worker at times when in an area where a potentially toxic, above the capacity of a gas mask, atmosphere exists. Areas to which high-pressure gases are frequently piped fall into this category.

A work mask type of apparatus is used for planned maintenance work. It is not an emergency apparatus as, for use, it requires setting up some auxiliary equipment and has a connecting hose which limits the free movement of the wearer to the length of the hose. The apparatus is actually a twofold unit.

1. A self-contained apparatus with which the wearer can move freely about by using a small cylinder to enter or leave the toxic atmosphere — either routinely or in the failure of the main breathing supply.

2. There is a provision for connecting a host to the demand regulator so that the wearer can breathe during the work period from a larger cylinder which has a pressure-reducing valve attached to provide low pressure for the connecting hose. Manifold assemblies can be provided to permit more than one worker to work from a common breathing source. It is imperative that the small emergency cylinder be used with this type of apparatus; otherwise, a failure of the supply from the large cylinder would expose the wearer to the toxic atmosphere.

Pressure-demand apparatus, using the same principle as described previously under pressure-demand air-line respirators, is approved and used where the toxicity is such that the potential back leak of demand apparatus is not tolerable.

Under the current Approval Schedule 13E, all demand-type apparatus must satisfy tests conducted at —25°F, although low-temperature components, such as nose cups to reduce fogging, may be added if needed.

The wearer of demand apparatus depends upon watching a pressure gauge to learn when the breathing supply has dropped to a point where return to fresh air is necessary. A self-actuating warning device is required for USBM approval. The warning device can sound an alarm or increase breathing resistance to alert the wearer to return to fresh air.

Units of 15- and 30-min duration are currently approved by USBM under Approval Schedule 13 for entry into and escape from irrespirable atmospheres; 3- or 10-min duration units can be approved for escape only. Combination apparatus, combining a 3- or 10-min escape unit with an air-line respirator are approved also.

The self-generating type apparatus is approved by USBM for 1 hr. This apparatus differs from conventional cylinder rebreathing apparatus in that it utilizes a chemical canister which evolves oxygen and removes the exhaled carbon dioxide in accordance with breathing requirements. It eliminates high-pressure cylinders, regulating valves, and other mechanical components.

The canister, which contains potassium superoxide, evolves oxygen when contacted by the moisture and carbon dioxide in the exhaled breath, and retains the carbon dioxide and moisture. Retaining moisture is important as it aids in preventing lens fogging.

In use, the self-generating unit operates as other rebreathing apparatus except that the wearer, using the canister, makes his or her own oxygen instead of drawing it from a compressed gas cylinder. The outstanding features of this type are its simplicity of construction and use, and lesser need for maintenance when compared with high-pressure apparatus.

STANDARDS

ANSI Standard Z88.2-1969, Practices for Respiratory Protection, is excerpted here to point out the various procedures this standard requires. The Z88 Standards Committee was organized and first met on September 24, 1963, to review the section of Z2.1-1959 dealing with respiratory protection. The committee decided to revise and write that section. The fifth draft of the proposed standard was submitted to letter ballot and accepted by the Z88 Committee on May 2, 1969. It was approved as an American National Standard on August 11, 1969 (ANSI Standards are updated every 5 years).

NOT IMMEDIATELY DANGEROUS ATMOSPHERE

If immediately dangerous atmospheres are not present or will not occur, then the consequences of respirator failure are lessened and emphasis can be placed on other factors such as long-term protection, convenience, cost, comfort, and wearer acceptance. These factors should be weighed one against another since they are not always compatible. However, long-term protection should be given priority over all other factors.

Long-term protection is determined primarily by the amount of inward leakage of atmospheric contaminants during normal usage of the respirator. The various types of respirators are grouped below according to the increasing amount of inward leakage one might expect during their routing use.

Group 1

- Supplied-air suit
- Pressure-demand, full face piece, open-circuit, and air-line devices
- Pressure-type, full face piece, closed-circuit, self-contained breathing apparatus
- Continuous-flow, full face piece, air-line respirator
- Air-line respirator with loose-fitting hood gathered around the waist
- Hose mask with blower and full face piece mask

Group 2

- Demand-type, full face piece, open-circuit, self-contained breathing apparatus
- Demand-type, full face piece air-line respirator
- Pressure-demand, half-mask air-line respirator

Group 3

- Continuous-flow, half-mask air-line respirator
- Demand-type, half-mask air-line respirator
- Air-line respirator with loose-fitting hood gathered around the neck

Group 4

- Air-purifying respirator with blower and with full face piece or hood gathered around the waist

Group 5

- Air-purifying respirator with blower and with half-mask face piece or hood gathered around the neck

Group 6

- Hose mask without blower
- Air-purifying full face piece respirator without blower

Group 7

- Air-purifying half-mask respirator without blower
- Mouth piece respirator

The negative pressure produced in the face piece of many respirators (some in Groups 2 and 3 and all in Groups 6 and 7) during inhalation will cause inward leakage of ambient air if leaks exist. The amount of inward leakage expected for various types of respirators is vaguely understood and is related to the face piece-to-face seal and leakage of various components including air-purifying units. For Group 1, inward leakage will normally be infinitesimal or nonexistent. For Groups 2, 3, 4, and 5, inward leakage will normally be minute. Group 6, inward leakage with a good face piece-to-face seal and with ideal wearing conditions may be as low as 0.1%. In practice, however, this low-level leakage is usually not attained due to poor face piece-to-face fit and adverse wearing conditions such as face movements and beard growth. For Group 7, inward leakage with good face piece-to-face fit and ideal wearing conditions may be as low as 1%. This leakage level, however, is seldom attained for reasons similar to those cited for devices in Group 6. It is therefore important (particularly for devices in Groups 6 and 7) that due consideration be given to potential inward leakage in selecting devices.

Experience has demonstrated that not all persons can obtain a satisfactory face piece-to-face seal with a single full face piece or half-mask air-purifying respirator. Therefore, to provide an adequate face piece-to-face seal on a variety of facial features, it may be necessary that two or more models of full face piece or half-mask respirators be available. When an individual can be fitted with two or more respirators, he or she should be permitted to select the most comfortable device.

It should be brought out that when using a full-face respirator, the individual should not wear contact lenses. Contact lenses ride or cover the cornea (lens) of the eye. If chemicals got under the face piece it would be very difficult to remove the contact lenses. Also, many people wear soft contact lenses rather than hard lenses. Thus, a constant exchange of air is needed for the lenses (especially extended wear). When this does not occur, the contact lenses adhere to the cornea and can literally be torn off in sections when removing them after a 4-hr shift. Presently the ANSI Standards Committee is attempting to address the concern of the use of soft contact lenses and full-face respirators. It is strongly suggested this agency be contacted for their recommendation.

USE OF RESPIRATORS

Operating Procedures

Standard procedures shall be developed for respirator use. These should include all information and guidance necessary for their proper selection, use, and care. Possible emergency and routine uses for respirators should also be anticipated and planned for. For examples of various selections and contaminants see Table 7.4.

RESPIRATORY PROTECTION

Hazards to the lungs are not always easy to detect. Some of the most common hazards are the lack of oxygen and the presence of harmful dust, fogs, smokes, mists, fumes, gases, vapors, or sprays including substances that may cause cancer, lung impairment, other diseases, or death. Respirators prevent the entry of harmful substances into the lungs during breathing. Some respirators also provide a separate supply of breathable air so work can be performed where there is inadequate oxygen or where greater protection is needed.

The prevention of atmospheric contamination at the work site generally should be accomplished as far as feasible by engineering control measures, such as enclosing or confining the contaminant-producing operation, exhausting the contaminant, or substituting with less toxic materials. Historically, the industrial hygiene profession has sought to control hazardous air contamination through engineering means. However, when effective engineering controls are not feasible, while those controls are being installed, or during cleanup operations, appropriate respirators must be used. The user should be aware that respirators have their limitations and are not a substitute for effective engineering controls. Where respirators are necessary for health protection, specific procedures are necessary to overcome any potential deficiencies and to assure the effectiveness of the equipment.

Employers are responsible for establishing an effective respiratory program — different hazards require different respirators — and employees are responsible for wearing the respirator and complying with the program.

The plant or company department of industrial hygiene, health physicians, safety engineering, or fire prevention should administer the program in liaison with the plant medical department. In small plants that have no such departments,

Table 7.4. Selection of Respirators[a]

Hazard	Respirator
1. Oxygen deficiency	
Immediately dangerous to life or health[b]	Any positive-pressure SCBA[c] Combination positive-pressure SAR[d] with auxiliary self-contained air supply
Not immediately dangerous to life or health	Any positive-pressure SCBA or SAR
2. Gas and vapor contaminants	
Immediately dangerous to life or health	Positive-pressure SCBA Combination positive-pressure SAR with auxiliary self-contained air supply
Not immediately dangerous to life or health	Any positive-pressure SAR; gas mask; chemical cartridge respirator
3. Particulate contaminants	Any positive-pressure SAR including abrasive blasting respirator; powered air-purifying respirator equipped with high-efficiency filters; any air-purifying respirator with a specific particulate filter
4. Gaseous and particulate contaminants	
Immediately dangerous to life or health	Positive-pressure SCBA; combination positive-pressure SAR with auxiliary self-contained air supply
Not immediately dangerous to life or health	Any positive-pressure supplied-air respirator; gas mask; chemical-cartridge respirator
5. Escape from contaminated atmosphere that may be immediately dangerous to life or health	Any positive-pressure SCBA; gas mask; combination positive-pressure SAR with escape SCBA
6. Firefighting	Any positive-pressure SCBA

[a] This table presents a simplified version of characteristics and factors used for respirator selection. It does not specify the contaminant concentrations or particle size.
[b] "Immediately dangerous to life or health" is any condition that poses either an immediate threat to life or health or an immediate threat of severe exposure to contaminants, such as radioactive materials, which are likely to have adverse delayed effects on health.
[c] SCBA — self-contained breathing apparatus.
[d] SAR — supplied air respirator.

the respirator program should be administered by an upper-level superintendent, foreman, or qualified person responsible to the principal manager. Responsibility for the program should rest with one person. The administrator must have sufficient knowledge of the subject to supervise the program properly.

Any respirator program should stress thorough training of all participants, especially the users who need to wear the respirators. Employees must be aware that the equipment does not eliminate the hazard. If the equipment fails, overexposure will occur. To reduce the possibility of failure, equipment must fit properly and be maintained in a clean and serviceable condition.

Employers and employees must understand the equipment's purpose and its limitations. The equipment must not be altered or removed from the wearer even for a short time, despite the fact the wearer may find it uncomfortable.

An effective respirator program should include the following:

- Written standard operating procedures
- Program evaluation
- Selection
- Training
- Fit testing
- Inspection, cleaning, maintenance, and storage
- Medical examinations
- Work area surveillance
- Air quality standards
- Approved respirators

WRITTEN STANDARD OPERATING PROCEDURES

In workplaces where respirators are used in potentially hazardous atmospheres present during normal operations or emergency situations, employers are required to have written operating procedures for the safe and proper use of respirators. Users must be familiar with these procedures as well as with available respirators and their limitations.

PROGRAM EVALUATION

The effectiveness of a company's respirator program should be evaluated regularly — at least annually — and the written operating procedure modified as necessary to reflect the evaluation results. The use of a labor-management team may be effective for the periodic evaluation.

Written procedures shall be prepared covering safe use of respirators in dangerous atmospheres that might be encountered in normal operations or in emergencies. Personnel shall be familiar with these procedures and the available respirators.

In areas where the wearer, with failure of the respirator, could be overcome by toxic or oxygen-deficient atmosphere, at least one additional man shall be present. Communications (visual, voice, or signal line) shall be maintained between both or all individuals present. Planning shall be such that one individual will be unaffected by any likely incident and have the proper rescue equipment to be able to assist the other(s) in case of emergency.

TRAINING

For safe use of any respirator, it is essential that the user be properly instructed in its selection, use, and maintenance. Both supervisors and workers shall be instructed by competent persons as required in CFR 1910.120 and TSCA 717.

All employees required to use respiratory protective equipment must be instructed in the proper use of the equipment and its limitations. Those employees who will be required to use respiratory protective equipment in atmospheres immediately dangerous to life or health should be trained in rescue procedures.

Minimum training shall include the following:

1. Instruction in the nature of the hazard, whether acute, chronic, or both, and an honest appraisal of what may happen if the respirator is not used.
2. Explanation of why more positive control is not immediately feasible. This shall include recognition that every reasonable effort is being made to reduce or eliminate the need for respirators.
3. A discussion of why this is the proper type of respirator for the particular purpose.
4. A discussion of the respirator's capabilities and limitations.
5. Instruction and training in actual use of the respirator (especially a respirator for emergency use) and close and frequent supervision to assure that it continues to be properly used.
6. Classroom and field training to recognize and cope with emergency situations.
7. Other special training as needed for special use.

Training shall provide the individuals an opportunity to handle the respirator, have it fitted properly, test its face piece-to-face seal, wear it in normal air for a long familiarity period, and finally, to wear it in a test atmosphere.

Training should include an explanation of the following:

1. Nature of the respiratory hazard and what may happen if the respirator is not used properly.
2. Engineering and administrative controls being used and the need for the respirator as added protection.
3. Reason(s) for selection of a particular type of respirator.
4. Limitations of the selected respirator.

5. Methods of donning the respirator and checking its fit and operation.
6. Proper wear of the respirator.
7. Respirator maintenance and storage.
8. Proper method for handling emergency situations.

Users should know the improper respirator use or maintenance may cause overexposure. They should know that continued use of poorly fitted and maintained respirators can cause chronic disease or death from overexposure to air contaminants.

FIT TESTING

Full face pieces, half masks, quarter masks, and even the different brands of the same type of respirator marketed have different fit characteristics. No one respirator will fit everyone. Employers will find it advantageous to purchase several brands of each type in various sizes to assure proper fit for all workers who must wear one.

Corrective glasses worn by employees also present a problem when fitting respirators. Special mountings are available to hold corrective lenses inside full face pieces. If corrective lenses are needed, the face piece and lenses must be fitted by a qualified individual to provide good vision, comfort, and proper sealing.

The user must receive fitting instructions including demonstrations and practice in how to wear the respirator, how to adjust it, and how to determine if it fits properly.

Although respirators are designed for maximum efficiency, they cannot provide protection without a tight seal between the face piece and wearer. Consequently, beards and other facial hair can substantially reduce the effectiveness of a respirator. The absence of dentures can seriously affect the fit of a face piece. To assure proper respiratory protection, a face piece must be checked each time the respirator is worn. This can be accomplished by performing either a positive- or negative-pressure check. Detailed instructions for performing these tests can be found in the ANSI Standards.

The effectiveness of the fit of the face piece can be tested two ways — qualitatively and quantitatively.

Qualitative fit testing involves the introduction of a harmless odorous or irritating substance into the breathing zone around the respirator being worn. If no odor or irritation is detected, a proper fit is indicated.

Quantitative fit testing offers the most accurate, detailed information on respirator fit. It involves the introduction of a harmless aerosol to the wearer while he or she is in a test chamber. While the wearer performs exercises that could induce face piece leakage, the air inside and outside the face piece is then measured for the presence of the harmless aerosol to determine any leakage into the respirator.

MAINTENANCE AND CARE OF RESPIRATORS

General

A program for maintenance and care of respirators shall be adjusted to the type of plant, working conditions, and hazards involved, and shall include the following basic services:

1. Inspection for defects (including a leak check)
2. Cleaning and disinfecting
3. Repair
4. Storage

Equipment shall be properly maintained to keep original effectiveness.

Inspection

All respirators shall be inspected routinely before and after each use. A respirator that is not routinely used but is kept ready for emergency use shall be inspected after each use and at least monthly to assure that it is in satisfactory working condition.

Self-contained breathing apparatus shall be inspected at least monthly. Air and oxygen cylinders shall be fully charged according to the manufacturer's instructions. It shall be determined that the regulator and warning devices function properly.

Respirator inspection shall include a check of the tightness of connections and the condition for the face piece, head bands, valves, connecting tube, and canister. Rubber of elastomer parts shall be inspected for pliability and signs of deterioration. Stretching and manipulating rubber or elastomer parts with a massaging action will keep them pliable and flexible and prevent them from taking a set during storage. A respirator inspection must include a check of the tightness of the connections.

Chemical cartridges and gas mask canisters should be replaced as necessary to provide complete protection. The manufacturer's recommendation should be followed. Mechanical filters must be replaced as necessary to avoid high resistance to breathing.

Repairs must be made only by experienced persons using parts specifically designed for the respirator. The manufacturer's instruction should be consulted for any repair and no attempt should be made to repair or replace components or make adjustments or repairs beyond the manufacturer's recommendations.

A respirator that has been used must be cleaned and disinfected before it is issued. Emergency-use rescue equipment must be cleaned and disinfected immediately after each use. Records must be kept of inspection dates and findings. The records must be maintained and kept for 30 years (CFR 1910.20).

Respirators must be stored to protect against dust, sunlight, heat, extreme cold, excessive moisture, or damaging chemicals. Protection against any mechanical damage should also be provided. Respirators should be stored so that face pieces and exhalation values will rest in a normal position to prevent the rubber or plastic from reforming into an abnormal shape.

Cleaning and Disinfection

Routinely used respirators shall be collected, cleaned, and disinfected as frequently as necessary to ensure that proper protection is provided for the wearer. Each worker should be briefed on the cleaning procedure and be assured that he or she will always receive a clean and disinfected respirator. Such assurances are of greatest significance when respirators are not individually assigned to workers. Respirators maintained for emergency use shall be cleaned and disinfected after each use.

The following procedure is recommended for cleaning and disinfecting respirators:

1. Remove any filters, cartridges, or canisters.
2. Wash face piece and breathing tube in cleaner-disinfectant or detergent solution (see following paragraphs). Use a hand brush to facilitate removal of dirt.
3. Rinse completely in clean, warm water.
4. Air dry in a clean area.
5. Clean other respirator parts as recommended by manufacturer.
6. Inspect valves, head straps, and other parts; replace with new parts if defective.
7. Insert new filters, cartridges, or canisters; make sure seal is tight.
8. Place in plastic bag or container for storage.

Cleaner-disinfectant solutions are available that effectively clean the respirator and contain a bactericidal agent. The bactericidal agent is generally a quaternary ammonium compound. The respirator may be immersed in the solution, rinsed in clean, warm water, and air dried.

Alternatively, respirators may be washed in a liquid detergent solution, then immersed in: (1) a hypochlorite solution (50 ppm of chlorine) for 2 min; (2) an aqueous iodine solution (50 ppm of iodine) for 2 min; or (3) a quaternary ammonium solution (200 ppm of quaternary ammonium compounds in water with less than 500 ppm total hardness). These disinfectants will kill the AIDS virus, too.

Different concentrations of quaternary ammonium salts are required to achieve a disinfecting solution with waters of varying hardness. Also, dermatitis may occur if the quaternary ammonium compounds are not completely rinsed from the respirator. The hypochlorite and iodine solutions are not stable; they age rubber parts, and are corrosive to metallic parts. Therefore, immersion times should not be extended, and the disinfectants shall be thoroughly rinsed from the respirator parts.

Strong cleaning and disinfecting agents can damage respirator parts. Temperatures above 185°F and vigorous mechanical agitation should not be used. Solvents which affect elastomer or rubber parts should be used with caution. Thus, do not put respirator in dishwashers unless temperature is regulated.

Respirators may be contaminated with toxic materials (that is, organic phosphate pesticides and radionuclides). If the contamination is light, normal cleaning procedures should provide satisfactory decontamination; if heavy, a separate decontamination step may be required before cleaning. Also, when heavy contamination is suspected, dispose of all wash and rinse waters as stated in RCRA and CWA requirements.

For complete decontamination against phosphate pesticides, the respirator should be washed with alkaline soap and rinsed with 50% alcohol (ethyl or isopropyl).

Respirators used to protect against radioactive contaminants should be decontaminated to levels not exceeding 100 disintegrations per minute per 100 cm^2 fixed alpha and 0.2 mrad/hr of beta-gamma above background at contact. There should be not detectable removable activity using standard swipe techniques.

Repair

Replacement or repairs shall be done only by experienced persons with parts designed for the respirator. No attempt shall be made to replace components or to make adjustment or repairs beyond the manufacturer's recommendations. Reducing or admission valves or regulators shall be returned to the manufacturer or to a trained technician for adjustment or repair.

Storage

After inspection, cleaning, and necessary repair, respirators shall be stored to protect against dust, sunlight, heat, extreme cold, excessive moisture, or damaging chemicals. Respirators placed at stations and work areas for emergency use should be stored in compartments built for the purpose, be quickly accessible at all times, and be clearly marked. Routinely used respirators, such as dust respirators, may be placed in plastic bags (i.e., freezer bags). Respirators should not be stored in such places as lockers or tool boxes unless they are in carrying cases or cartons.

Respirators should be packed or stored so that the face piece and exhalation valve will rest in a normal position and function will not be impaired by the elastomer setting in an abnormal position.

Instructions for proper storage of emergency respirators, such as gas masks and self-contained breathing apparatus, are found in "use and care" instructions, which are usually mounted inside the carrying case lid.

MEDICAL EXAMINATIONS

Persons assigned to tasks that require the use of a respirator must be physically able to perform the work while using the respirator. The respirator user's medical status must then be reviewed periodically. The local physician must determine what health and physical conditions are pertinent.

When respirators are worn in toxic atmospheres, the individual should be provided appropriate laboratory tests. These may include urine, blood, or fecal analysis and other techniques to determine the intake and excretion of toxic substances. The findings of these tests, when correlated with other exposure data, such as air sampling data for wearers of such equipment, can serve as an indication of the effectiveness of the program. Positive evidence of exposure should be followed up with appropriate surveillance of work area conditions to determine if there is any relationship to inadequate respiratory protection or a need for additional engineering controls.

WORK AREA SURVEILLANCE

Surveillance must be maintained of the conditions in the work area and of the degree of worker exposure or stress (combination of work rate, environmental conditions, and physiological burdens of wearing a respirator).

Changes in operating procedures, temperature, air movement, humidity, and work practices may influence the concentration of a substance in the work area atmosphere. These factors necessitate periodic monitoring of the air contaminant concentration. Testing should continue to assure that the contaminant exposure has not risen above the maximum protective capability of the respirators being used.

Employees using self-containing breathing apparatus or supplied air respirators with auxiliary self-contained breathing apparatus in confined spaces, where the environment is or may be immediately dangerous to life or health, must wear safety harnesses and lifelines. A second person equipped with complete protective gear must be standing by ready to help if the first worker gets into trouble. Communications (visual, voice, or signal line) must be maintained with all persons present. Precautions must be taken so that in the event of an accident, one person will be unaffected and have the proper rescue equipment to be able to assist the others in an emergency situation.

AIR QUALITY STANDARDS

Compressed air, compressed oxygen, liquid air, and liquid oxygen used for respiration must be of high purity. Oxygen must meet the requirements of the U.S. Pharmacopoeia for medical or breathable oxygen. Breathable air must meet at least the requirement for Grade D breathable air described in Compressed Gas Association (CGA) Commodity Specification G-7.1-1966. (A later edition of this specification, G-7.1-1973, has been issued by CGA.) Compressed oxygen must not be used in open circuit self-contained breathing apparatus or supplied air respirators that have previously used compressed air. Oxygen must never be used with air-line respirators.

Breathable air may be supplied to respirators from cylinders or air compressors. For testing cylinders, see Shipping Container Specifications of the DOT (49 CFR Part 178).

Containers of breathable gas containers must be clearly marked. (See ANSI Method of Marking Portable Compressed Gas Containers to identify the Material Contained, Z48.1-1954; Federal Specification BB-A-1034a, June 21, 1968, "Air, Compressed for Breathing Purposes;" or Interim Federal Specification GG-B-00675b, April 27, 1965, "Breathing Apparatus Self Contained.") Further details on the sources of compressed air and its safe use will be found in the CGA's pamphlet G-7-1968.

The compressor for supplying air must be equipped with the necessary safety devices and alarms. Compressors must be constructed and situated to avoid any entry of contaminated air into the system and must be equipped with suitable in-line, air-purifying sorbent beds and filters installed to assure air quality. The system must also have a receiver of sufficient capacity to enable the wearer to escape from a contaminated atmosphere in the event of compressor failure and alarms to indicate compressor failure and overheating. If an oil-lubricated compressor is used, it must have a high-temperature or carbon monoxide alarm, or both. If only the high-temperature alarm is used, the air from the compressor must be tested frequently for carbon monoxide.

Air-line couplings must be incompatible with outlets for other gas systems to prevent accidental servicing of air-line respirators with nonrespirable gases or oxygen.

SUMMARY

When planning a program to control occupational illness caused by breathing air contaminated with harmful dusts, fogs, fumes, mists, gases, smokes, vapors, or sprays, the primary objective should be to prevent such atmospheric contamination. Respirators must be used while effective engineering controls, if they are feasible, are being installed. If engineering controls are not feasible, employers must provide respirators and employees must wear them when it is necessary to protect the health of the employee. The equipment issued to the employee must be properly selected, used, and maintained for a particular work environment and contaminant.

REFERENCES

1. Patty, F. A. General Principals of Industrial Hygiene, Respirators and Respiratory Protective Equipment, chap.XI.
2. Lundin, A.M. "Respiratory Protective Equipment," Fundamentals of Industrial Hygiene (Chicago: National Safety Council, 1971).

3. OSHA 3079. Respiratory Protection, U.S. Department of Labor, Occupational Safety and Health Administration (revised), Government Printing Office, Washington, DC (1988).
4. NIOSH/OSHA/USCG/EPA. Occupational Safety and Health Guidance Manual for Hazardous Waste Site Activities. Superintendent of Documents, U.S. Government Printing Office, Washington, DC, DHHS (NIOSH) Publ. No. 85-115 (October 1985).
5. Pinkham, J. Occupational Health and Safety Magazine. "Good Rapport, Education are Keys to Getting Workers to Use Respirators," pp. 45–46, Waco, TX (August 1988).
6. Ginnold, R. Occupational Health and Safety Magazine. "Grain Fumigation Hazard Invoke Nationwide Protective Measures," pp. 52–56, Waco, TX (October 1987).
7. CFR as stated below. Government Printing Office, Washington DC (1989) for Respiratory Protection. CFR 29 Part 1910.134. CFR 29 Part 1926.103. CFR 29 Part 1915.152. CFR 29 Part 1918.102.
8. OSHA 2254 (Revised). Training Requirements in OSHA Standards and Training Guidelines, U.S. Department of Labor, Occupational Safety and Health Administration. U.S. Government Printing Office, Washington DC (1987), pp. 27, 47, and 51.
9. Code of Federal Regulation (30 CFR 11).
10. U.S. Bureau of Mines Approval Section 19.

CHAPTER 8

Personal Protective Equipment

INTRODUCTION

In order for a chemical to do harm to the body, it must first gain entrance. The skin, which includes the respiratory tract, is nature's way of preventing harmful substances from entering the body. On a hazardous waste site or industrial workplace this may not prove adequate. Personal protective equipment provides a barrier between the person and the environment, thus preventing harmful effects.

The proper selection of equipment is important in preventing exposures. The person doing the selection will have to take into consideration several factors. The level of protection type and kind of equipment selected depends on the hazardous conditions and, in some cases, cost, availability, compatibility with other equipment and performance. An accurate assessment of all these factors must be made before work can be safety carried out.

CONTROL MEASURES FOR SPECIAL OPERATIONS

Exposure to the causes of industrial skin diseases varies considerably from one industry to another. Although general preventive measures can be outlined for many types of exposure, it is advisable to solve each problem individually after complete information on the conditions surrounding the job has been obtained. A thorough understanding of the basic methods of control will enable a safety professional to arrive at the simplest solution, which often is the most effective.

For example, on some wet polishing operations it has proven practical to install barriers to prevent skin irritants from splashing on operators. Before this simple solution was conceived, the operators were expected to wear protective clothing which covered a large part of their body. Since the clothing was uncomfortable.

this regulation was difficult to enforce. The barriers not only furnished protection against the hazard, but also eliminated a supervisory problem.

The type and quantity of skin irritants used in various industrial processes affect the degree of control which can be obtained, but the primary objective in every case should be complete elimination of skin contact through change of process or use of protective measures.

Some compounds can be successfully used when the percentage of the irritant in the compound is reduced. In other cases, a less irritating or nonirritating material can be substituted for a skin irritant. The supplier should be consulted since he may be able to provide a closely related and generally satisfactory substitute for the irritating material.

When no substitute can be found for agents that produce skin irritation, they can best be prevented from entering the work room air by installation of a complete enclosure for the process along with an exhaust system to remove the irritating materials to a collecting unit or to the outside air. Many operations involving the use of acids or caustics, vapors, dusts, and mists can be totally enclosed.

Where complete enclosure is not practicable, the irritants may be locally exhausted at the point of operation to prevent amounts sufficient to cause irritation from being dispersed into the work room air.

Exposure to irritants is often minimized by having workers wear personal protective equipment, such as coveralls or coats, gloves, sleevelets, aprons, caps, face shields, acid hoods, special shoes or boots, and other articles of protective attire that is impervious to the exposure. Provision of clean clothing daily, such as long-sleeved and long-legged coveralls which fasten at the neck, are recommended for many exposures.

On some jobs, protective ointments may be used successfully to prevent contact with skin irritants. Such ointments must be selected on the basis of the physiochemical characteristics of the specific irritants against which they are to be used. Otherwise, they may be useless as a means of preventing contact.

Before a particular protective ointment is purchased, its composition should be ascertained and its protective ability evaluated. The supplier of the ointment should be asked to furnish proof that the product does protect against the particular skin hazard and will not contaminate the product that is to be manufactured when handling it.

Satisfactory protection against some exposures may be provided by simple agents, such as cold creams, lanolin, petrolatum, and many other bland oils, or by compounds of these materials.

Where it is not possible to eliminate contact with solvents, it has been found expedient to have the employee, at the end of the shift, use some material which puts back into the skin the oil which has been removed by the solvents. This procedure should be adopted only where the plant physician or other competent person has passed on the advisability of using material. Some substances that may restore oil to the skin are themselves irritants.

Remember UV light protection is available in many sun block products for the skin. They should be used when the employee is working in the sunny outdoors, black light operation, or UV light overhead when no diffusers are used.

PROTECTIVE CLOTHING — CRITERIA FOR SELECTION

Chemicals

The most important factors in selecting personal protective clothing is the determination of what chemicals the employee will be exposed to. On a hazardous waste site this may range from a few to several hundred, and in an industrial workplace only a handful. Determining the exact chemical or group of chemicals (certain groups tend to require similar protection, e.g., aliphatic chlorinated hydrocarbons) can be done by taking air samples or using preliminary soil samples for extrapolation. Both of these methods require a person to enter the site in order to do the sampling.

Not all chemicals are absorbed through the intact skin. The ACGIH TLV booklet uses a skin notation in its listing. The EPA has composed a list, from "Oil and Hazardous Materials Technical Assistance Data System" (OHMTADS), of common chemicals that have been identified as having the potential for rapid absorption through the intact skin. Since both the ACGIH and OHMTADS lists contain only about 1200 chemicals, they may be of limited use. Other reference sources should also be consulted.

When selecting personal protective equipment (PPE) one must always consider the chemical they are using in addition to the type of work they are doing. Then the most appropriate PPE can be selected.

Protective materials are made of several different substances for protection against specific chemicals. There is *no universal protective material.* All will decompose, be permeated by, or otherwise fail to protect under given circumstances. Most manufacturers make guides to products. These guides are usually for gloves and only measure rate of degradation (failing to maintain structure, see Table 8.1). Therefore, a protective material may not degrade, but at the same time allows the chemical to pass through. Some manufacturers put disclaimers on their charts. For example, Edmont states:

> SPECIAL NOTE: The Guide indicates only the degrading effect of these chemicals on product material. Permeability of the product materials by these chemicals is not considered.

Therefore, these guides must be used with caution. When permeation tables are available, they should be used in conjunction with degradation guides.

When a chemical comes into contact with a barrier, such as a glove, the chemical seeks equilibrium between both sides of the barrier. In the case of a

Table 8.1. EPA Protective Ensembles Level of Protection Equipment A to D

Level of Protection	Equipment	Protection Provided	Should Be Used When:	Limiting Criteria
A	Recommended Pressure-demand, full-face piece SCBA[a] or pressure-demand supplied-air respirator with escape SCBA Fully encapsulating, chemical-resistant suit Inner chemical-resistant gloves Chemical-resistant safety boots/shoes Two-way radio communications Optional Cooling unit Coveralls Long cotton underwear Hard hat Disposable gloves and boot covers	The highest available level of respiratory, skin, and eye protection.	The chemical substance has been identified and requires the highest level of protection for skin, eyes, and the respiratory system based on either measured (or potential for) high concentration of atmospheric vapors, gases, or particulates or site operations and work functions involving a high potential for splash, immersion, or exposure to unexpected vapors, gases, or particulates of materials that are harmful to skin or capable of being absorbed through the intact skin Substances with a high degree of hazard to the skin are known or suspected to be present, and skin contact is possible Operations must be conducted in confined, poorly ventilated areas until the absence of conditions requiring Level A protection is determined	Fully encapsulating suit material must be compatible with the substances involved
B	Recommended Pressure-demand, full-face piece SCBA or pressure-demand supplied-air respirator with escape SCBA	The same level of respiratory protection, but less skin protection than Level A	The type and atmospheric concentration of substances have been identified and require a high level of respiratory protection, but less skin protection; this involves atmospheres	Use only when the vapor or gases present are not suspected of containing high

	Chemical-resistant clothing (overalls and long-sleeved jacket; hooded, one- or two-piece chemical splash suit; disposable chemical-resistant one-piece suit) Inner and outer chemical-resistant gloves Chemical-resistant safety boots/shoes Hard hat Two-way radio communication Optional Coveralls Disposable boot covers Face shield Long cotton underwear	It is the minimum level recommended for initial site entries until the hazards have been further identified	with IDLH concentrations of specific substances that do not represent a severe skin hazard or that do not meet the criteria for use of air-purifying respirators Atmosphere contains less than 19.5% oxygen Presence of incompletely identified vapors or gases is indicated by direct-reading organic vapor detection instrument, but vapors and gases are not suspected of containing high levels of chemicals harmful to skin or capable of being absorbed through the intact skin	concentrations of chemicals that are harmful to skin or capable of being absorbed through the intact skin. Use only when it is highly unlikely that the work being done will generate either high concentrations of vapors, gases, or particulates or splashes of material that will affect exposed skin
C	Recommended Full-face piece, air-purifying canister-equipped respirator Chemical-resistant clothing (overalls and long-sleeved jacket; hooded, one- or two-piece chemical splash suit; disposable chemical-resistant one-piece suit) Inner and outer chemical-resistant gloves Chemical-resistant safety boots/shoes Hard hat Two-way radio communications	The same level of skin protection as Level B, but a lower level of respiratory protection	The atmospheric contaminants, liquid splashes, or other direct contact will not adversely affect any exposed skin The types of air contaminants have been identified, concentrations measured, and a canister is available that can remove the contaminant All criteria for the use of air-purifying respirators are met	Atmospheric concentration of chemicals must not exceed IDLH levels The atmosphere must contain at least 19.5% oxygen

Table 8.1. (Continued)

Level of Protection	Equipment	Protection Provided	Should Be Used When:	Limiting Criteria
	Optional Coveralls Disposable boot covers Face shield Escape mask Long cotton underwear			
D	Recommended Coveralls Safety boots/shoes Safety glasses or chemical splash goggles Hard hat Optional Gloves Escape mask Face shield	No respiratory protection Minimal skin protection	The atmosphere contains no known hazard Work functions preclude splashes, immersion, or the potential for unexpected inhalation of or contact with hazardous levels of any chemicals	This level should not be worn in the Exclusion Zone The atmosphere must contain at least 19.5% oxygen

Source: NIOSH/OSHA/USCG/EPA. "Occupational Safety and Health Guidance Manual for Hazardous Waste Site Activities (#85-115)," Superintendent of Documents, U.S. Government Printing Office, Washington, DC (1985).

[a] SCBA — self-contained breathing apparatus.

vapor it is called vapor pressure. This process of seeking equilibrium causes the chemical to permeate the barrier. This process will continue until the concentration on both sides are equal. Breakthrough, or the time it takes the chemical to reach the opposite side of the barrier, can usually be plotted as an S-shaped curve with concentration of breakthrough vs time. The slope of the curve is dependent on the concentration of the challenging chemical.

Several private companies do testing for permeation. One such study was printed in the *American Industrial Hygiene Association Journal*, March 1981, by G. Nelson of Miller-Nelson Research, Inc., Dublin, California. The testing was performed by placing a piece of the material between two metal plates with the testing chemical being placed on one side. Air samples were taken from the other side. Initial breakthrough was defined as detection limits of the Miran 1A being used. He noted that most chemicals exhibit a typical S curve (permeation rate vs time). Breakthrough was dependent on the test chemical and the type and thickness of the material being tested.

Another source of data is the USCG. Comparing the USCG Chart to Nelson's data, some discrepancies appear. For example, the USCG Chart states that polyvinyl chloride is excellent protection against alcohol. However, Nelson's study using a Pioneer glove breakthrough was 3 to 20 min depending on material thickness; for comparison, natural latex had a resistance time of greater than 60 min.

The USCG rated both as excellent. This illustrates, without actual data on the chemical vs material and breakthrough times, the selection charts may be misleading.

However, it should be noted that all testing is done with a pure material, unless otherwise stated, and continuous contact until breakthrough occurs. During most site work, chemicals are usually in mixed combinations and the protective material is not in continuous contact for long periods of time. This may not be true in industrial workplaces where a potpourri of pure chemicals may be used. Example: isopropyl alcohol to wash down a part and then add an epoxy resin. Therefore, the selected material may not be the "best" protecting material. Glove selection should be dependent upon the most hazardous chemical based on skin hazard and concentration. Sometimes, layering — using several different layers of protective materials — affords the best protection. This is especially true when not just one material will protect an individual.

Concentration of the Chemical(s)

As previously stated, one of the major criteria for selecting protective material is the concentration of the chemicals in air, liquid, or solid state. At low concentrations the S curve will move away from the axis and flatten, thus producing a longer period of time until breakthrough occurs. Air-borne levels should be compared to the OSHA standards or ACGIH guidelines to determine the level of skin protection needed. The standards are not designed for skin exposed directly to the liquid, solid or gas. For example, the air-borne levels of PCB at a site may be low because it is not very volatile, but the liquid coming into direct contact with the skin may cause an overexposure.

Physical State

The hazard characteristics of a chemical depends on the physical state. Inorganic lead in soil would not be considered toxic to site personnel unless it became airborne, since it is not absorbed through intact skin. Organic lead in a liquid could be readily absorbed. Soil is frequently contaminated with hazardous materials. Concentrations will vary from a few parts per million to nearly 100%. The degree of hazard is dependent on the type of soil (i.e., amount of clay) and concentration of the chemical. In order for the chemical to permeate the material, it must first attach to the surface of the glove in either a solid, liquid, or vapor state, thus limiting the amount which is available to breakthrough. Generally speaking, "dry" soils do not cause hazard to site personnel if they take minimal precautions such as wearing some type of light-weight, nonporous gloves.

Length of Exposure

The length of time a material is exposed to a chemical increases the probability of breakthrough. Determinations of actual breakthrough times for short-term exposures indicate that several different materials can be used which would be considered to be inadequate under long-term exposures. It should be kept in mind that during testing, usually a pure (100% composition) liquid is placed in direct contact with the material producing a worst-case situation.

Abrasion

When selecting protective clothing, the job in which the employee is engaged in must be taken into consideration. Persons moving drums will need added abrasion protection for their hands, lower chest, and thighs. The use of leather gloves and a heavy apron over their regular protective clothing will prevent damage and, thus, exposure. If the leather gloves become exposed to a chemical, it should be discarded, because leather will absorb the chemical and become a reservoir for toxic chemical exposure.

Dexterity

Abrasion protection must be weighted against the ability to function, which must be maintained. A person cannot be expected to do work that requires fine dexterity if he/she has to wear a 20-mil glove.

Ability to Decontaminate

If disposable clothing cannot be used, the ability to decontaminate the materials selected must be taken into consideration. Once a chemical is absorbed by the material, it must be cleaned before it can be reused. (See chapter on Decontamination.) If the chemical has completely permeated the material, it is unlikely that the clothing can be adequately decontaminated.

Climatic Conditions

The human body was designed to work with few restraints from clothing. Protective clothing adds a burden by placing weight and restricting movement as well as preventing the natural cooling process. All protective clothing adds to this burden. In severe situations, a modified work program must be used. This program should be set up by the safety officer in consultation with someone who is familiar with dealing with hot and cold conditions. Some materials act differently when they are very hot and very cold. For example, PVC becomes almost brittle in very cold (0°F) temperatures. If there is any question, the manufacturer should be contacted.

Work Load

Like climatic conditions, the amount of protective materials a person has to wear will affect his/her ability to perform certain tasks. A person in a total encapsulating suit, even at 72°F cannot work for more than a short period of time without requiring a break. The work schedule has to be adjusted to maintain health of the employees. Special consideration should be made to selecting clothing that both protects and adds the least burden.

Reflective Clothing

Reflective clothing is available for individuals who work in extraordinary hot environments, where the glove temperature is high. Coats, pants, coveralls, and helmets are made of aluminized fabric. They are highly reflective, although care is needed to keep them from wearing out or getting dirty or creased. As a rule, they are not porous; hence, water vapor from the body cannot "breathe" through them. This deficiency, however, becomes minor compared with the benefit that comes from reflecting a radiant heat load that would otherwise be overwhelming. A full suit of this material is amazingly light in weight, as compared with the more conventional kevlar garments, but it could not be worn continuously because of its lack of porosity.

Extreme conditions of high ambient air temperature plus large sources of radiated heat have been met by a suit of quilted material faced with aluminized cloth, a helmet with an infrared-reflective glass face plate, and a self-contained air supply of compressed air. With such an outfit, individuals have made emergency repairs in kilns, remaining for a considerable period under temperatures up to 1000°F. For brief periods, they can work in temperatures of 1500°F, and momentary exposures to environments of 2000°F.

Relief to employees performing routine maintenance or inspection work in areas of high ambient temperature has been obtained with refrigerated suits. One type carries a supply of dry ice in the top of the helmet. Another gets refrigerated air from a portable refrigeration unit, this air being distributed about the wearer's body through air channels. Tests in 150 to 200°F ambient air temperature have been reported in which the air inside the suits was maintained at 90 to 100°F for extended periods. Other methods include use of liquid air, portable heat exchangers, or vortex devices.

The vortex tube (such as is commercially available for cooling a suit) is fed compressed air at 75 to 100 psi at the side and discharges a stream of cold air from one end and hot air from the other. There are no moving parts. Pressure of the air is the sole source of energy. A typical model operating on 100 psi compressed air consumes 25 cfm and is capable of producing a cold stream of 20 cfm of air at a temperature 55°F below the temperature of the supply air.

In less extreme and more common exposures, the body should be covered as much as possible, to protect against radiant heat. Clothing should be light in weight, loose-fitting, and porous. One-piece coveralls are better than two-piece clothing. Except for psychological reasons, the color is unimportant, because in the infrared region of the spectrum white and black are equally absorptive. A lightweight, porous hat should be worn, such as a broad-brimmed tropical helmet covered with aluminum foil, because radiant heat is particularly uncomfortable when it falls on top of the head.

If there is a high radiant heat exposure on one arm or leg, on the back, or some other particular area, aluminum foil can be pasted over this spot of the clothing to give local protection. Some companies keep a supply of aluminum foil on hand, and use it liberally wherever radiation screening is needed. When heat can be "felt" in some direction, that is a sign that radiation screening will help.

PROTECTIVE CLOTHING — CONSIDERATIONS FOR SELECTION

The choice of protective clothing is a function of hazard potential and task requirements. The following factors are taken into consideration:

1. *Performance requirements.* Clothing must be able to withstand a variety of physical abuses. The advantages and disadvantages of disposable vs reusable clothing must be considered.
2. *Construction requirements.* The construction requirements of any garment depend on the intended use of the garment. The material that the garment is made of is selected because of its effectiveness as a barrier against specific hazards — there is no such thing as "universal" protection.
   a. The physical construction of the garment must prevent penetration (e.g., location of seams and zippers, size of clothing).
   b. The material that the garment is constructed of must resist penetration at the molecular level. In many instances, it may be necessary to layer protective clothing to achieve the desired protection.
3. *Permeation rate.* Permeation rate is affected by a combination of the base material, the nature of the chemicals to which the material is exposed, and the duration and nature of exposure. Most materials allow some degree of permeation.

4. *Ease and cost of decontamination.* Considerations that should be made upon purchasing garments are the ability and degree to which the garment can be decontaminated and the cost of decontamination. Disposable clothing may be advantageous in some situations; however, such clothing is rather expensive in the long run. In most instances, field personnel will use a combination of disposable and reusable clothing.

TYPES OF PROTECTIVE MATERIALS

The following materials are generally available for a variety of garments:

1. Cellulose of paper
2. Natural and synthetic fabrics
   a. Tyvek — "spunbonded olefin"
   b. Nomex — fire resistant
3. Elastomers
   a. Polyethylene
   b. Neoprene
   c. Butyl rubber
   d. Chlorapel
   e. Viton
4. Thermal plastics films
   a. Polyvinyl chloride
   b. Saran-Dow-Product

Charts are available which indicate the effectiveness of materials to resist degradation (see Table 8.2 through Table 8.4). Note that degradation is not an indicating of the material's permeability.

TYPES OF PPE

The selection of appropriate gear is based on the level of protection needed. The following gear forms the basis of the protective clothing scheme.

1. *Head protection.* Hard hats must meet ANSI Z89.1-1969 specifications for puncture resistance. They come in several different shapes as well as colors and composition. Selection is based on personal preference. Some manufacturers have adapters so ear protectors and face shields may be attached. Most allow for liners for use in cold weather. Regulation: Hard hats. Regulated by 29 CFR 1910.135; specified in ANSI Z89.1, Safety Requirements for Industrial Head Protection (1969).
2. *Face shield.* When there is a possibility that splashing may occur, a face shield

User Precautions

Protective gloves and other protective apparel selection must be based on the user's assessment of the workplace hazards. Apparel materials do not provide unlimited protection against all chemicals, and the user must determine before use that the apparel will resist permeation and degradation by the chemicals (including chemical mixtures) in the environment of intended use.

To obtain maximum life, protective gloves and other protective apparel should have chemicals removed from the surface by washing or other appropriate methods after each use. The apparel should be stored away from the contaminating atmosphere.

Punctured, torn or otherwise ruptured apparel must be removed from service; unservicable apparel may be disposed of only in accordance with applicable waste disposal regulations.

Key to Degradation and Permeation Ratings

E Excellent .... Fluid has no effect
G Good .... Fluid has minor effect
F Fair .... Fluid has moderate effect
P Poor .... Fluid has severe effect, ranging from moderate to complete destruction
ND .... None detected
ID .... Insufficient data, data not available or conflicting data

Physical Performance Chart

Physical Characteristics	Silver Shield®	Viton†	Butyl	Nitrile
Abrasion Resistance	F	G	G	E
Cut Resistance	P	G	G	E
Puncture (Snag) Resistance	F	G	G	E
Flexibility	E	G	G	E
Heat Resistance	F	G	G	G
Ozone Resistance	E	E	G	F
Tensile Strength	E	G	G	E
Low Gas Permeability	E	E	E	P

Note: Products in these categories vary in capabilities. Laboratory tests are necessary for specific recommendations.

†Viton is a Registered Trademark of E.I. DuPont.

Permeation Chart Color Key

Good for Total Immersion

Good for Accidental Splash/ Intermittent Exposure

Not Recommended

	Silver Shield (4 Mil)			Viton (9 Mil)			Butyl (17 Mil)			Nitrile Latex (11 Mil)		
CHEMICAL	D	BT	PR	D	BT	PR	D	BT	PR	D	BT	PR
Acetaldehyde	E	>6 hrs	ND	P	0 min	281.9	E	9.6 hrs	0.066	F	4 min	161
Acetone	E	>6 hrs	ND	P	ID	ID	E	>17 hrs	ND	P	ID	ID
Acetonitrile	E	>8 hrs	ND	ID	ID	ID	E	>8 hrs	ND	ID	ID	ID
Acrylic Acid	ID	ID	ID	G	5.9 hrs	0.23	E	>8 hrs	ND	F	ID	ID
Acrylonitrile	E	ID	ID	F	1 min	176	G	3.1 hrs	0.000048	P	3 min	176
Aldehyde	E	>6 hrs	ND	P	0 min	281.9	E	9.5 hrs	0.0665	P	4 min	161

Aniline	E	> 8 hrs	ND	G	10 min	18.7	F	> 8 hrs	ND	P	1.1 hrs	45.0
Benzaldehyde	ID	ID	ID	F	9.9 hrs	4.0	E	9 hrs	ND	P	ID	ID
Benzene	E	> 8 hrs	ND	G	6 hrs	0.012	P	31 min	32.3	P	ID	ID
Benzoyl Chloride	ID	ID	ID	E	> 8 hrs	ND	F	6.2 hrs	16.6	P	ID	ID
Bromobenzene	E	ID	ID	E	8 hrs	ND	P	32 min	39.8	P	13min	9.1
Butyl Acetate	E	> 6 hrs	ND	P	ID	ID	G	1.9 hrs	7.61	P	29 min	54.4
p-t Butyltoluene	E	> 8 hrs	ND	E	> 8 hrs	ND	G	1.7 hrs	8.0	P	ID	ID
Butyraldehyde	E	ID	ID	P	54 min	9.0	E	> 15 hrs	ND	P	ID	ID
Carbon Disulfide	G	> 8 hrs	ND	E	> 8 hrs	ND	P	7 min	98.0	P	1 min	51
Carbon Tetrachloride	E	> 6 hrs	ND	E	> 13 hrs	ND	P	ID	ID	G	3.4 hrs	5
Cellosolve	G	> 6 hrs	ND	F	ID	ID	G	ID	ID	P	ID	ID
Chlorobenzene	E	ID	ID	E	> 8 hrs	ND	P	35 min	308	P	ID	ID
Chloroform	P	10 min	0.009	E	9.5 hrs	0.46	P	ID	ID	P	4 min	352
Chloronaphthalene	E	> 8hrs	ND	E	> 16 hrs	ND	P	ID	ID	P	2.9 hrs	>1.32
Chloroprene	ID	ID	ID	ID	> 8 hrs	ND	P	28 min	18	ID	ID	ID
Cyclohexane	E	> 6 hrs	ND	E	> 7 hrs	ND	P	1.1 hrs	20.3	P	ID	ID
Cyclohexanol	E	> 6 hrs	ND	E	> 8 hrs	ND	E	> 11 hrs	ND	E	> 16 hrs	ND
Cyclohexanone	E	> 6 hrs	ND	P	29 min	86.3	E	> 16 hrs	ND	P	ID	ID
Dibutylphthalate	E	> 6 hrs	ND	E	> 8 hrs	ND	E	> 16 hrs	ND	E	> 16 hrs	ND
1,1, Dichloroethane	ID	2.4 hrs	6	G	1.5 hrs	31	ID	ID	ID	P	ID	ID
1,2, Dichloroethane	E	> 6 hrs	ND	E	6.9 hrs	0.81	P	2 hrs	53	P	8 min	311
Diethylamine	E	> 8 hrs	ND	P	35 min	852	P	47 min	46	F	ID	ID
Diethylaminoethanol	E	ID	ID	E	> 8 hrs	ND	E	> 8 hrs	ND	E	> 8 hrs	ND
1,4-Diethylene Dioxide	ID	> 8 hrs	ND	P	23 min	26.8	E	> 20 hrs	ND	P	28 min	77.1
Diethylenetriamine	ID	ID	ID	E	> 8 hrs	ND	E	> 8 hrs	ND	P	ID	ID
Diisobutyl Ketone 80%	E	> 6 hrs	ND	F	1.2 hrs	90.6	G	3.3 hrs	41.2	F	3 hrs	48.9
Dimethyl Acetamide	ID	1.5 hrs	0.728	P	25 min	3	ID	> 8 hrs	ND	ID	ID	ID
Dimethyl Formamide	E	> 8 hrs	ND	P	8 min	6.5	E	> 8 hrs	ND	F	1 min	> 15
Dimethysulfoxide	G	ID	ID	F	1.5 hrs	5	E	> 8 hrs	ND	F	ID	ID
Dioxane	E	> 8 hrs	ND	F	23 min	26.8	E	> 20 hrs	ND	P	28 min	77.1
Divinyl Benzene	E	> 8 hrs	ND	E	> 17 hrs	ND	F	2.2 hrs	238	P	ID	ID
Epichlorohydrin	ID	ID	ID	P	2 hrs	4	G	> 8 hrs	ND	P	ID	ID
Ether	ID	> 6 hrs	ND	P	12 min	21.5	P	8 min	92.2	P	14 min	21.8
Ethyl Acetate	E	> 6 hrs	ND	P	ID	ID	G	7.6 hrs	3.4	P	8 min	145
Ethyl Ether	ID	> 6 hrs	ND	P	12 min	21.5	P	8 min	92.2	P	14 min	21.8

D = Degradation, BT = Breakthrough Time, PR = Permeation Rate, ND = None Detected, ID = Insufficient Data

	Silver Shield (4 Mil)			Viton (9 Mil)			Butyl (17 Mil)			Nitrile Latex (11 Mil)		
CHEMICAL	D	BT	PR	D	BT	PR	D	BT	PR	D	BT	PR
Ethylamine (70% in water)	E	47 min	7.64	P	ID	ID	E	> 12 hrs	ND	F	1.1 hrs	30.1
Ethylene Dibromide	E	ID	ID	E	> 8 hrs	ND	F	3.3 hrs	6.0	P	ID	ID
Formaldehyde (37% in water)	E	> 6 hrs	ND	E	> 16 hrs	ND	E	16 hrs	ND	E	> 21 hrs	ND
Furan	ID	ID	ID	P	20 min	23	P	1.3 hrs	10	P	ID	ID
Furfural	E	> 8 hrs	ND	F	3.6 hrs	14.8	E	> 16 hrs	ND	P	28 min	265
Glutaraldehyde	E	ID	ID	E	> 8 hrs	ND	E	> 8 hrs	ND	P	ID	ID

	Silver Shield (4 Mil)			Viton (9 Mil)			Butyl (17 Mil)			Nitrile Latex (11 Mil)		
CHEMICAL	D	BT	PR	D	BT	PR	D	BT	PR	D	BT	PR
n-Hexane	E	> 6 hrs	ND	ID	> 11 hrs	ND	P	ID	ID	E	ID	ID
Hydrazine (70% in water)	G	> 6 hrs	ND	P	ID	ID	E	> 8 hrs	ND	G	> 8 hrs	ND
Hydrochloric Acid (37%)	E	> 6 hrs	ND	E	ID	ID	E	ID	ID	P	ID	ID
Hydrofluoric Acid (50%)	G	> 6 hrs	ND	G	ID	ID	F	ID	ID	P	ID	ID
Isobutyl Alcohol	E	ID	ID	E	> 8 hrs	ND	E	> 8 hrs	ND	G	> 8 hrs	ND
Isobutyraldehyde	E	ID	ID	P	4 min	11.5	E	> 8hrs	ND	P	ID	ID
Methacrylic Acid	ID	ID	ID	F	> 8 hrs	ND	G	> 8 hrs	ND	P	1.7 hrs	23
Methacrylonitrile	E	ID	ID	F	4 min	462	G	6.8 hrs	0.001	P	7 min	560
Methyl Chloroform	ID	> 6 hrs	ND	E	> 15 hrs	ND	P	ID	ID	P	41 min	76.4
Methyl Cyanide	ID	> 8 hrs	ND	ID	ID	ID	E	> 8 hrs	ND	ID	ID	ID
Methyl Ethyl Ketone	E	> 24 hrs	ND	P	ID	ID	E	> 8 hrs	ND	P	ID	ID
Methyl Isocyanate	ID	ID	ID	P	4 min	121	P	1.1 hrs	9.0	P	ID	ID
Methylamine (40% in water)	F	1.9 hrs	2.0	E	> 16 hrs	ND	E	> 15 hrs	ND	G	> 8 hrs	ND
Methylene Chloride	G	> 8 hrs	ND	F	1 hr	7.32	P	24 min	133	P	4 min	766
Methylene Dianiline	E	> 24 hrs	ND	E	> 8 hrs	ND	E	> 24 hrs	ND	F	ID	ID
Methylene Dichloride	ID	1.9 hrs	0.002	G	1 hr	7.32	P	ID	ID	P	4 min	766
Morpholine	E	> 8 hrs	ND	G	1.9 hrs	97	E	> 16 hrs	ND	P	48 min	206
Nitric Acid (3 Molar)	E	> 6 hrs	ND	G	ID	ID	F	ID	ID	P	ID	ID
Nitrobenzene	E	> 8 hrs	ND	E	> 8 hrs	ND	E	> 23 hrs	ND	F	33 min	1.7
Nitropropane	E	> 8 hrs	ND	P	21 min	26.1	E	> 8 hrs	ND	P	16 min	29.5
Oxalic Acid	E	> 8 hrs	ND	E	> 8 hrs	ND	E	> 8 hrs	ND	G	ID	ID
PCB (Aroclor 1254 (50%))	E	> 8 hrs	ND	E	> 8 hrs	ND	P	ID	ID	F	ID	ID

Pentachlorophenol (1% in kerosene)	E	> 8 hrs	ND	ID	> 13 hrs	ND	P	ID	ID	E	> 13 hrs	ND
n-Pentane	E	> 6 hrs	ND	E	> 8 hrs	ND	P	ID	ID	E	ID	ID
Perchlorethylene	E	> 6 hrs	ND	E	> 17 hrs	ND	P	ID	ID	F	> 1.3 hrs	5.5
Phenol (85% in water)	G	> 6 hrs	ND	E	> 15 hrs	ND	E	> 20 hrs	ND	P	39 min	> 1500
Propyl Acetate	E	> 6 hrs	ND	P	ID	ID	G	2.7 hrs	2.86	P	17 min	72.5
n-Propyl Acetate	ID	> 6 hrs	ND	P	ID	ID	G	2.7 hrs	2.86	P	17 min	72.5
Propylenediamine	ID	ID	ID	E	> 8 hrs	ND	E	> 8 hrs	ND	F	ID	ID
Pyridine	ID	ID	ID	P	38 min	74.0	G	> 8 hrs	ND	P	ID	ID
Red Fuming Nitric Acid	P	35 min	ID	P	ID	ID	P	ID	ID	P	ID	ID
Sodium Hydroxide 50%	E	> 6 hrs	ND	G	ID	ID	E	ID	ID	G	ID	ID
Styrene	G	> 4 hrs	ND	G	ID	ID	P	ID	ID	P	ID	ID
Sulfuric Acid (3 Molar)	E	> 6 hrs	ND	E	ID	ID	G	ID	ID	P	ID	ID
Tetrachloroethylene	E	> 6 hrs	ND	E	> 17 hrs	ND	P	ID	ID	F	1.3 hrs	5.5
Tetraethylenepentamine	ID	ID	ID	E	> 8 hrs	ND	E	> 8 hrs	ND	F	ID	ID
Tetrafluoroethylene	E	ID	ID	E	> 8 hrs	ND	E	> 8 hrs	ND	ID	ID	ID
Tetrahydrofuran	E	> 8 hrs	ND	P	4 min	327	F	31 min	112	P	4 min	167
Thiophene	ID	ID	ID	E	> 8 hrs	ND	P	1.8 hrs	17	P	ID	ID
Toluene	E	> 6 hrs	ND	E	> 16 hrs	ND	F	21 min	22.1	P	11 min	68.1
Toluene Diisocyanate	E	> 8 hrs	ND	E	> 16 hrs	ND	E	> 8 hrs	ND	G	3.7 hrs	1.8
Trichlorethylene	E	> 6 hrs	ND	G	7.4 hrs	0.24	P	18 min	550	P	8 min	283
1,1,1 Trichloroethane	E	> 6 hrs	ND	E	> 15 hrs	ND	P	ID	ID	F	41 min	76.4
1,1,2 Trichloroethane	ID	ID	ID	E	> 8 hrs	ND	P	5.7 hrs	7.0	P	ID	ID
Triethylamine	ID	ID	ID	E	> 8 hrs	ND	P	ID	ID	E	> 8 hrs	ND
Vinyl Chloride	E	> 8 hrs	ND	G	4.4 hrs	0.098	P	ID	ID	G	5.7 hrs	0.14
Xylene	E	> 24 hrs	ND	E	> 8 hrs	ND	P	ID	ID	P	ID	ID

D = Degradation, BT = Breakthrough Time, PR = Permeation Rate, ND = None Detected, ID = Insufficient Data

Table 8.3. Chemical Protection of Clothing Materials by Generic Class

Generic Class	Butyl Rubber	Polyvinyl Chloride	Neoprene	Natural Rubber
Alcohols	E[a]	E	E	E
Aldehydes	E-G	G-F	E-G	E-F
Amines	E-F	G-F	E-G	G-F
Esters	G-F	P	G	F-P
Ethers	G-F	G	E-G	G-F
Fuels	F-P	G-P	E-G	F-P
Halogenated hydrocarbons	G-P	G-P	G-F	F-P
Hydrocarbons	F-P	F	G-F	F-P
Inorganic acids	G-F	E	E-G	F-P
Inorganic bases and salts	E	E	E	E
Ketones	E	P	G-F	E-F
Natural fats and oils	G-F	G	E-G	G-F
Organic acids	E	E	E	E

[a] Key: E = excellent; F = fair; G = good; P = poor.

Source: "Survey of Personnel Protective Clothing and Respiratory Apparata...," DOT, USCG, Office of Research and Development (September 1974).

should be worn (except with a self-contained breathing apparatus). The shield should cover the entire face and the chemical cartridges on the respirator. The shield should fit firmly against the brim of the hat in order to prevent any liquid from running down the face in the event of a splash.

3. *Safety goggles.* Safety goggles should be used when eye protection is needed, such as when a half-mask respirator is being used. Safety goggles are good against particulants, but may not be adequate for some vapors.
4. *Safety glasses.* When visual correction is needed or when eye protection is desired, only approved safety glasses should be used on a site. Regulation: Face shields and safety glasses. Regulated by 29 CFR 1910.133(a); specified in ANSI Z87.1, Eye and Face Protection (1968). Glasses must be approved for industrial use. Glasses used inside the face piece of a respirator should, but are not required to, meet those standards.
5. *Hearing protectors.* Hearing protectors are required in areas where the TWA is above 90 dB. It is recommended that they be provided to any employee who is exposed to noise at 85 dB and ones who request them. Either ear plugs or ear muffs meeting ANSI specifications for hearing protection and may be used, providing they meet the necessary attenuation requirements.
6. *Body protection.* Whole body protection ranges from cotton coveralls to totally encapsulating suits. As previously discussed, an evaluation must be made to determine if the site contains chemicals which by their nature of concentration create a hazard to the skin. Each job must be assessed to determine the degree of protection necessary. For example, a person operating a drill rig may need only cotton or Tyvek coveralls. Opening drums may require PVC overalls with a butyl rubber apron. Like gloves, body protection comes in a variety of

Table 8.4. Physical Characteristics of CPC Materials[a]

(Designation in Matrices)	Abrasion Resis-tance	Cut Resis-tance	Flexi-bility	Heat Resis-tance	Ozone Resis-tance	Punc-ture Resis-tance	Tear Resis-tance	Relative Cost
Butyl rubber (butyl)	F	G	G	E	E	G	G	High
Natural rubber (nat. rub.)	E	E	E	F	P	E	E	Medium
Neoprene (neop.)	E	E	G	G	E	G	G	Medium
Neoprene/Styrene-butadiene rubber (neop./SBR)	G	G	G	G	G	G	G	Medium
Neoprene/natural rubber (neop./nat. rub.)	E	E	E	G	G	G	G	Medium
Nitrile rubber (nitrile)	E	E	E	G	F	E	G	Medium
Nitrile rubber/polyvinyl Chloride (nitrile/PVC)	G	G	G	F	E	G	G	Medium
Polyethylene (PE)	F	F	G	F	F	P	F	Low
Chlorinated polyethylene (CPE)	E	G	G	G	E	G	G	Low
Polyurethane (PU)	E	G	E	G	G	G	G	High
Polyvinyl alcohol (PVA)	F	F	P	G	E	F	G	Very
high								
Polyvinyl chloride (PVC)	G	P	F	P	E	G	G	Low
Styrene-butadiene rubber (SBR)	E	G	G	G	F	F	F	Low
Viton	G	G	G	G	E	G	G	Very
high								

[a] Ratings are subject to variation depending on formulation, thickness, and whether the material is supported by fabric. E = excellent; G = good; F = fair; P = poor.

materials. Permeation data for gloves are applicable to body protection made of the same material. Unlike gloves, coveralls have seams, zippers, or buttons which pose penetration problems. When selecting coveralls, a person should look for double-stitched overlapping seams, buttons only on splash suits not meant to protect from vapors, and zippers which are covered by another layer of protective material. If the suit becomes ripped, its barrier properties are lost. Damaged suits should be disposed of to prevent inadvertent use. Regulation: Body protection (e.g., fully encapsulating suits, aprons) is not specifically regulated, except in 29 CFR 1910.120.

7. *Hand protection.* As previously discussed, there are several different materials with which gloves are made. When making glove selection, the chemical(s), type of material, includes the length and type of cuffs. The use of disposable gloves should be encouraged when practical, unless cost and chemical hazards prohibit it. Regulation: Hand protection. Not specifically regulated, except sited in 29 CFR 1910.120.
8. *Foot protection.* In many cases, the part of the body which has the closest contact with the site is the feet. Minimum standards require the use of steel toes and shanks. This rule applies to all foot protection, from leather work boots to "rubber" boots. Selection of rubber boots should be similar to gloves. The chemicals should be identified and the concentration, physical state, length of exposure, and terrain must be considered before making selection. Because boots are so hard to decontaminate, booties that fit over the regular boots should be worn. When they start to deteriorate, they can be discarded. Many are designed with a reinforced sole so they do not twist sideways and provide for longer use. They come in various thicknesses and materials. (It must be noted that leather does absorb chemicals. A protective disposable bootie or rubber boot should be worn in addition to the leather boot.) Regulation: Foot protection is regulated by 29 CFR 1910.136; specified in ANSI Z41.1, Safety Toe Footwear (1967).

LEVELS OF PROTECTIVE CLOTHING

Personnel must wear protective equipment when response activities involve known or suspected atmospheric contamination, when vapors, gases, or particulates may be generated, or when direct contact with skin-affecting substances may occur. Respirators can protect the lungs, gastrointestinal tract, and the eyes against air toxicants. Chemical-resistant clothing can protect the skin from contact with skin-destructive and -absorbable chemicals. Good personal hygiene limits or prevents ingestion of material.

The level of protection selected should be based primarily on

1. Type(s) and measured concentration(s) of the chemical substance(s) in the atmosphere and the toxicity of which.

2. Potential or measured exposure to substances in the air, splashes of liquids, or other direct contact with material due to work being performed.

In situations where the type(s) of chemical(s), concentration(s), and possibilities of contact are not known, the appropriate level of protection must be selected on the basis of professional experience and judgment until the hazards can be better characterized.

While PPE reduces the potential for contact with harmful substances; ensuring the health and safety of response personnel requires safe work practices, decontamination, sit entry protocols, and other safety considerations. Together, these protocols establish a combined approach for reducing potential harm to workers.

A protective clothing scheme providing various levels of protection is a convenient means of selecting protective clothing. Levels A, B, C, and D are used by EPA to denote degrees of respiratory protection; Level A also specifies protective clothing (i.e., fully encapsulating chemical-resistant suits). In order to distinguish respiratory protection clothing levels below the fully encapsulating suit, this guide will discuss protective clothing in terms of Levels 1, 2, and 3. Most hazardous waste sites require a variety of levels of protection. A generalization that usually holds true is that on a given site respiratory protection is "area specific," while protective clothing is "task specific." There are occasions — sampling an unknown liquid — where, as a precaution, a self-contained breathing apparatus might be worn even though instruments show the area to be "safe." In general, however, respiratory protection levels will be designated for a site on the basis of air monitoring and historical data, while the clothing will be chosen to carry out a specific task in that area.

Level A Protection

Criteria for Selection

Level A protection (a fully encapsulating suit) is used when percutaneous hazards exist or when there is no known data that positively rule out percutaneous hazards. Since Level A protection is extremely psychologically stressful, the decision to use this protection must be carefully considered. The following conditions suggest a need for Level A protection:

1. Confined facilities where probability of skin contact is high.
2. Sites containing known percutaneous hazards.
3. Sites with no established history to rule out percutaneous hazards.
4. Atmosphere immediately dangerous to life and health — skin absorption route.
5. Site exhibiting signs of acute mammalian toxicity (e.g., dead animals, illnesses associated with past entry into site by humans).
6. Sites at which sealed drums of unknown materials must be opened.
7. Total atmospheric readings of the PID, FID, and similar instruments indicate 500 to 1000 ppm of unidentified substances.

8. Extremely hazardous substances (for example: dioxin, cyanide compounds, concentrated pesticides, DOT Poison "A" materials, suspected carcinogens, and infectious substances) are known or suspected to be present, and skin contact is possible (see Table 8-1).

PPE

The following items constitute level A protection:

1. Pressure-demand, self-contained breathing apparatus, approved by the Mine Safety and Health Administration (MSHA) and the National Institute of Occupational Safety and Health (NIOSH).
2. Fully encapsulating chemical-resistant suit.
3. Cotton overalls.
4. Cotton underwear.
5. Inner (surgical) gloves.
6. Boots, chemical-resistant, steel toe and shank. (Depending on suit construction, worn over or under suit boot.)
7. Booties, butyl rubber or polyvinyl chloride.
8. Gloves, disposable* (additional pair).
9. Booties, disposable* (additional pair).
10. Hard hat* (under suit).
11. Two-way radio communications (intrinsically safe).

Guidance on Selection Criteria

The fully encapsulated suit provides the highest degree of protection to skin, eyes, and respiratory system if the suit material is resistant to the chemical(s) of concern during the time the suit is worn and/or at the measured or anticipated concentrations. While Level A provides maximum protection, the suit material may be rapidly permeated and penetrated by certain chemicals from extremely high air concentrations, splashes, or immersion of boots or gloves in concentrated liquids or sludges. These limitations should be recognized when specifying the type of chemical-resistant garment. Whenever possible, the suit material should be matched with the substance it is used to protect against.

The use of Level A protection and other chemical-resistant clothing requires evaluating the problems of physical stress, in particular, heat stress associated with the wearing of impermeable protective clothing. Response personnel must be carefully monitored for physical tolerance.

Protective equipment, being heavy and cumbersome, decreases dexterity, agility, visual acuity, etc., and so increases the probability of accidents. This probability decreases as less protective equipment is required. Thus, increased probability of accidents should be considered when selecting a level of protection. Also mandatory is the full time presence of a second team member.

Many toxic substances are difficult to detect or measure in the field. When such substances (especially those readily absorbed by or destructive to the skin) are

---

* Optional

known or suspected to be present and personnel contact is unavoidable, Level A obtained. Ordinarily, Level A protection is not necessary.

Level of Protection Below Fully Encapsulating Suit

Whereas Level A protection encompasses both respiratory protection and protective clothing, the lower levels of protection — Level B, Level C, and Level D — primarily encompass respiratory protection. Therefore, the selection of protective clothing at these levels depends upon conditions on site and the tasks to be accomplished. The levels used for selecting protective clothing are referred to as Levels 1, 2, and 3 and are described below.

Level 1 Protective Clothing

This level of protection is used when there is a likelihood of being splashed with an unknown chemical or with a chemical known to be dermally active. It is used when sampling drums that have been open or when there is reason to believe that the vapors are not percutaneous.

Personal protection equipment for Level 1:

1. Butyl rubber apron, ankle length with sleeves*
2. Chemical-resistant coveralls, e.g., Saranex
3. Cotton underwear
4. Neoprene boots with steel toe and shank
5. Butyl rubber booties
6. Butyl rubber, neoprene, or nitrile gloves
7. Surgical gloves
8. Hard hat with face shield
9. Face and eye covering (usually when this level of protection is chosen, it will be in conjunction with Level B or C respiratory protection, which provides face and eye protection)
10. 5-min escape pack (respiratory)
11. Protective glasses
12. Hard hat with optional face shield*

Additional Considerations

As previously stated, a hazardous waste site will contain a variety of situations involving many chemicals in several combinations and concentrations. The following are some work practices which will help avoid some problems.

1. Safe foot protection is needed on a site. If the ground surface to be worked on is contaminated with liquid and it is necessary to walk in the chemicals, some sort of protective "booties" must be worn over the boots. This cuts down on

* Optional

decontamination requirements. They are designed with soles to help prevent them from slipping around in the boot. For example, the PVC bootie with elastic tops are available from several manufacturers. If the ground contains any sharp objects, the use of booties is questionable. Boots should be worn with either cotton or wool socks to help absorb the perspiration.

2. If the site situation requires the use of hard hats, chin straps should be used if a person will be stooping over where his/her hat may fall off. Respirator straps should not be placed over the hard hats. This could effect the fit of the respirator.
3. Some types of protective materials conduct heat and cold readily. In cold conditions, natural material clothing should be worn under the protective clothing. Protective clothing should be removed prior to allowing a person "to get warm." Applying heat, such as a space heater, to the outside of the protective clothing may drive the contaminants through. In hot weather, underclothing will absorb sweat. It is recommended that hazardous waste site workers use all cotton undergarments.
4. Body protection should be worn over the boots and taped. In the case of a splash, this will prevent anything from running into the top of the boot. Gloves should be worn outside the sleeves and *taped*. This will prevent substances from entering the top of the glove. Duct tape is preferred, but masking can be used. When aprons are used, they should be taped across the back for added protection. However, this should be done in such a way that the person should be able to have mobility and be able to get out of his or her protective wear. (Leave a little tab before you seal down the tape.)
5. Atmospheric conditions such as temperature, wind direction, wind velocity, and pressure determine the behavior of contaminants in air or the potential for volatile material getting in air. These parameters should be considered in determining the need for the level of protection required.
6. A program must be established for periodic monitoring of the air during site operations. Without an air monitoring program, any changes could go undetected and jeopardize response personnel. Monitoring can be done with various types of air pumps and filtering devices followed by analysis of filtering media; portable real time monitoring instruments located strategically onsite; personal dosimeter; and periodic walk-throughs by personnel carrying survey instruments.
7. For operations in the onsite exclusion zone (area of potential contamination), different levels of protection may be selected, and various types of chemical-resistant clothing may be worn. This selection would be based not only on measured air concentrations, but also on the job function or reason for being in the area of the potential for skin contact or inhalation of the materials present.
8. Escape masks should be made readily available when levels of respiratory protection do not include a self-contained breathing apparatus. Their use can be made on a case-by-case basis. Escape masks could also be strategically located at the site in areas that have higher possibilities of vapors, gases, or particulates.

Hearing Protection

Where noise level surveys show exposures above the acceptable level for damage to the hearing system, risk criteria, the exposure can be reduced by:

1. Environmental control
   a. Reducing the amount of noise produced at the source
   b. Reducing the amount of noise transmitted through air and building struc tures
   c. Revising operational procedures
   d. Isolation of noise or receiver
2. Personal protection
   a. Plugs
   b. Muffs

The most satisfactory method of environmental control is to control the noise at the source. Unfortunately, this is not always possible. When the amount of noise produced at the source cannot be sufficiently reduced, a combination of control methods may be required to conserve hearing.

Environmental Control

The engineering control of noise has not been sufficiently employed in industry or at hazardous work sites (drill rigs). Few companies include noise level specifications when they purchase new equipment. Those who do, however, have found that suppliers can, in fact, often make machines much quieter. In many instances, the supplier does not realize his equipment is so noisy.

Noisy products are not only of concern to the purchaser — because the doctrines of product liability are extending to noise, compensation insurance carriers are, in turn, making claims against equipment manufacturers.

Engineering control of noise is sometimes fairly simple, particularly where machinery can be completely isolated. However, there are many instances where engineering control to limit the noise to safe levels is impractical. In such instances, hearing protection in the form of ear muffs or plugs can be used alone or in combination with partial environmental control methods.

Personal Protection

There was a time when muff-type noise attenuators were recommended in preference to plugs, because muffs offer a few more decibels of protection than plugs. However, experience in industrial hearing protection programs has indicated that overprotection can be the enemy of success, and that there are such strong psychological objections to wearing any hearing protection that sometimes a little less attenuation is a good trade-off for improved comfort and increased acceptability.

Here are the principal types of hearing protection devices, with comments on their relative merits and appropriate uses.

Muffs

Muffs offer the highest attenuation. They should be mandatory for exposure to high-intensity (above 95 dB) noise. Fitting is not a problem and they are easily adjusted. They cover the whole pinna, or ear flap, providing protection from noise that might be conducted through the ear canal. Disadvantages: some workers complain of discomfort, heat, headaches; standard models cannot be worn with safety hats, welding shields, etc.

Plugs

Since, when properly fitted, plugs are more comfortable than muffs, they are to be preferred where there is an 8-hr/day exposure to moderate (80 to 95 dB) noise levels. Plugs are available in a variety of types — solid, sponge, or air filled. Materials are rubber, neoprene, and plastics. All come in a variety of sizes. Disadvantages: not sufficient attenuation for really high-intensity exposures; require expert initial fit by medical department and proper daily insertion by the employee for maximum comfort and effectiveness.

Although disposable malleable plugs have been frowned on in the past, recent studies have indicated that certain types of moldable materials have attenuation almost equal to the best plugs, and they have the advantage of being comfortable, providing a good fit, and having good employee reception. These include the waxed cotton type and a new product known as "Swedish wool." The former was developed initially by an aircraft firm and is now available under several trademarks. The latter was developed by a Swedish automotive company and consists of extremely fine "glass down" fibers. Both types are especially good for problem employees who complain of discomfort with standard plugs. Disadvantages: slightly higher cost than plugs (since they are disposed of after a single use) and slightly less attenuation than some plugs.

Motivation

There seems to be more resistance to wearing hearing protection than any other type of personal equipment. The two most common reasons:

1. Workers may not be convinced that they need the protection — hearing loss is so gradual, even in intense exposures, that it does not make a dramatic impression.
2. Many find hearing protectors "uncomfortable."

Evidence of these worker attitudes can be found in such practices as "springing" muffs so they don't seal against the head, clipping off the inner end of plugs and leaving only the outer end tab to fool the supervisor, and indifferent molding and insertion of malleable-type plug materials.

Some proven techniques for overcoming these attitudes have been found and include:

1. Acquainting new workers with the hearing conservation program when they are being briefed on health insurance, first aid facilities, safety glasses, safety hats, etc. In this way, they will tend to view wearing the hearing protective devices as a part of the job.
2. Using preemployment audiograms to show the worker the present status of his or her hearing. A graphic presentation of the minor loss he/she may already have will help to convince that hearing loss is real, and encourage the worker to conserve the balance.
3. On-the-job fitting: usually, fitting is done in a quiet dispensary or other part of the plant remote from the noisy exposures, but at a hazardous waste site this may not be true. Consequently, the worker gets no immediate idea of the protective value of the muff, plug, or malleable insert. The reason for dispensary fitting is obvious — a great many types of plugs and sizes are needed to find the combination that suits each individual, and these are not easy to transport about the plant or in the field. However, if on-the-job demonstrations can be arranged, they usually make a strong and lasting impression.
4. Group audiometric examinations which serve not to obtain audiograms, but to demonstrate to a group of workers the different degrees of hearing loss. Invariably such demonstrations will produce several older men in noisy jobs who cannot hear the increasing signals until long after the other men have. There may be some who won't hear a 4000-Hz signal at all. This dramatizes hearing loss.

REFERENCES

1. OSHA Instruction DFO CPL 2.70, January 29, 1986, Special Emphasis Program: Hazardous Waste Sites.
2. OSHA Instruction DFO CPL 2-2.37A, January 26, 1986, Technical Assistance and Guidelines for Superfund and Other Hazardous Waste Site Activities.
3. "Occupational Safety and Health Guidance Manual for Hazardous Waste Site Activities," NIOSH, OSHA, USCG, and EPA (October 1985).
4. "Protecting Health and Safety at Hazardous Waste Sites: An Overview," U.S. Environmental Protection Agency, EPA/625/9-85/006 (September 1985).
5. "Personal Protective Equipment for Hazardous Materials Incidents: A Selection Guide," U.S. Department of Health and Human Services, Public Health Service, Centers for Disease Control, National Institute for Occupational Safety and Health (October 1984).
6. "Personal Protective Equipment," U.S. Department of Labor, Occupational Safety and Health Administration, OSHA 3077, U.S. Printing Office, Washington, DC (1987).

7. TLV book.
8. USEPA, OHMTADS (Section).
9. AIHA *Journal*, G. Nelson of Miller-Nelson Research, Inc., Dublin, CA (March 1981).
10. ANSI Z89.1-1969, Head Protection.
11. 29CFR1910.135.
12. 29CFR1910.136.
13. Z41.1 Safety Toe Footwear (1967).

CHAPTER 9

Drum Storage and Handling

INTRODUCTION

Accidents may occur during handling of drums and other hazardous waste containers. Hazards include detonations, fires, explosions, vapor generation, and physical injury resulting from moving heavy containers by hand and working around stacked drums, heavy equipment, and deteriorated drums. While these hazards are always present, proper work practices — such as minimizing handling and using equipment and procedures that isolate workers from hazardous substance — can minimize the risks to site personnel.

Whether in the field as a project team leader or in a plant as a maintenance/ hazardous waste supervisor, each should be aware of all pertinent regulations. OSHA regulations (29 CFR Parts 1910 and 1926) include general requirements and standards for storing, containing, and handling chemicals and containers, and for maintaining equipment used for handling materials. EPA regulations (40 CFR Part 265) stipulate requirements for types of containers, maintenance of containers and containment structures, and design and maintenance storage areas. DOT regulations (49 CFR Parts 171 through 178) also stipulate requirements for containers and procedures for shipment of hazardous wastes. Containers are handled during characterization and removal of their contents and during other operations. A flow chart showing one set of possible procedures for drum handling is given in Figure 9.1.

CONTAINERS

Many industrial plants and commercial establishments purchase flammable liquids in 55-gal drums and from them fill small containers, 5 gal or less in capacity, for routine use.

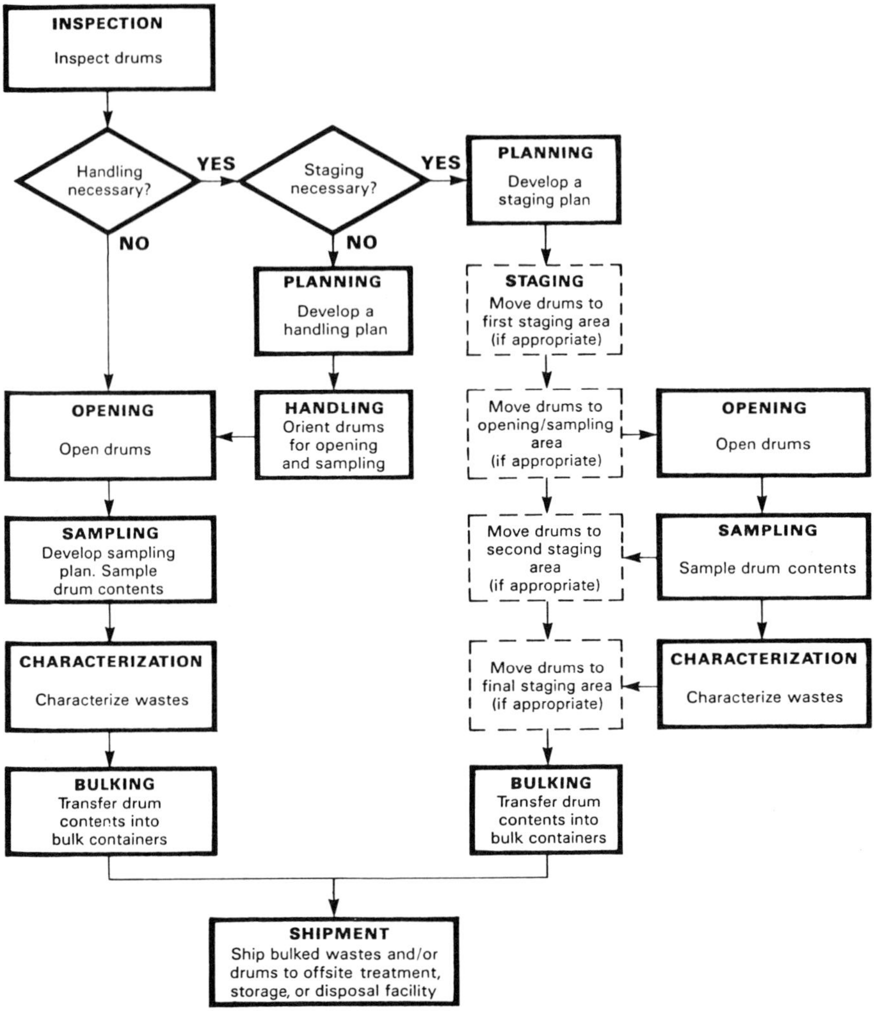

Figure 9.1. Flow chart for drum handling. Dashed boxes indicate optional steps. Number of staging areas necessary is site specific.

The drums must meet rigid DOT standards to qualify as shipping containers for flammable liquids. However, such standards are not intended to qualify the drums as long-time storage containers.

Many users assume that it is safe to store sealed drums exactly as they are received. To be safe for storage or dispensing, a drum must be protected against exposure to fire and explosion hazards. Outside storage is preferred to inside storage. However, drums should be protected from the direct rays of the sun, other sources of heat, cold, and becoming punctured. The bung should be replaced with a pressure and vacuum relief vent as soon as the sealed drum is opened.

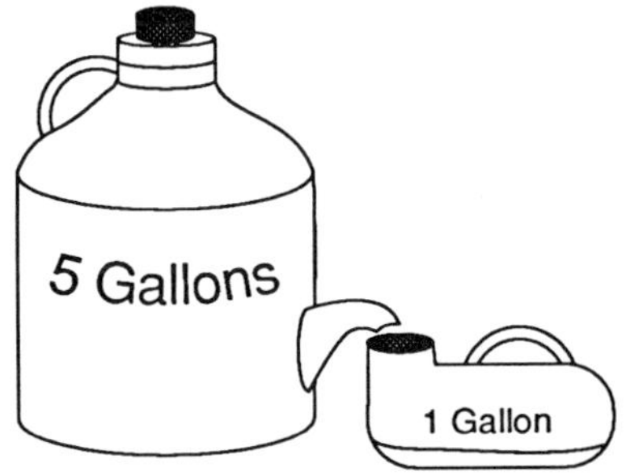

Figure 9.2. Two safety containers for viscous flammable liquids. The larger is a 5-gal point-of-use storage can with gate-valve dispensing outlet and automatic pressure-relief assembly to fill cap. The smaller point-of-use dispensing can has an automatic pressure-relief feature in its spring-closure spout.

A pressure and vacuum relief vent should be installed in a sealed drum of a flammable liquid if there is any chance that it will be exposed to the direct rays of the sun or in any other way subjected to considerable variations in temperature. If a drum leaks or is otherwise damaged, its contents should be immediately transferred to a sound container(s) that is clean or that previously held the same liquid. The replacement container should be one that meets the necessary safety requirements.

Each drum should be checked for proper labeling to avoid confusing the contents with other flammable, combustible, or nonflammable liquids, and also to facilitate safe disposal.

Probably the most common piece of equipment for handling small quantities of a flammable liquid is the portable safety can, ranging in size from 1 pt to 5 gal. Safety cans are made in numerous styles with faucets, pouring spouts, or dispensing hoses. A tilt-type safety container mounted on a frame, which permits manual tipping of the can for dispensing contents, is available. Special safety cans are made for viscous liquids such as rubber cement and heavy oils (Figure 9.2). End-use containers also are manufactured in many different styles for many different applications.

Only FM- or UL-recognized safety containers should be considered acceptable for handling flammable liquids ... containers approved or listed specifically for the purpose, whether it is storing, carrying, dispensing or end-use. Containers generally should be painted red and should bear clearly legible labels that identify contents and indicate the hazards. Stainless steel or unpainted containers may be used for corrosive or paint-dissolving liquids as applicable (see Figure 9.3 through Figure 9.5).

Flammable liquids are often purchased in small, light-weight containers with screw caps — such as 1-gal cans. Although these vessels must meet rigid standards to qualify as shipping containers, they do not necessarily afford the fire protection required of a container for storing, carrying, or dispensing flammable and combustible liquids. It is therefore, recommended in each instance where a

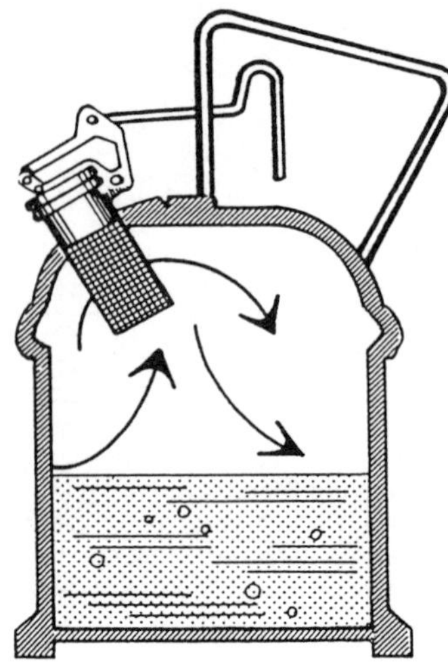

Figure 9.3. Properly designed spring-action cover relieves excess vapor pressure. Cover should not appreciably leak liquid itself, even if container is inverted. Other picture shows dispensing station for withdrawal of small amounts of flammable liquids into portable cans.

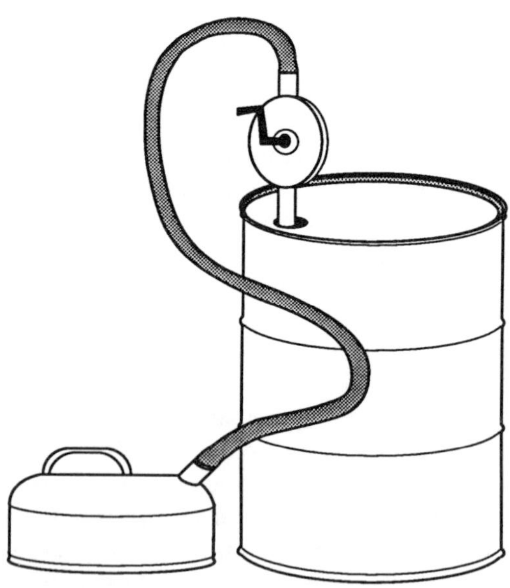

Figure 9.4. Laboratory use of flammable liquids often impose a need for chemical purity as well as safety. Use reaction-resistant metal (like stainless steel) or lined cans. Note tilting dispenser and flammable liquid waste can.

Figure 9.5. Rotary-type transfer pump threaded into end bung of a 55-gal drum. Flammable liquid is dispensed from grounded drum to portable safety container by means of conductive hose designed to electrically bond drum to container during transfer.

greater degree of safety can be attained that all flammable liquids be transferred to "recognized" containers as soon as the sealed shipping containers are opened.

STORAGE

Rooms in which drums containing flammable liquids are stored or dispensed should be considered potentially hazardous areas, but consistent application of basic fire and accident prevention rules should result in elimination or control of hazards. Special attention should be given to the location of the room, its arrangement, its ventilation, the equipment used, and storage procedures.

Drums stored outdoors or indoors should have floors that are impervious to chemical spills and proper diking around them.

A storage and dispensing room located inside a building should be protected by an automatic fire extinguishing system (automatic sprinklers, carbon dioxide, dry chemical, or foam) as well as by strategically located portable fire extinguishers (preferably outside the room, but near an entrance to it) designed for use on Class B fires. It should have ventilation to the outside air, special trapped floor drainage facilities, and explosion relief in the form of pressure-opening windows or a weak wall, providing 1 ft^2 of venting for every 50 ft^3 of room volume. For Class I flammable liquids, electrical equipment should conform to the requirements of the N.F.P.A. Standard No. 70, National Electrical Code, for Class I hazardous locations.

Design features, storage limitations, and other precautionary measures should meet the specifications given in the N.F.P.A. Standard No. 30, Flammable and

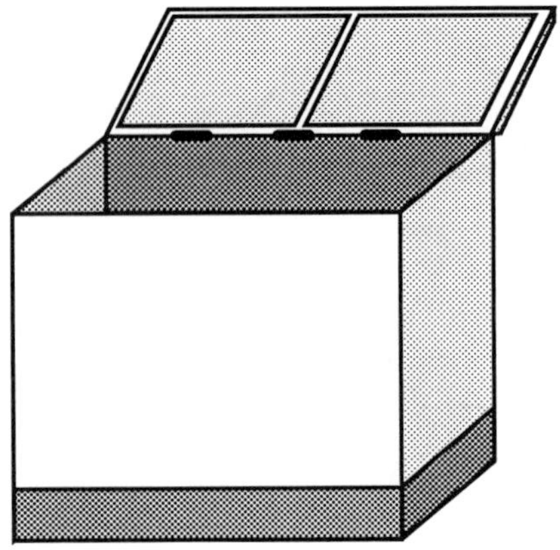

Figure 9.6. Metal storage cabinet for flammable liquid containers (maximum of 5 gal capacity each).

Combustible Liquids Code. In addition, the local fire ordinances should be consulted.

Storage cabinets (Figure 9.6) that are of a recognized type for holding containers of flammable and combustible liquids can be installed near the areas where the liquids are used.

Flammable liquids should not be stored in domestic-type refrigerators unless properly modified to remove ignition sources. Special refrigerators for this service are listed by Underwriters' Laboratories.

BONDING AND GROUNDING

The action of transferring a liquid from one container to another may produce voltage potentials that can result in static sparks capable of igniting flammable vapors. Because static charges can generate from this source and others, it is important that flammable liquid dispensing and receiving containers be bonded together before pouring. It is also important for large containers, such as drums, to be connected to an adequate electrical ground when used as dispensing or receiving vessels.

To be effective, all grounding and bonding connections must be metal to metal. Therefore, all dirt, paint, rust, or corrosion should be removed from points of contact before such connections are made.

The practice of placing the nozzle of the dispensing container in contact with the opening of the receiving container is used to establish an electrical bond between the vessels. This method may be satisfactory, providing a good electrical contact is established and maintained during the entire dispensing operation (Figure 9.7).

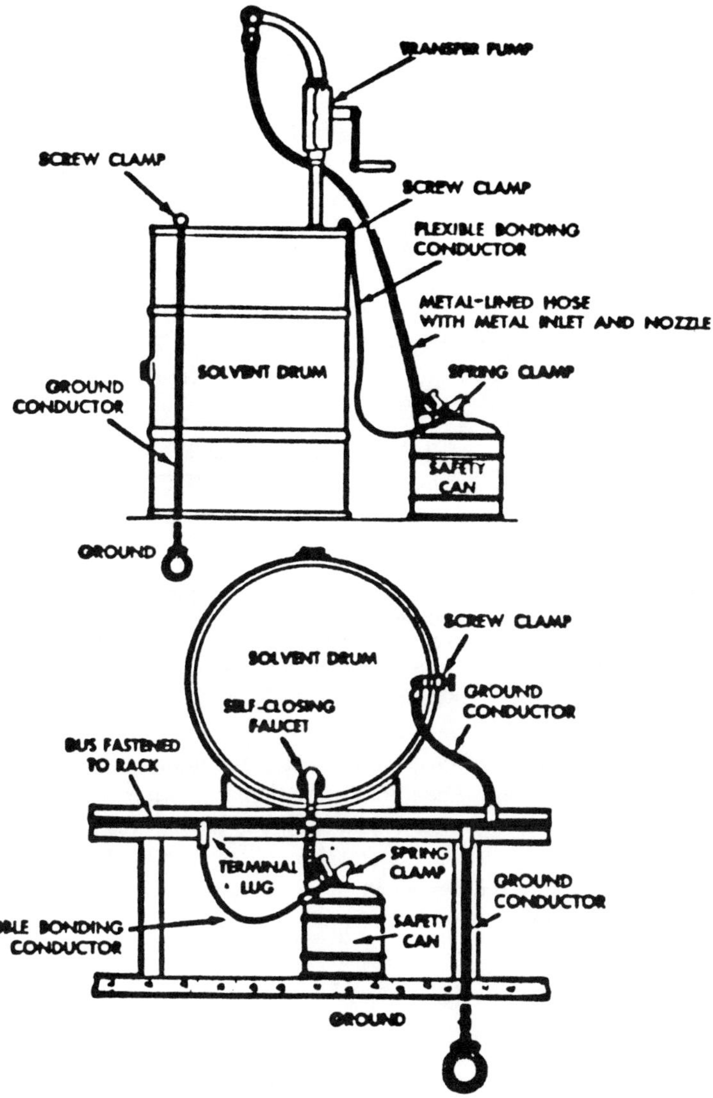

Figure 9.7. Two methods of dispensing flammable liquid from 55-gal drum to portable container. (Top) drum upright; transfer pump threaded into bung; metal-lined delivery hose with nozzle in contact with container. (Bottom) Drum horizontal; metal self-closing valve threaded into bung; same hose setup. If continuous electrical bond cannot be maintained this way, bond vessels by means of a flexible conductor (shown).

However, prevention of static spark while transferring a flammable liquid from one container to another is best accomplished by prior bonding of the containers with a wire having alligator clamps.

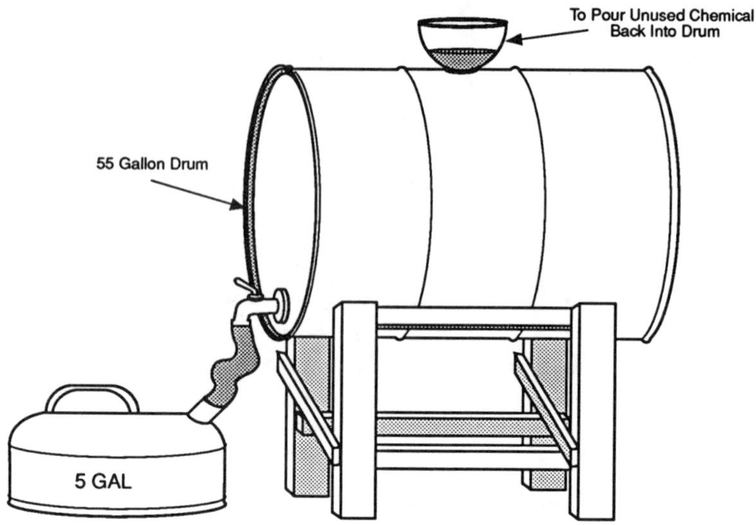

Figure 9.8. Hand-operated, self-closing metal valve is threaded into end bung of drum mounted on wheeled rack. When dispensing liquid, hose nozzle is kept in contact with exposed metal of container spout, aided by spring pressure of cover. A funnel with flame arrester is shown inserted into side bung of drum to facilitate pouring unused liquid back into the drum.

A drip return pan attached to the dispensing pump may afford effective bonding when the receiving container is placed on it preliminary to and during the transfer. Likewise, a recognized dispensing hose having a metal connector and nozzle joined by a sealed-in conductor can establish a continuous bond if the nozzle is kept in contact with exposed metal of the receiving container during the entire transfer (Figures 9.7 and 9.8).

WASTE DISPOSAL

Cloth, paper, and other solid wastes that have been soaked with a flammable or combustible liquid should be placed in a recognized disposal container. Recognized containers are made of metal and are shipped with self-closing covers. Such containers should be clearly labeled as to the type of waste they are to receive (Figure 9.9).

The containers should be emptied at the end of each working day and their contents removed to a safe location and disposed of as hazardous or non-hazardous waste per RCRA regulations. Incinerators can be built to burn solid waste with a minimum of air pollution when properly designed.

Most laws and ordinances strictly prohibit the pouring of flammable and combustible liquids into sinks or floor drains which connect with sanitary or storm sewer facilities. Standard 30- or 55-gal drums or smaller recognized containers

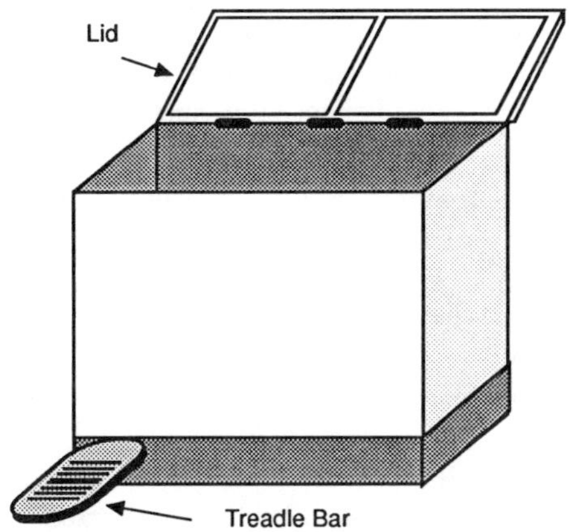

Figure 9.9. Waste container for discarded flammable solids. Gravity-activated; self-closing cover is opened by pressure on treadle bar.

should be used for disposal of flammable and combustible waste liquids, and they should be emptied periodically. They should be safeguarded against fire according to all applicable provisions that are specified for dispensing containers. The liquids should be disposed of by burning of a solvent burner or delivered to a waste liquids collection agency. In many cities, there are firms that collect waste liquids that are recyclable. In some instances, the original supplier still picks up used liquids for reclaiming and resale.

CONTROL OF FIRE AND EXPLOSION HAZARDS

Regardless of the quantities involved, each flammable liquid to be used should be analyzed to determine the extent of its flammable and health hazards so that appropriate control measures can be taken. To control these hazards, the nature of the specific liquid, the amounts of vapor released, the ignition sources, the kinds of operations, the usage temperature, ventilation rates, and the type of building construction all must be taken into consideration.

Since the relative degree of hazard varies widely, competent qualified advice should be obtained regarding types of electrical equipment, ventilation requirements, the need for eliminating sparks, open flames, and other sources of ignition, requisite handling procedures, and other factors relating to the maintenance of a safe environment.

TRAINING OF PERSONNEL

Safe practices on the part of employees who handle flammable and combustible liquids are essential in the prevention of fire and explosion hazards. Before being permitted to undertake jobs that require the use of such liquids, workers should be fully instructed in their characteristics, hazards, and methods of control. Supervisors should make frequent checks to assure that the required safe practices are being followed consistently.

Among the basic safe practices employees should be trained to observe the following:

1. Use only recognized containers in good condition and keep them closed when they are not in use.
2. Never use a container for any liquid other than that for which it is intended and so marked.
3. Keep at the job site only that quantity of liquid needed during the shift. At the end of the shift, return any unused liquid to designated storage area.
4. Cleanup spills of liquids immediately. Dispose of cleanup rags or absorbent pads into recognized containers. Never use sawdust or deodorized kitty litter to absorb a spill.
5. Never smoke, use open flames, or strike sparks where there is a possibility of igniting a flammable or combustible liquid.
6. Check bonding and grounding connections for electrical continuity.

Only specially trained personnel should be assigned to work in storage and dispensing rooms. Periodic checks should be made for adherence to the prescribed safe practices, and refresher training should be given annually as required by 29 CFR 1910.120 and OSHA training document.

Employees should be kept constantly aware that flammable liquids are hazardous and must be handled with particular care. Posters and signs are among the means that can be used effectively. No smoking signs should be posted conspicuously in buildings and areas where smoking is prohibited because of the presence of flammable liquids.

REFERENCES

1. HHE 78-049-553. National Steel Drum Company. "Hazard Evaluation and Technical Assistance Reports," National Steel Drum Company, Philadelphia, PA, U.S. Government Printing Office, Washington, DC (1978).
2. Contract Report 210-80-0042. "Safety Information Profile — Hazards of Flammable and Combustible Liquids Stored in Plastic Containers," U.S. Government Printing Office, Washington, DC (1980).
3. Mayhew, J. J., G. M. Sodear, and D. W. Carroll. A *Hazardous Waste Site Management Plan*. Washington, DC: Chemical Manufacturer's Association, Inc., (1982).
4. deVero, E. R., B. P. Simmons, R. D. Stephens, and D. L. Storm. "Samples and Sampling Procedures for Hazardous Waste Streams," EPA-600/2-80-018, U.S. Environmental Protection Agency, Cincinnati, OH (1980).

5. U.S. EPA. "Characterization of Hazardous Waste Sites — A Methods Manual," Vol. II, Available Sampling Methods, 2nd ed., EPA 600/4/84-076 (1984).
6. NIOSH. "Criteria for a Recommended Standard: Working in Confined Spaces," NIOSH No. 80-106, Also available from U.S. Government Printing Office (#017-033-00353-0) and National Technical Information Service (PB-80-183015) (1979).
7. 29CFR1910.120 Government Printing Office, Washington, DC (1988).
8. OSHA Document, OSHA Required Training Courses, Department of Labor, U.S. Government Printing Office, Washington, DC.

CHAPTER 10

Principles of Air Monitoring Equipment

DIRECT-READING PHYSICAL PROPERTY INSTRUMENTS

The physical properties of aerosols, gases, and vapors are used in the design of direct-reading physical instruments for quantative estimations of contaminants in the atmosphere. The principles upon which these instruments are based are presented (alphabetically) in the following discussion.

Operating Principles

Aerosol Photometry (Light Scattering)

The principle of aerosol photometry is the generation of an electrical pulse by a photocell which detects the light scattered by a particle. A pulse height analyzer estimates the effective particle diameter. The number of electronic pulses is related to the number of particles counted per unit flow rate of the sampled gaseous medium. Calibrations may be made using a reference standard such as polystyrene spheres about which the diameters and refractive index are known. The aerosol under study, however, is the reference standard of choice because of the unique effects of shape factor, angle of scatter, and refractive index, as well as particle size. Whereas certain instruments are designed to give a size analysis based upon the above principles, there are others which use a forward light scattering principle to provide an integrated measurement of total particle concentration in an illuminated volume. The latter are used in monitoring particulate concentrations in experimental rooms and exposure chambers.

Aerosol photometry can usually provide only an approximate analysis of particulates because of the impracticality of calibrating the instrument to each type of particulate suspension which is to be measured. The great variations in shape, size, degrees of agglomeration, and refractive indices of the mixture of chemical components in a given dust or fume suspension make such a calibration exceedingly difficult. Aerosol photometry can, therefore, provide an indication of the particulate concentration in different particle size ranges of interest. It may still be necessary to perform size distribution analyses by microsieving and microscopic procedures for greatest accuracy.

Chemiluminescence

Chemiluminescence is a phenomenon in which there is a distinctly colored glow which accompanies a reaction, much as the oxidation of certain decaying wood, of luciferin in fireflies, and of yellow phosphorus. Recently, this phenomenon has been measurable in the reaction of ozone with such other gases as ethylene and nitric oxide for the measurement of ozone or nitrogen oxides in ambient atmospheres.

The chemiluminescence principle has been incorporated into continuous ambient air monitors. The monitors are selective for ozone or for NO_2-NO_x. Measurements of ozone at concentrations extending from 0.001 to 1 ppm in ambient atmospheres are based upon the (photometric) photomultiplier tube detection. The chemiluminescence is produced by the flameless gas phase reaction of the ozone in the air sample stream with an ethylene gas stream, the flow of which from a bottled supply is regulated through a calibrated capillary tubing to the reactor chamber. Similarly, NO measurements from 0.01 to 5000 ppm are based upon the chemiluminescent reaction of NO and ozone to produce NO_2 and O_2. The ozone is produced from bottled oxygen by an in-line ozone generator. Monitoring of NO_x is accomplished by means of an NO_2 to NO catalytic converter which operates in a bypass line on a timed sequence basis. Thus, NO_2 measurements can be obtained by difference.

The selectivity of these instruments is enhanced by the use of narrow-band optical filters to provide negligible interference effects from other atmospheric contaminants. Although designed for ambient air studies, these instruments may be used advantageously as fixed-station monitors for in-plant atmospheres.

Combustion

A combustible gas or vapor mixture is passed over a filament heated above the ignition temperature of the substance of analytical interest. If the filament is part of the bridge circuit, the resulting heat of combustion changes the resistance of the filament, and the measurement of the imbalance is related to the concentration of the gas or vapor in the sample mixture. The method is basically nonspecific, but it may be made more selective by choosing appropriate filament temperatures for individual gases or vapors or by using an oxidation catalyst for a desired reaction such as Hopcalite for carbon monoxide.

Combustible gas indicators must be calibrated in the laboratory for their response to the anticipated individual test gases and vapors, such as benzene, toluene, hexane (for hydrocarbons in general), carbon monoxide, acetone, and styrene. These instruments are definitely portable and they are valuable survey meters in the industrial hygienist's collection of field instruments. Readings are in terms of 0 to 1000 ppm or 0 to 1.0 lower explosive limit (LEL). However, it is essential to recognize that industrial atmospheres rarely contain one gaseous contaminant and that these indicators will respond to all the combustible gases present. Hence, supplementary sampling and analytical techniques should be used for a complete definition of hazardous environmental conditions.

Conductivity, Electrical

A gas-air mixture is drawn through an aqueous solution. Those gases which form electrolytes produce a change in the electroconductivity as a summation of the effects of all ions thus produced. The method is nonspecific. If the concentrations of all other ionizable gases are either constant or insignificant, then the resulting changes in conductivity may be related to the gaseous substance of interest. Temperature control is extremely critical in conductance measurements. If thermostated units are not used, then electrical compensation must enter into the measurements to allow for the 2% per degree Celsius conductivity temperature coefficient average for many gases.

The electrical conductivity method has found its greatest application in the continuous monitoring of sulfur dioxide in ambient atmospheres. However, a lightweight portable analyzer which uses a peroxide absorber to convert SO_2 to H_2SO_4 is now available; this battery-operated instrument can provide within 1 min an integrated reading of the SO concentration over the 0- to 1-ppm range. A larger portable model which may be operated off a 12-V automobile battery is also available for the higher concentration ranges of SO_2 encountered in field sampling.

Conductivity, Thermal

The specific heat of conductance of a gas or vapor is a measure of its concentration in a carrier gas such as air, argon, helium, hydrogen, or nitrogen. However, thermal conductivity measurements are nonspecific, and the method finds its greatest usefulness in estimating the concentration of the separately deluted components from a gas chromatographic column. The method operates by virtue of the loss of heat from a hot filament to a single component of a flowing gas stream, the loss being registered as a decrease in electrical resistance measured by a Wheatstone bridge circuit. The applications of this method are limited mostly to binary gas mixtures and are based upon the electrical unbalance produced in the bridge circuit by the difference in the filament resistances of the sample and reference gases passed through the separate cavities in the thermal conductivity cell.

Coulometry

Coulometry is the precise measurement of the quantity of electricity passing through a solution during an electrochemical reaction. The substance of analytical interest is oxidized or reduced at one electrode in a primary coulometric analysis or it may react stoichiometrically (in a secondary coulometric analysis) with one of the electrolytic products. The method is capable of a high degree of precision. It is used in the automatic monitoring of part per billion to part per million concentrations of reactive inorganic gases present in ambient atmospheres; air samples are drawn through the electrolytic cell in which the reactant is generated in controlled quantities to meet the concentration requirements.

The method is basically nonspecific; it is made more selective for specific atmospheric oxidants by adjusting the concentration, pH, and composition of the electrolyte used in the reaction. In certain instances, a chemical filter or a selective membrane is used to remove serious interferents from the sampled gas stream. Both portable and fixed monitor types of instruments, based upon this principle, are used to monitor ozone, nitrogen dioxide, and sulfur dioxide concentrations.

Flame Ionization

The hydrogen flame ionization detector (FID) is basically a stainless steel burner in which hydrogen is mixed with the sample gas stream in the base of the unit; combustion air or oxygen is fed axially and diffused around the jet through which the hydrogen-gas mixture flows to the cathode tip where ignition occurs. A loop of platinum serves as the collector electrode, which is set about 6 mm above the tip of the burner. The current carried across the electrode gap is proportional to the number of ions generated during the burning of the sample; the detector responds to all organic compounds, except formic acid, but its response is greatest with hydrocarbons and diminishes with increasing substitution of other elements, notably oxygen, sulfur, and chlorine. Its low noise level of 10 to 12 A provides a high sensitivity of detection, and it is capable of the wider linear dynamic range of 10^7. Its usefulness is enhanced by its insensitivity to water, the permanent gases, and most inorganic compounds, thus simplifying the analysis of aqueous solutions and atmospheric samples. It is used to great advantage in both laboratory and field models of gas chromatographs as well as in hydrocarbon analyzers which are set up as fixed station monitors of ambient atmospheres in the laboratory or field.

Hydrocarbon analyzers, operating with an FID detector, are carbon counters; their response to a given quantity of a typical C_6 hydrocarbon is six times to that of methane at a fixed flow rate of the sample stream. Thus, the instrumental characteristics such as sensitivity are usually given as methane equivalent. In addition to hydrocarbons, these analyzers respond to alcohols, aldehydes, amines, and other compounds, which will produce an ionized carbon atom in the hydrogen flame. The electronic stability of at least one model is within 1% over 24 hr of operation; this instrument is equipped with electronic span calibration to improve the accuracy of the data.

Gas Chromatography

Gas chromatography is a physical process for separating the components of complex mixtures. A gas chromatograph consists of

1. A carrier gas supply complete with a pressure regulator and flow meter.
2. An injection system for the introduction of a gas or vaporizable sample into a port at the front end of the separation column.

3. A stainless steel, copper, or glass separation column containing a stationary inert material, such as diatomaceous earth, used as in gas-solid chromatography (GSC) or as a support for a thin layer of a liquid substrate, such as silicone oils, in gas-liquid chromatography (GLC).
4. A heater and oven assembly to control the temperature of the column(s), injection port, and detector unit.
5. A detector.
6. A recorder for the chromatograms produced during the separations.

The separations are based upon the varied affinities of the sample components for the packing materials of a particular column, the rate of carrier gas flow, and the operating temperatures of the column. Improved separations are made possible by the use of temperature programming. The components of the sample(s), as a consequence of their varied affinities for a given column, are eluted sequentially and thus evoke separate responses by the detection system, the signal of which is amplified to produce a peak on the strip chart recorder. The height and area of the peak are proportional to the concentration of the eluted sample component. Calibrations are made using known mixtures of the pure substance in a gas-air mixture prepared in a 50- to 100-L Saran bag or other suitable container. The time of retention on the column and supporting analytical techniques (infrared spectrophotometry, for example) are used in the identification of the individual peaks of a chromatogram. The method is capable of providing extremely clean-cut separations and is one of the most useful techniques in the field of organic analysis. It is sensitive to fractional part per million concentrations of organic substances. The most commonly used detectors include flame ionization, thermal conductivity, and electron capture.

Rugged, battery operated, portable gas chromatographs have been refined to the point where they are considered practical for many field study applications. These instruments may now be obtained with a choice of thermistor-type thermal conductivity, flame ionization, and electron capture detectors, and in some instances, the latter two types are interchangeable. Complete with gas sampling valve, rechargeable batteries, appropriate columns, and self-contained supplies of gases, these chromatographs have much to offer to the industrial hygienist engaged in on-site analyses of trace quantities of organic compounds and the permanent gases. The gas bottles provide 8 to 20 hr of operation dependent on the flow rates and must be recharged using high-pressure gas regulators. The retention times of the compounds of analytical interest must be determined in the laboratory for a given type of column, as is true for the laboratory-type chromatographs.

Photometry (Colorimetry)

Photometry is the measurement of the power of a beam of radiant energy, in the visible, ultraviolet, or infrared region of the electromagnetic spectrum, which has been attenuated as a result of passing through a solution, a gas-air mixture containing a substance such as mercury vapor, ozone, or benzene vapor, a suspension of solid or liquid particulates in air or other gaseous medium, or a photographic image of a spectral line or an x-ray diffraction pattern on a photographic film or plate. Photometers used for the indicated types of applications contain

1. A lamp or other generating source of energy.
2. An optical filter to limit the band width of the incident beam.
3. An optical system to collimate the filtered beam, which is then passed through.
4. The sample system contained in a cuvette or gas cell to
5. A photocell, bolometer, thermocouple, or pressure sensor type of detector where the signal is amplified and fed to a
6. Readout meter or to a strip chart recorder.

The more sophisticated technique is termed spectrophotometry, which makes use of prisms made of glass (visible region), quartz (ultraviolet), and sodium chloride or potassium bromide (infrared) or of diffraction gratings, instead of optical filters, to provide essential monochromatic radiation as a "purer" source of energy. Spectrophotometers are used mostly in laboratories for highly specific and precise analytical determinations.

Most field-type calorimetric analyzers have been designed to function as fixed-station monitors for the active gases such as oxides of nitrogen, sulfur dioxide, "total oxidant," ammonia, aldehydes, chlorine, hydrogen fluoride, and hydrogen sulfide. These instruments require frequent calibration with zero and span gases at the sampling site to assure the provisions of reliable data. However, built-in automated calibration systems, which standardize regularly zero and span controls against pure air and a calibrated optical filter, are now available from one source of colorimetric analyzers for the nitrogen oxides, sulfur dioxide, and aldehydes. A further advantage is the 0- to 10-ppm working range of the instruments with the optional capability of extending the upper limit to 10,000 ppm.

Another recent advance is the provision of a portable, calorimetric analyzer for NO2-NO_x by another manufacturer. This instrument, which uses dual photometric cells, is designed for rugged field use, may be operated from a 12-V automobile battery, and operates over the 0- to 2.0-ppm range, with higher ranges available. It may also be used as a field monitor, if desired.

Polarography

Polarographic analysis is based on the electrolysis of a sample solution using an easily polarized microelectrode (the indicator electrode) and a large nonpolarizable reference electrode. In laboratory instruments, the indicator electrode is a noble metal, usually a dropping mercury electrode for reduction reactions and a platinum electrode for oxidation reactions. The reference electrode may be a pool of mercury on the bottom of the electrolysis vessel or a saturated calomel electrode.

The method provides both qualitative and quantitative information. As the increasing voltage to the polarographic cell is applied at a steady rate, the decomposition potentials of electroeducible (or electrooxidizable) ions are reached, in turn. At the decomposition potential of a given ion, the current increases rapidly and then levels off to a limiting current, thus producing an S-shaped curve or "wave." The value of the half-wave potential is characteristic of the discharged ion in a given electrolyte. The height of the wave, i.e., the rise in the current, is proportional to the concentration of the discharged ion species in the sample.

The method has been used largely for the analysis of metallic ions and organic species. Manufacturers of modern field instruments, however, are taking advantage of advanced, compact, long-lived polarographic sensors to provide continuous monitors for oxygen in such diversified environments as furnace atmospheres, flue gases, auto exhausts, space vehicles, manholes, and physiology test chambers.

The portable instruments may be easily calibrated for gaseous oxygen by exposing the sensor to ambient air and adjusting the calibration pot to provide a meter reading of 20.9 on the 0 to 25% scale. These instruments provide a rapid response to changes in the concentration of oxygen and should prove valuable to the industrial hygienist who may encounter oxygen-deficient atmospheres during surveys.

Radioactivity

Radioactive substances emit three principal types of radiation, i.e., alpha (α), beta (β), and gamma (γ). Radioactive particles and gases may be monitored manually or automatically in gas streams, in ambient atmospheres, in process water, or in solid process materials or products by means of ionization chambers, scintillation detectors, or Geiger-Mueller counters. Choice of detectors for α, β-γ or α-β-γ monitors is determined by the isotopes of interest. Electronic recorders are available for graphic presentation of monitoring data.

Portable, light-weight air monitors for γ, β-γ, and α-β-γ radiation are available with interchangeable probes to survey work areas for the different types of radiation encountered. One such instrument thus provides the capability of measuring α- and weak β- radiation using one probe and γ-radiation up to 10 or 60 mrad/hr with the others. Other portable survey meters include the "Cutie Pie," with ranges of 50, 500, and 5000 mrad/hr and a fast neutron survey meter designed with tissue equivalent response in making health hazard surveys in the range of 0.2 to 14 MeV around reactors and neutron generators.

A summary of the operating characteristics of the current (1987) commercially available direct-reading physical instruments is presented in Table 10.1. The information provided in this table is based upon that given in the manufacturer's literature; in certain instances, e.g., repeatability, a range given by more than one source of supply or different applications of an operating principle to the estimation of multiple chemical entities. In other cases, the information has not been provided in a manufacturer's list of specifications for an instrument, and this has been so noted in Table 10.1. The material presented under "Applications and Remarks" provide information on the individual substances which may be analyzed directly by the stipulated technique along with specified interferences and other important considerations. This tabulation is not an official certified list; it is intended as a useful guide in selecting direct-reading physical instruments on the basis of desired operating parameters.

Table 10.1. Direct Reading Physical Instruments

Principle of Operation	Applications & Remarks	Code[a]	Range	Repeat-ability (Precision)	Sensitivity	Response Time
Aerosol photometry	Measures, records, and controls particulates continuously in areas requiring sensitive detection of aerosol levels; detection of 0.05- to 40-μm diameter particles; computer interface equipment is available	A & B	10^{-2} to 10^{2} μg per liter	Not given	10^{-3} μg per liter (for 0.3 μm DOP)	Not given
Chemilumin-escence	Measurement of NO in ambient air selectivity and NO_2 after conversion to NO by hot catalyst; specific measurement of O_3; no atmospheric interferences	B	0 to 10,000 ppm	±0.5 to ±3%	Varies: 0.1 ppb to 0.1 ppm	ca 0.7 sec NO mode and 1 sec NO_x mode; longer period when switching ranges
Colorimetry	Measurement and separate recording of NO_2-NO_x, SO_2, total oxidants, H_2S, HF, NH_3, Cl_2, and aldehydes in ambient air	A & B	ppb & ppm	±1 to ±5%	0.01 ppm (NO_2, SO_2)	30 sec to 90% of full scale
Combustion	Detects and analyzes combustible gases in terms of percent LEL on graduated scale; available with alarm set at $^1/_3$ LEL	A	ppm to 100%	—	ppm	<30 sec
Conductivity, electrical	Records SO_2 concentrations in ambient air; some operate off a 12-V car battery; operates unattended for periods up to 30 days	A & B	0 to 2 ppm	<±1 to ±10%	0.01 ppm	1 to 15 sec (lag)
Coulometry	Continuous monitoring of NO, NO_2, O_3, and SO_2 in ambient air; provided with strip chart recorders; some require attention only once a month	A & B	Selective: 0 to 1.0 ppm overall, or to 100 ppm (optional)	±4% of full scale	Varies: 4 to 100 ppb dependent on instrument range setting	<10 min to 90% of full scale

Flame ionization (with gas chromato-graph)	Continuous determination & recording of methane, total hydrocarbons, and carbon monoxide in air; catalytic conversion of CO to CH_4; operates up to 3 days unattended	B	Selective: 0 to 1 ppm; 0 to 100 ppm	±1% of full scale	Not given	5 min (cycle time)
Same as above	Separate model for continuous monitoring of SO_2, H_2S, and total sulfur in air; unattended operation up to 3 days	B	0–20 ppm	±4% of full scale	0.005 ppm (H_2S); 0.01 ppm (SO_2)	5 min (cycle time)
Flame ionization (hydrocarbon analyzer)	Continuous monitoring of total hydro-carbons in ambient air; potentiometric or optional current outputs compatible with any recorder; electronic stability from 32 to 110°F	B	0 to 1 ppm as CH_4, X1, X10, X100, X1000 with continuous span adjustment	±1% of full scale	1 ppm to 2% full scale as CH_4, 4 ppm to 10% as mixed fuel	<0.5 sec to 90% of full scale
Gas chromatograph (portable)	On-site determination of fixed gases, solvent vapors, nitro and halogenated compounds, and light hydrocarbons; instruments available with choice of flame ionization, electron capture, or thermal con-ductivity detectors and appropriate columns for desired analyses; rechargeable batteries	A	Depends on detector	Not given	<1 ppb (SF_6) with electron capture detector; <1 ppm (HCs)	—
Infrared analyzer (photometry)	Continuous determination of a given component in a gaseous or liquid stream by measuring amount of infrared energy absorbed by component of interest using pressure sensor technique; wise variety of applications include CO, CO_2, Freons, hydrocarbons, nitrous oxide, NH_3, SO_2, and water vapor	B	From ppm to 100% depending on application	±1% of full scale	0.5% of full scale	0.5 sec to 90% of full scale
Photometry, ultraviolet (tuned to 253.7 μm)	Direct readout of mercury vapor; calibration filter is built into the meter; other gases or vapors which interfere include acetone, aniline, benzene, ozone, and others which absorb radiation at 253.7 μm	A	0.005 to 0.1 and 0.03 to 1 mg/m^3	±10% of meter read-ing or ± min-imum scale division, which-ever is larger	0.005 mg/m^3	Not given

Table 10.1. (Continued)

Principle of Operation	Applications & Remarks	Code[a]	Range	Repeatability (Precision)	Sensitivity	Response Time
Photometry, visible (narrow-centered 394-μm band pass)	Continuous monitoring of SO_2, SO_3, H_2S, mercaptans, and total sulfur compounds in ambient air; operates more than 3 days unattended	B	1 to 3,000 ppm (with air flow dilution)	±2%	0.01 to 10 ppm	<30 sec to 90% of full scale
Particle counting (near-forward scattering)	Reads and prints directly particle concentrations at 1 of 3 preset time intervals of 100, 1,000, or 10,000 sec corresponding to 0.01, 0.1, and 1 ft^3 of sampled air	B	Preset (by selector switch); particle size ranges: 0.3, 0.5, 1.0, 2.0, 3.0, 5.0, and 10.0 μm; counts up to 10^7 particles/ft^3 (35×10^3/L)	±0.05% (probability of coincidence)	—	Not given
Polarography	Monitor gaseous oxygen in flue gases, auto exhausts, hazardous environments, and in food storage atmospheres and dissolved oxygen in wastewater samples; battery operated, portable, sample temperature 32 to 110°F, up to 95% relative humidity; potentiometric recorder output; maximum distance between sensor and amplifier is 1000 ft	A	0–5 and 0–25%	±1% of reading at constant sample temperature	Not given	20 sec to 90% of full scale

Radioactivity	Continuous monitoring of ambient γ- and X-radiation by measurement of ion chamber currents, averaging or integrating over a constant recycling time interval, sample temperature limits 32 to 120°F; 0 to 95% relative humidity (weatherproof detector); up to 1,000 ft remote sensing capability; recorder and computer outputs; complete with alert, scram, and failure alarm systems; all solid-state circuitry	B	0.1 to 10^7 mR/hr	±10% (decade accuracy)	—	<1 sec
Radioactivity	Continuous monitoring of β- or γ-emitting radioactive materials within gaseous or liquid effluents; either a thin-wall Geiger-Mueller tube or a gamma scintillation crystal detector is selected depending on the isotope of interest; gaseous effluent flow — 4 cfm; effluent sample temperature limits 32 to 120°F using scintillation detector and –65 to 165°F using G-M detector; complete with high radiation alert and failure alarms	B	10 to 10^4 cpm	±2% full scale (rate-meter accuracy)	$<10^{-7}$ μCi of ^{131}I per cc of air and 10^{-7} μCi of ^{137}Cs per cc of water using a scintillation detector	0.2 sec at 10^4 cpm (rate meter)
Radioactivity	Continuous monitoring of radioactive airborn particulates collected on a filter tape transport system; rate of air flow — 10 SCFM; scintillation and G-M detectors, optional but a β-sensitive plastic scintillator is provided to reduce shielding requirements and offer greater sensitivity; air sample temperature limits 32 to 120°F; weight 550 lb; complete with high- and low-flow alarm and a filter failure alarm	B	10 to 10^4 cpm	±2% of full scale (rate-meter accuracy)	10^{12} μCi of Cs-137 per cc of air using a scintillation detector	0.2 sec at 10^4 cpm (rate-meter)

Source: Draeger Detector Tube Handbook, Draegerwerk, Lubeck, West Germany (1970), pp. 33–71.

[a] Code: A — portable instruments; B — fixed monitor of "transportable" instruments.

DIRECT-READING COLORIMETRIC DEVICES

Operating Principles

Direct-reading colorimetric devices utilize the chemical properties of an atmospheric contaminant for the reaction of that substance with a color-producing reagent. Reagents used in detector kits may be in either a liquid or a solid phase or provided in the form of chemically treated papers. The liquid and solid reagents are generally supported in sampling devices through which a measured amount of contaminated air is drawn. On the other hand, chemically treated papers are usually exposed to the atmosphere and the reaction time noted for a color change to occur.

Liquid Reagents

Liquid reagents may be supplied in sealed ampoules or in tubes for field use. Such preparations are provided in a concentrated or a solid form for easy dilution or dissolution at the sampling site. Representative of this type of reagent are the ortho-tolidine and the Griess-Ilsovay kits for chlorine and nitrogen dioxide, respectively. Although the glassware needed for these applications may be somewhat inconvenient to transport to the field, methods based on the use of liquid reagents are more accurate than those which use solid reactants. This is due to the inherently greater reproducibility and accuracy of color measurements made in a liquid system.

Chemically Treated Papers

Papers impregnated with chemical reagents have found wide applications for many years for the detection of toxic substances in air. Examples include the use of mercuric bromide papers for the detection of arsine, lead acetate for hydrogen sulfide, and a freshly impregnated mixture of o-tolidine and cupric acetate for hydrogen cyanide. When a specific paper is exposed to an atmosphere containing the contaminant in question, the observed time of reaction provides an indication of the concentration of that substance. Thus, in the case of hydrogen cyanide, a 5-sec response time by the o-tolidine-cupric acetate paper is indicative of a concentration of 10 ppm of HCN in the tested atmosphere.

Similarly, sensitive detector crayons have been devised for the preparation of a reagent smear on a test paper, the response of which to a specific toxic substance in a suspect atmosphere may then be timed to obtain an estimation of the atmospheric concentration of a contaminant. Crayons for phosgene, hydrogen cyanide, cyanogen chloride, and Lewisite (ethyl dichloroarsine) have been formulated for this purpose.

Colorimetric Indicator Tubes

Colorimetric indicating tubes containing solid reagent chemicals provide compact direct-reading devices, which are convenient to use for the detection and semiquantitative estimation of gases and vapors in atmospheric environments.

There are tubes for nearly 200 atmospheric contaminants on the market, and 7 U.S. companies manufacture and/or distribute these devices currently in this country. Whereas it is true that the operating procedures for these tubes are simple, rapid, and convenient, there are distinct limitations and potential errors inherent in this method of assessing atmospheric concentrations of toxic gases and vapors. Therefore, dangerously misleading results may be obtained with these devices unless they are used under the supervision of an adequately trained industrial hygienist who

1. Enforces rigidly.
   a. The periodic (as required) calibration of individual batches of each specific type of tube for its response to known concentrations of the contaminant.
   b. The refrigeration storage of all tubes to minimize their rate of deterioration.
2. Inform their staff of the physical and chemical nature of extent of interferences to which a given type of tube is subject and limits the usage of the tube accordingly.
3. Stipulates how and when other independent sampling and analytical procedures will be employed to derive needed quantitative data.

Colorimetric indicating tubes are filled with a solid granular material, such as silica gel or aluminum oxide, which has been impregnated with an appropriate chemical reagent. The ends of the glass tubes are sealed during manufacture. When a tube is to be used, its end tips are broken off, the tube is placed in the manufacturer's holder, and the recommended volume of air is drawn through the tube by means of the air-moving device provided by the manufacturer. This device may be one of several types, such as a positive displacement pump, a simple squeeze bulb, or a small electronically operated pump with an attached flow meter. Each air-moving device must be calibrated after each usage or after sampling 100 tubes as an arbitrary rule or more often if there are reasons to suspect changes due to effects of a corrosive action from contaminants in tested atmospheres. An acceptable pump should be correct +5% by volume; with use, its flow characteristics may change. It should also be checked for leakage and plugging of the inlet after every ten samples.

In most cases, a fixed volume of air is drawn through the detector tube, although with some systems, varied amounts of air may be sampled. The operator compares either an absolute length of strain produced in the column of the indicator gel or a ratio of the length of strain to the total gel length against a calibration chart to obtain an indication of the atmospheric concentration of the contaminant that reacted with the reagent. In another type of tube, a progressive change in color intensity is compared with a chart of color tints in making the estimation. In a third type of detector, the volume of sampled air required to produce an immediate color change is noted; it is intended that this air volume should be inversely proportional to the concentration of the atmospheric contaminant. The remainder of this chapter is devoted to the direct-reading, colorimetric indicating detector tube systems because of the widespread use of these devices.

Detector Tube Characteristics

Reagents and Interferences

Complete information on the formulations of the chemical reagents used in the manufacture of the commercial devices is not available, owing to the understandably competitive nature of this enterprise. However, there is sufficient knowledge of the chemical nature of certain solid reactants, commonly used for this purpose to provide the limited information on chemical reactants, products, color changes, and stated interferences given in Table 10.2. From this table it is apparent that the lack of specificity encountered with certain detector tubes is due to the use of common reagents for numerous compounds. Thus, the use of a hexavalent chromium compound for the oxidation of a wide variety of organic substances with the production of a green chronic reaction product is a completely nonspecific procedure for these substances. This type of detector tube must be used with the realization that all readily oxidizable substances may affect the indication and efforts should, therefore, be taken to ascertain the chemical nature of the various associated contaminants in each exposure situation. Further, aromatic hydrocarbons, hydrides, halides, and chlorinated hydrocarbons provide other examples of class compounds for which single formulations in detector tube reactants may have been used. Such formulations limit the usefulness of these tubes in mixed exposure areas, as the estimations are then based on the results of the reaction of a mixture of contaminants. Furthermore, errors of estimation may be positive or negative depending upon whether a stain is intensified (or lengthened) by the presence of a similarly reacting contaminant or bleached by a differing chemical contaminant present in the sampled air stream. This general situation has evoked a great deal of criticism of detector tube systems by the industrial hygiene profession, which recognizes the need for improved devices of this type.

Efforts to overcome this cross-sensitivity have been attempted by

1. Use of the prelayer to remove certain known interferences from the air stream before they reach the indicating layer.
2. Use of a conversion layer to transform the desired gaseous substance to a different chemical compound which can then react with the indicating layer.

Examples of the incorporation of a prelayer in the formulation of a tube include the Drager Arsine 0.05/a and the Benzene 0.05 tubes. The former uses a copper compound to retain in the prelayer such interferences as hydrogen sulfide, hydrogen selenide, mercaptans, ammonia, and hydrogen chloride, but allows arsine to pass through to the indicating layer containing a gold compound with which the arsine reacts to yield a weak violet-gray color. The benzene 0.05 tube has an acid-aldehyde reagent to remove toluene, xylene, and naphthalene; benzene passes into the indicating layer, where it produces a brown color from its reaction with formaldehyde and sulfuric acid.

Another difficulty arising from detector tube reagents is the catalyzed reaction of gaseous contaminants with one another due to the active contact surface

Table 10.2. Select List of Detecting Reactions in Colorimetric Indicating Tubes

Test Gas or Vapor	Reagents in: Pre-cleanse Layer	Reagents in: Ampoule or Conversion Layer	Reagents in: Indicating Layer	Product(s)	Color Change	Stated Interferences
Acetone	None	None	2,4-Dinitro-phenylhydrazine	A hydrazone	Pale yellow to yellow	Other ketones and aldehydes, alcohols, esters
Acrylo-nitrile	None	(1) Chromate (VI) compound	(2) Mercuric chloride (3) Methyl red	(1) Hydrogen cyanide (2) Hydrogen chloride	Yellow to red	HCl, HCN, organic CN compounds, aromatic
solvents				(3) Red form of indicator		
Alcohol ated	Drying agent	None	Chromate (VI)	Chromic (III)	Yellow to	Other oxygen-
			compound	compound	green	compounds
Ammonia	None	None	(1) Acid (2) Bromophenol blue	(1) Ammonium salt (2) Blue form of indicator	Orange to dark blue	Amines, hydrazines
Aniline	None	(1) Furfural	(2) Acid	(1) Schiff base (2) Dianiline derivative	White to red	Ammonia
Arsine	Copper com-pound (to retain reduced S com-pounds, H_2Se, NH_2, and HCl)	None	Gold compound	Colloidal gold	White to weak violet gray	Phosphine, stibine

Table 10.2. (Continued)

	Reagents in					
Test Gas or Vapor	**Pre-cleanse Layer**	**Ampoule or Conversion Layer**	**Indicating Layer**	**Product(s)**	**Color Change**	**Stated Interferences**
Benzene	Acid and aldehyde (to retain other aromatics)	None	(1) Formaldehyde (2) Sulfuric acid	(1) Diphenyl-methane (2) p-Quinoid compound	White to brown	None affect the indication
Carbon dioxide	None	None	(1) Hydrazine (2) Crystal violet	(1) Carbonic acid mono-hydrazine (2) Blue form of indicator	White to blue	None affect the indication
Carbon disulfide	Copper compound (to retain H_2S)	None	(1) Copper compound (2) Amine	Copper dialkyldi-thiocarbamate	Pale blue to yellowish green	None affect the indication (except H_2S)
Carbon monoxide	Chromate (VI) compound (to retain H_2S, C_4H_4, petroleum compounds)	None	(1) Iodine pentoxide (2) Selenium dioxide (3) Fuming sulfuric acid	Iodine (and carbon dioxide)	White to brownish green	Acetylene and easily cleaved halogenated hydrocarbons
Carbon tetra-chloride	None	(1) Fuming Sulfuric acid	(2) Dimethyl-amino-benzaldehyde (3) Dimethyl-aniline	(1) Phosgene (2) "Blue reaction product"	Yellow to blue	Fluorochloro-methane compounds

Chlorine	Drying agent	None	o-Tolidine	"Yellow reaction product"	White to yellow	Bromine, chlorine dioxide; discoloration by nitrogen dioxide
Chloro-prene	None	None	Permanganate	"Yellowish brown reaction product"	Violet to yellowish brown	Other organic compounds with carbon-carbon double bonds
Cyanogen chloride	None	(1) Pyridine (2) Water	(3) Barbituric acid	(1) & (2) Glutacon-aldehyde cyanamide (3) "Pink reaction product"	White to pink	Cryanogen bromide
Ethyl acetate	Drying agent	None	Chromosulfuric acid	Chromic (III) compound	Orange to brown green	Easily oxidized organic com-pounds including other acetates
Formalde-hyde	None	(1) Xylene vapor	(2) Sulfuric acid	(1) Dixylyl methane (2) "Pink quinoid compound"	White to pink	Other aldehydes and styrene
Hydrazine amines	None	None	(1) Acid (2) Bromophenol blue	(1) Hydra-zinium slat (2) Blue form of indicator	Yellow to blue	1,1-Dimethyl-hydrazine, ammonia,
Hydro-chloric acid	Drying agent	None	Bromophenol blue	Yellow form of indicator	Blue to yellow	Chlorine; R.H. >80%

Table 10.2. (Continued)

	Reagents in					
Test Gas or Vapor	Pre-cleanse Layer	Ampoule or Conversion Layer	Indicating Layer	Product(s)	Color Change	Stated Interferences
Hydro-cyanic acid	Lead compound (to retain H_2S, HCl, SO_2, NO_2, and NH_3)	None	(1) Mercuric chloride (2) Methyl red	(1) Hydrogen chloride (2) Red form of indicator	Yellow to red	None affect the indication (except those not retained in precleanse layer)
Hydrogen fluoride	None	None	Zirconium-alizarin lake	Alizarin	Pale violet to pale yellow	High humidity
Hydrogen sulfide	None	None	Lead compound	Lead sulfide	White to light brown	None affect the indication
Mercaptan	None	(2) Sulfur solution	(1) Copper compound	(1) Copper mercaptide (2) Yellowish brown copper compound	White to yellowish brown	Other mercaptans, hydrogen sulfide, ammonia, amines
Methyl bromide	None	(1) Sulfur trioxide (2) Perman-ganate	(3) o-Dianisidine	(1) & (2) Bromine (3) "Brown reaction product"	White to brown	Halogens, halides, halogenated hydrocarbons
Mono-styrene	Drying agent	None	Sulfuric acid	"Yellow reaction product"	White to yellow	Butadiene and other polymer-tending organic compounds

Nickel tetracar-bonyl	None	(2) Dioxime	(1) Iodine	(1) Nickel iodide (2) Nickel dioxime complex	Pale brown to pink	Iron pentacarbonyl, hydrogen sulfide, sulfur dioxide
Nitrogen dioxide (NO_2)	Drying agent	None	N,N′-di-phenylbenzidine	"Bluish grey reaction product"	White to bluish gray	Ozone, chlorine
Nitrous fumes (NO +NO_2)	None	(1) Chromium (VI) compound	(2) N,N′-di-phenylbenzidine	(1) Nitrogen dioxide (2) "Bluish grey reaction product"	White to bluish gray	Ozone, chlorine
Ozone	None	None	Indigo	Isatine	Pale blue to white	Chlorine, nitrogen dioxide
Perchlor-oethylene halides,	None	(1) Perman-ganate	(2) N,N′-di phenylbenzidine	(1) Chlorine (2) "Greyish blue reaction product"	White to grayish blue	Halogens, hydrogen easily cleaved halogenated hydrocarbons, petroleum vapor
Phenol	None	(1) 2,6-di-bromoquinone chlorimide	(2) Activated silica gel	(1) & (2) Indophenol dye	White to blue	Other aromatic hydroxy compounds quinones; ammonia and amines discolor, indicating layer
Phosgene	Drying agent	None	(1) DImethyl-aminobenz-aldehyde (2) Diethyl-aniline	Bluish green complex	Yellow to bluish green	Carbonyl bromide

Table 10.2. (Continued)

	Reagents in					
Test Gas or Vapor	**Pre-cleanse Layer**	**Ampoule or Conversion Layer**	**Indicating Layer**	**Product(s)**	**Color Change**	**Stated Interferences**
Phosphine	Copper compound (to retain reduced sulfur compounds, NH_3, H_2S, H_2Se)	None	Gold compound	Colloidal gold	White to weak grayish violet	Arsine, stibine
Sulfur dioxide	Copper compound (to retain H_2S)	None	Iodine/starch	Sulfuric acid	Blue to white	Nitrogen dioxide
Toluene benzene	Drying agent	None	(1) Iodine pentoxide (2) Dilute sulfuric acid	(1) & (2) Iodine	White to brown	Xylenes,
Trichloroethane	None	(1) Sulfuric acid (2) Oxidizing agent	(3) o-Dianisidine	(1) & (2) Chlorine (3) "Brownish red reaction product"	Pale grey to brownish red	Some other halogenated hydrocarbons; petroleum hydrocarbons & aromatic compounds >1,000 ppm

Trichloro-ethylene	None	(1) Perman-ganate	(2) o-Tolidine	(1) Chlorine (2) "Orange reaction product"	White to orange	Halogens, halides, easily cleaved halogenated hydrocarbons, petroleum distillate vapors
Vinyl chloride	None	None	Permanganate	Light brown manganese compound	Violet to light brown	Organic compounds with carbon-carbon double bond

provided by the column of chemicals in the detector tube. This phenomenon has been observed with sulfur dioxide and nitrogen dioxide when these gases are brought into closer molecular contact in detector tubes than that prevailing in ambient atmospheres, where their reaction rate with each other is generally low.

Quality Control

Rigid control of the purity of reagents, grain size of the gel, method of packing the tubes, moisture content of the gel, uniformity of tube diameter, and proper storage precautions is required for the optimal performance of any detector tube. The lack of such controls will lead to an inferior product, the use of which may produce disastrous results in the evaluation of health-related problems. Thus, an incompletely responsive tube could cause a hazardous situation to remain uncorrected. Conversely, an overly reactive response may cause undue concern, wasted effort, and needless expense for unwarranted corrections.

Failure of certain manufacturers to maintain rigid control over manufacturing practices has been observed in the past. Although it has been established that an elevated moisture content above 20% will cause a rapid deterioration of the silica gel used in the potassium pallado-sulfite formulation for a carbon monoxide tube, one observer found during a visit to a detector tube manufacturing facility that there was a lapse of many hours following the packing but preceding the hermetical sealing of the tubes. Indeed, such a practice permits the entry of not only moisture, but also other contaminating vaporous substances which might produce a deleterious effect on the future response of the reagent to the analysis substance.

Another critical parameter is the tube diameter. Manufacturing variations in the internal diameter of the narrower tubes can produce appreciably different cross-sectional areas. Saltzman has pointed out that this situation leads to calibration errors as high as 50% due to a varying volume of air sample per unit cross-sectional area as compared with those provided under standard test conditions. At least one manufacturer minimizes the effect of this source of error by loading equal amounts of the indicating gel into each tube; the variations in cross-sectional areas are then compensated by corresponding variations in the filled lengths of the tubes. The manufacturer provides calibration scales which permit the positioning of each tube in accordance with its filled length when measuring the length of stain and thus reduce the overall error of measurement of the gas concentration.

Variations in the grain size of the gel, in the purity of the reagents, and the cleanliness of the air in the tube manufacturing facility can affect the properties of different batches of the indicating gel markedly. If not controlled carefully, these parameters can cause marked and unpredictable changes in the number of active centers on the solid surface of the gel and thus affect the reaction velocity of the indicating system.

The method of storage has a profound effect on the shelf life of an indicator tube. Deterioration of the tubes increases greatly at elevated temperatures, and storage under refrigeration by both the manufacturer and the user is mandatory to realize a useful shelf life, which may be approximately 2 years for some of the tubes. Multiple-layer tubes may have a shorter shelf life due to diffusion of chemicals between layers. For these reasons, a realistic expiration date for tubes stored under refrigeration should be stamped on each box of the tubes by the manufacturer.

Calibration

Saltzman has presented an excellent treatment of the theory of indicator tube calibration where he has developed a basic mathematical analysis of the relationships between the variables which affect the length of strain, i.e., the concentration of test gas, volume of air sample, sampling flow rate, grain size of gel, tube diameter, and other variables. This source should be consulted for a full appreciation of the complex interrelationships between the factors affecting the kinetics of indicator tube reactions. It is sufficient to point out in this chapter that the length of stain is proportional to the logarithm of the product of gas concentration and air sample volume as shown in the following equation:

$$L/H = \text{Ln}\,(CV) + (K/H)$$

where:

- L = the length of stain in centimeters,
- C = the gas concentration in parts per million,
- V = the air sample volume in cubic centimeters,
- K = a constant for a given type of indicator tube and test gas,
- H = a mass transfer proportionality factor having the dimensions of centimeters and known as the height of a mass transfer unit.

If this mathematical model is correct for a given indicator tube, a linear plot of L vs the logarithm of the CV product, for a fixed constant flow rate, will yield a straight line with slope = H. The significance of this equation is the implication that it is important to control the flow rate, which may produce a greater effect on the length of stain than does the concentration of the test gas. Therefore, in the optimal design of an indicator tube it is desirable that the reaction rate be sufficiently rapid to permit the establishment of equilibrium between the indicating gel and the test gas and thus produce a stoichiometric relationship between the volume of stained indicating gel and the quantity of the absorbed test gas. Such equilibrium conditions may be assumed to exist when stain lengths are directly proportional to the volume of sampled air and are not affected by the sampling flow rate. With this situation a log-log plot of stain length vs concentration for a fixed sample volume may be prepared in the calibration of a given batch of tubes.

From the preceding discussion of the complexity of the heterogeneous phase kinetics of indicator tube reactions, the quality control problems associated with their manufacture and storage, and the difficulties posed by interfering substances, it is obvious that frequent, periodic calibration of these devices should be made by the user. Dynamic dilution systems for the reliable preparation of low concentrations of a test gas or vapor are recommended for this purpose. Such calibrations should be performed before each use if there has been an appreciable period since the last calibration was performed.

EVALUATION OF PERFORMANCE

As of January 1, 1972, the results of the evaluation of gas detector tubes for five substances had been published by NIOSH.

The results are as follows

No. of Manufactured Tubes Meeting Approval Criterion Within

Substance	+25% (at 95% C.L.)[a]	±50% (at 95% C.L.)
Benzene	None	3
Carbon monoxide	None	8
Carbon tetrachloride	None	None
Perchloroethylene	3	—
Sulfur dioxide	4	—

[a] 95% confidence level.

Presently there are specific detector tubes that allow for air monitoring without the use of an air pump. They are cost effective and no additional analytical fee is necessary. The detector tube reactions are read in parts per million. An example of the new 15-min to 24-hr sampling detector tubes is Sensidyne Gastic Passive Dosi-Tube for formaldehyde and several other chemicals.

INSTRUMENTS FOR SOUND MEASUREMENT

There is probably a greater variety of instruments for measuring noise than for any other environmental factor of concern to industrial hygienists. Almost every measurement need can be satisfied with instrumentation commercially available today. Only those instruments more useful to the industrial hygienist will be discussed here.

SOUND LEVEL METERS

The standard sound level meter is the basic measuring instrument for the industrial hygienist. It consists of a microphone, an amplifier with calibrated volume control, and an indicating meter. It measures the root-mean-square (rms) sound pressure level in decibels, which is proportional to intensity or sound energy flow.

Sound level meters of the same type differ mainly in external shape, arrangement of controls, and other convenience features that frequently influence the selection made by a prospective user. A typical sound level meter is pictured in Figure 10.1.

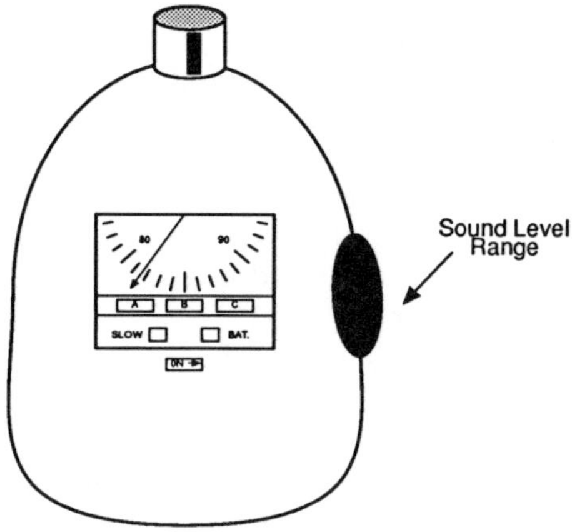

Figure 10.1. A standard sound level meter.

Table 10.3. Relative Response of Sound Level Meter Weighing Networks

Frequency Hz	Weighted Response (dB)		
	A	B	C
31.5	–39	–17	–3
63	–26	–9	–1
125	–16	–4	0
250	–9	–1	0
500	–3	0	0
1000	0	0	0
2000	1	0	0
4000	1	–1	–1
8000	–1	–3	–3

Source: *Handbook of Noise Measurement*, (West Concord, MA: General Radio Company, 1967).

Standards for sound level meters specify performance characteristics in order that all conforming instruments will yield consistent readings under identical circumstances. The more important characteristics specified are frequency response, signal averaging, and tolerances. Table 10.3 shows the relative response of sound level meter weighing networks.

Three weighing networks are provided on standard sound level meters in an attempt to duplicate the response of the human ear to various sounds. These weighing networks cause the sensitivity of the meter to vary with frequency and intensity of sound like the sensitivity of the human ear.

The relative responses of the three networks are shown in Table 10.3 where the A, B, and C weighings mimic ear response to low-, medium-, and high-intensity sounds, respectively. Entries in the table show relative readings of the meter for constant sound pressure level of variable frequency. These A, B, and C meter response curves correspond to the 40, 70, and 100 phon equal loudness contours. The D-weighing network is provided on some sound level meters for approximating the "perceived noise level" used in appraising the offensiveness of aircraft noise.

The A-weighing network is the most useful one on the sound level meter. It indicates the A-weighted sound level, often abbreviated dBA, from which most human responses can be predicted quite adequately.

Action of the indicating meter may be selected as "fast" or "slow." Relatively steady sounds are easily measured using the "fast" response. Unsteady sounds can be averaged with the more sluggish "slow" response to reduce meter needle swings.

The speed of meter response affects the readings obtained for transient sounds. For example, the level of a whistle toot lasting $^1/_5$ sec would be indicated no more than 2 dB low on the "fast" scale. On the "slow" scale, the level of a toot lasting $^1/_2$ sec would read 3 to 5 dB low.

American standard sound level meters are furnished in three types, offering varying degrees of precision. Designated Types 1, 2, and 3 in order of increasing tolerances, the Type 2 generally measures within 2 or 3 dB of true levels, which is satisfactory for most purposes. Errors are about half as large with the Type 1 of "precision" sound level meter and about twice as large with the Type 3 "survey" instrument. These errors can be reduced somewhat by careful calibration.

CALIBRATORS

The overall accuracy of sound measuring equipment may be checked by using an acoustical calibrator such as is shown in Figure 10.2. It consists of a small, stable sound source that fits over the microphone and generates a predetermined sound level within a fraction of a decibel. If the meter reading is found to vary from the known calibration level, the meter may be adjusted to eliminate this error. The acoustical calibration procedure supplements the electrical calibration incorporated in some meters to check the gain of all electronic components following the microphone. Sound level calibrators should be used only with the microphones for which they are intended in order to avoid errors and microphone damage.

IMPULSE METERS

The sound level meter is too sluggish to indicate peak levels of transient noise lasting a fraction of a second, such as those produced by hammer blows or punch press strokes. Such noise must be measured with a special meter that indicates the peak level.

Accessory impulse meters are available for connection to sound level meters and can be calibrated to indicate the peak level of the sound at the microphone. One of these is shown in Figure 10.3. In taking such readings, it is necessary to

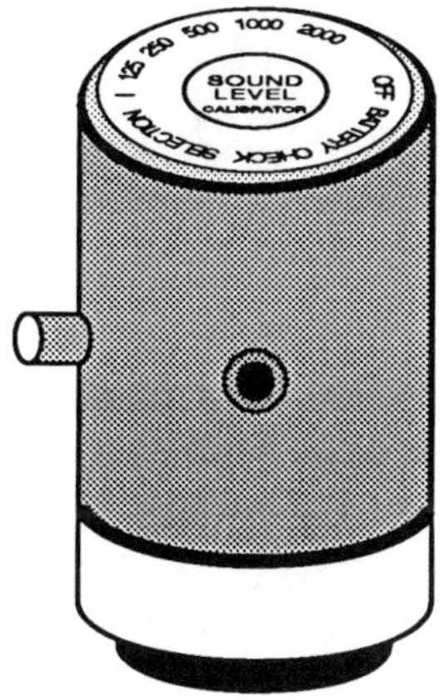

Figure 10.2. A sound level calibrator.

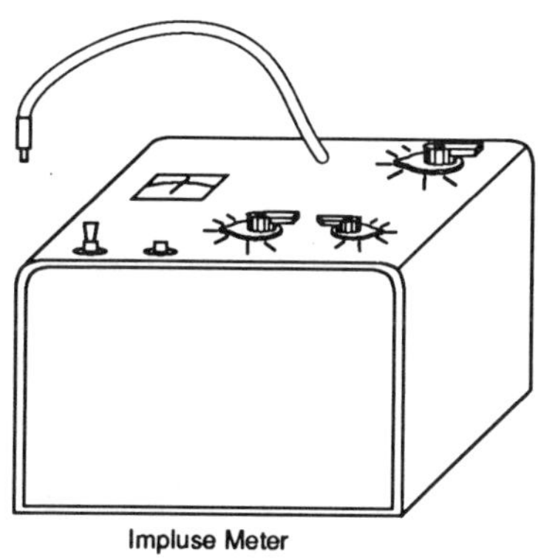

Figure 10.3. Illustration of an impulse meter.

be sure that the upper sound pressure limit of the microphone is not exceeded, as the indicated level will then be too low. The upper limit of the sound level meter can be extended by replacing the standard microphone with a less sensitive one that is linear to higher levels. Such microphone substitutions affect the calibration of the sound level meter and must be taken into account when interpreting readings.

The sound level meter in Figure 10.4 is equipped with a circuit for measuring impulse noise according to a German standard. It has a rather slow rise time in order that the indication of impulsive sound will correlate with the loudness perceived by the ear. For impulses that rise abruptly, the reading with this circuit is lower than would be found with a true peak meter. However, the instrument shown has been provided also with an instantaneous peak reading circuit that will indicate the true peak.

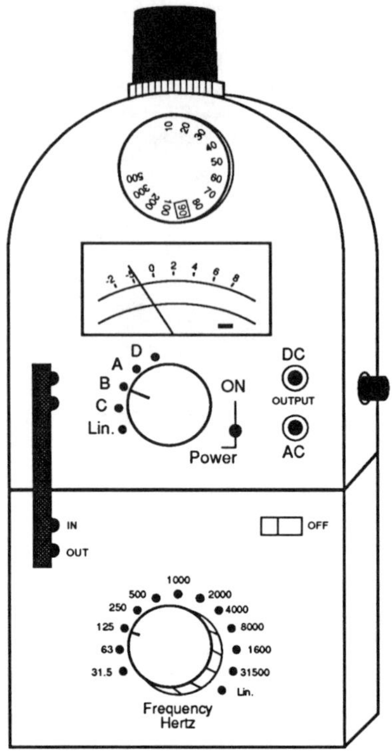

Figure 10.4. A precision sound level meter with an octave band filter attached to the base.

FREQUENCY ANALYZERS

It is often necessary to know the frequency distribution of the sound energy. It is important in noise abatement, for example, since the reduction afforded by control devices varies with frequency. Such information is provided by one of the several types of sound spectrum analyzers available. They may be connected to the sound level meter or other sound sensing systems. The electrical signal from the microphone is filtered by the analyzer circuitry so that only signals within a limited frequency range are transmitted to the indicating meter. Measurement of sound pressure level in contiguous frequency ranges provides data for a plot of sound pressure level vs frequency.

Octave band analyzers are the types most commonly encountered. An octave band filter set is shown attached to the lower end of a sound level meter in Figure 10.4. The frequency range of each band is such that the upper band limit is twice the lower band limit. Formerly, octave bands were described by these cut-off frequency limits such as 300 to 600, 600 to 1200, etc. Currently they are designated by the geometric mean of the cut-off frequencies, which are called center frequencies. Thus, when the 1000-Hz band is mentioned, it is understood that it extends from 710 to 1420 Hz. The center and cut-off frequencies of octave band filters in common use are shown in Table 10.4.

Table 10.4. Center Frequencies and Limits of Zoctave Bands

Traditional Bands (Hz)			Preferred Bands (Hz)		
Lower Limit	Center Freq.	Upper Limit	Lower Limit	Center Freq.	Upper Limit
19	27	38	22	31.5	44
38	53	75	44	63	88
75	106	150	88	125	177
150	212	300	177	250	355
300	425	600	355	500	710
600	850	1,200	710	1,000	1,420
1,200	1,700	2,400	1,420	2,000	2,840
2,400	3,400	4,800	2,840	4,000	5,680
4,800	6,800	9,600	5,680	8,000	11,360

Source: *Handbook of Noise Measurement*, (West Concord, MA: General Radio Company, 1967).

Often bands narrower than octaves are required for pinpointing the frequency of a tone. In such applications, the one-third and one-twenty octave analyzers are quite valuable. Some of these can be coupled to a graphic level recorder which tunes the analyzer through the frequency range and simultaneously plots its output on a moving paper chart. This equipment is very useful in determining sources of noise in machinery, since sound of a particular frequency must be generated by mechanical events such as the meshing of gear teeth, passing of fan blades, etc. which are repeated at the same rate.

ACCESSORY EQUIPMENT

As most sound is generated by vibration of some body, it is sometimes desirable to study the vibration as well as the sound. Accessory vibration pickups are available for most sound level meters to make such vibration analysis possible. They are generally accelerometers for which the electrical output is proportional to the acceleration of the surface to which they are attached. Types sensitive to the velocity or displacement of the surface are also available. Complete vibration meters are supplied, too, for greater convenience.

The vibration signal may be examined for frequency content using any of the sound analyzers discussed earlier. In this way, vibration at a particular frequency may be correlated with the sound of the same frequency it produces. Calibrators are available for checking the sensitivity of the vibration pickup to reduce errors in measuring the acceleration of a surface.

The cathode ray oscilloscope is useful for observing the waveform of a sound. It plots the sound pressure vs time on a television-type screen. From observation of the wave shape, it is sometimes possible to determine the mechanical process reasonable for the noise. It is also possible to observe the peak pressures of impulsive sounds. Cameras are available for photographing the face of the cathode ray tube to obtain a permanent record of the waveform.

A graphic record of sound level may be obtained by connecting a sound level meter to a graphic level recorder which plots the sound level on a moving paper chart. The pen speed of the recorder must be capable of following the fluctuations in sound level if an accurate record is to be obtained, or one can plug in the sound level meter and dosimeter to a data logger to get precise precalculated data in numeric values or historgrams.

The magnetic tape recorder can be used to store a sound for later analysis or replay, just as the data logger does for present sound measurements. An instrument of broadcast quality must be used to obtain high-fidelity reproduction of the sound on playback. For analysis on replay, the recorder output should be connected to the analyzer input by means of a patch cord. Too much distortion will occur if the recording is played back through a loud-speaker and picked up with a microphone. If only the frequency content is of interest, then no precautions need be taken to keep track of level during recording and playback. However, if it is necessary to determine the true levels of the original noise on playback, careful recording of calibration tones must be undertaken. Generally, it is better to measure sound directly in the field if possible.

SOUND MONITORS

When sound level varies erratically over a wide range, it is difficult to describe the noise by meter readings. Therefore, statistical analyzers have been developed to assist in this process. They indicate the percentage of time that the sound level lies in certain predetermined level ranges. From these data, the mean level, standard deviation, as well as other statistical indices may be calculated.

Another type of monitor evaluates noise exposures according to the rules established by the ACGIH, which will be described in a later section. These instruments may be exposed to varying noise for a work day and will indicate whether this exposure limit has been exceeded. One of the battery-powered, wearable types is shown in Figure 10.5.

A different type of noise hazard meter recently developed integrates the effects of noise like the ear does. When exposed to nonimpulsive noise of any duration, it indicates the amount of temporary shift in hearing threshold that a group of normal ears would experience in the same exposure. Its reading is interpreted according to the theory that noise exposures producing little temporary hearing loss are not likely to produce much permanent hearing loss even after many repetitions.

ACCEPTABILITY CRITERIA

Criteria for the acceptability of noise are dictated by the effects which are to be avoided. The most important of these is hearing damage resulting from prolonged exposure to excessive noise. Another undesirable effect is speech interfer-

Figure 10.5. A battery-powered noise dosimeter that can be worn by an employee.

ence or interruption of communications by noise. Annoyance is a third undesirable effect of noise more difficult to assess. There are also certain nonauditory effects of noise we are just beginning to recognize which are discussed later in this chapter.

HEARING DAMAGE

The damaging effect of noise on hearing depends on

1. The level and spectrum of the noise.
2. Duration of exposure.
3. How many times it occurs per day.
4. Over how many years daily exposure is repeated.
5. The effects on hearing regarded as damage.
6. Individual susceptibility to this type of injury.

All of these factors must be considered in establishing limits of acceptable exposures to dangerous noise. Table 10.5 illustrates the relationship between sound pressure level in decibels (dB) and sound pressure in Newtons per square meter (N/m2).

Noise Evaluation

Early in the study of the effects of noise on hearing, it was learned that noise frequency as well as intensity influenced the effect produced. High-frequency noise was found to be more damaging than low-frequency noise of the same sound

Table 10.5. Relationship Between Sound Pressure Level in Decibels (dB) and Sound Pressure in Newtons per Square Meter (N/m²)

Sound Pressure Level in dB RE 0.00002 N/m²	dB	N/m²	Sound Pressure
	120–	–20	
Pneumatic chipper (at 5 ft)		–10	Rock-n-roll band
	110–	–5	
Textile loom			
	100–	–2	Power lawn mower
Newspaper press		–1	(at operator's ear)
	90–		
Diesel truck 40 mph (at 50 ft)		–0.5	Milling Machine (at 4 ft)
	80–	–0.2	Garbage Disposal (at 5 ft)
		–0.1	
	70–		Vacuum Cleaner
Passanger car 50 mph (at 50 ft)		–0.05	
	60–	–0.02	Air conditioning window
Conversation (at 3 ft)		–0.01	unit (at 2 ft)
	50–		
		–0.005	
Quite room	40–	–0.002	
		–0.001	
	30–		
		–0.0005	
	20–	–0.0002	
		–0.0001	
	10–		
		–0.00005	
	0–	–0.00002	

Source: Dr. Loren Hatch, Retired NIOSH Medical Director, St. Augustine, FL. Developed in 1982 while working for NIOSH on loan to the University of South Florida, Tampa, FL, as technical advisor for the College of Public Health.

pressure level. Therefore, noise spectra were evaluated with standard octave band analyzers, which were the only portable spectrum analyzers than available.

As knowledge of noise effects grew, some investigators began to feel that octave band analysis was a needlessly complicated evaluation of noise that could be replaced with the A-weighted sound level measured using a standard sound level meter. It seemed that the A-weighing network made the meter less sensitive to low-frequency sounds to about the same extent that the ear is less susceptible to injury by these low-frequency sounds.

In these studies, the damage to be avoided was impairment of ability to understand "everyday speech" as defined by the medical profession. This medico-legal definition allows some observable change in hearing thresholds not sufficient to affect ability to understand everyday speech significantly.

Steady Noise

All-day exposure to steady noise has been investigated to determine the level at which hearing damage begins after many years of redundant exposure. Such

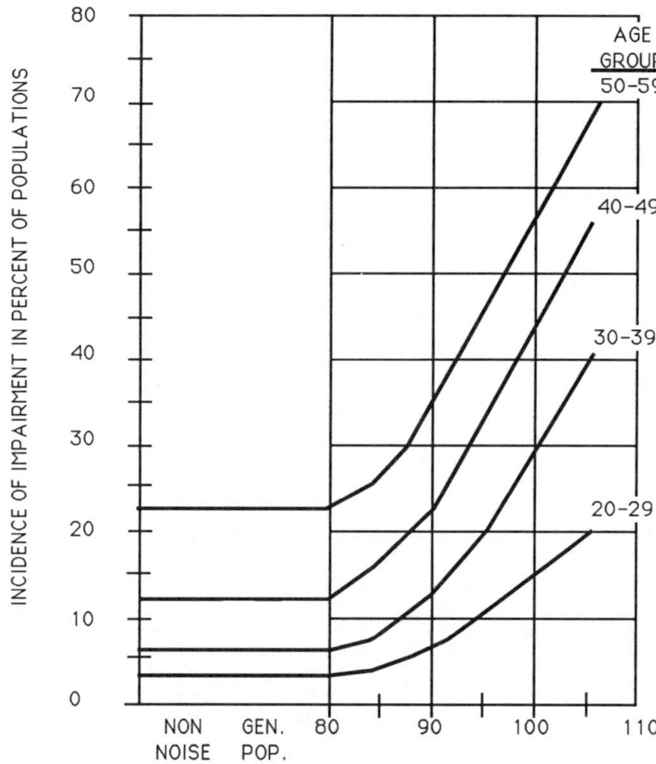

Figure 10.6. Data gathered on industry-impaired hearing.

studies are the basis for the curves in Figure 10.6, which indicate the risk of hearing impairment associated with exposure to a steady noise level at work. Each curve indicates on the vertical scale the percentage of workers that showed impaired hearing as defined by the medical profession after working continuously in the noise levels shown on the horizontal scale.

To interpret Figure 10.6, note that the upper curve shows that in a group of 100 men, aged 50 to 59 years exposed to 90 dB at work for 33 years, 33 men should show evidence of impaired hearing. However, note that the lower flat portion of the same curve indicates that the general population and others not exposed to dangerous noise at work exhibit 22 cases of impaired hearing out of every 100. Therefore, near-lifetime exposure to 90 dB at work seems to produce about 11 more cases of impaired hearing per 100 surviving than would otherwise have occurred. As the data generating the curves of Figure 10.6 are not so consistent as the precise lines would indicate, this difference of 11 percentage points is about the smallest that can be considered significant. For lower age groups exposed for shorter periods, the increase in prevalence of impaired hearing is much less

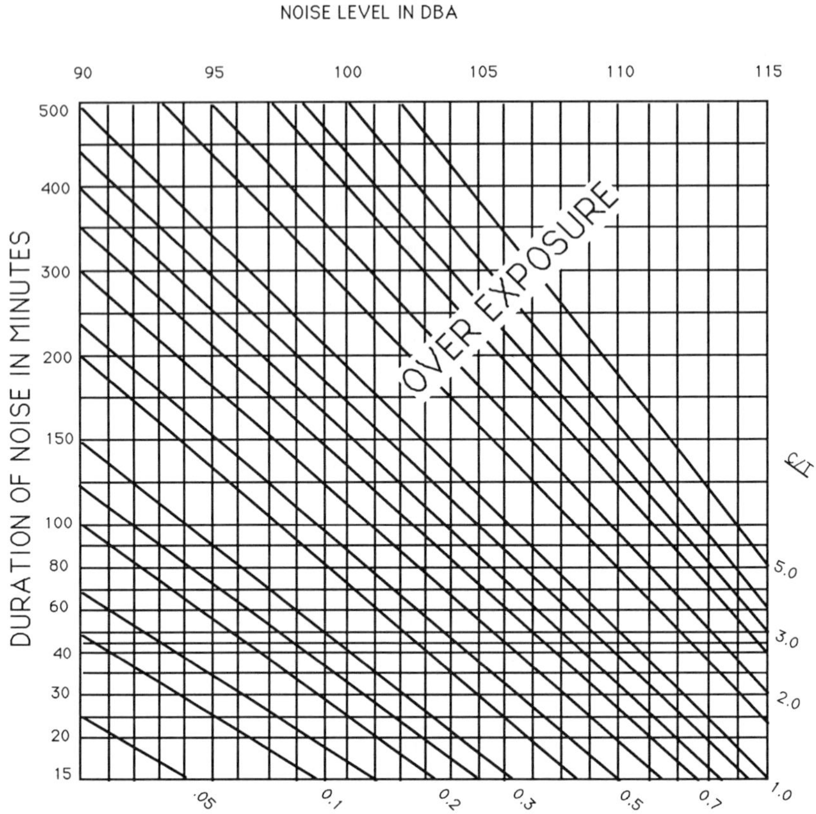

Figure 10.7. Graph of curves suggesting 90 dBA as one limit for steady exposure to continuous noise.

pronounced. The curves of Figure 10.7 suggest 90 dB as one limit for steady exposure to continuous noise, a limit that has become rather widely accepted. Future standards may lower this limit.

Intermittent Noise

Most occupational noises are intermittent rather than continuous. Interrupting harmful noise allows the ear to rest and recover, which reduces the likelihood of permanent damage. Such intermittent exposures have not been studied much because of the great complexities of exposure description. As a result, theories are relied upon to set limits for intermittent noise.

The theory most generally accepted postulates that the hazard of noise exposure increases in proportion to the average temporary hearing loss which the exposure would produce in a group of normal ears. This theory arises out of the observation that those noise exposures that ultimately produce permanent hearing loss also produce temporary hearing loss in normal ears. While the true relation between temporary and permanent hearing loss has not been established, it is logical to assume that those noise exposures that do not cause much temporary loss will not cause much permanent loss either. Any temporary threshold shift (TTS) that disappears before the next exposure to noise commences is considered acceptable.

On the basis of this assumption, results of TTS studies have been used to define safe limits for all day exposures to steady noise. These limits agree with those established by permanent threshold shift studies.

TTS studies have also indicated that intermittent noise is much less harmful than steady noise. The laws describing growth of TTS during exposure and recovery afterwards have been used to calculate exposures producing acceptably small amounts of TTS. Combinations of sound level, duration of exposure, and degree of repetition that are considered acceptable for personnel exposures at work are shown in Table 10.5. This method for appraising noise exposures was derived from the report describing hazardous exposure to intermittent and steady-state noise prepared by the National Academy of Science-National Research.

REFERENCES

1. "Specification for Sound Level Meters," S1, 13-1971 (New York: American National Standards Institute 1971).
2. "General Purpose Sound Level Meters," IEC/123 (New York: American National Standards Institute 1961).
3. Harrington, D. U.S. Bureau of Mines Information Circular No. 7072, Washington, DC (1939).
4. Timbrell, V. "The Inhalation of Fibrous Dists, Biological Effects of Asbestos," *Ann. N. Y. Acad. Sci.* 123:255–273 (December 31, 1965).
5. Witten, B. and A. Prostak. "Sensitive Detector Crayons for Phosgene, Hydrogen Cyanide, Cyanogen Chloride, and Lewisite," *Anal. Chem.*, 29:885–887 (1957).
6. Saltzman, B. E., Ph.D. "Direct Reading Colorimetric Indicators," in *Air Sampling Instruments*, Section S, 4th ed. (Cincinnati OH: American Conference of Governmental Industrial Hygienists, 1972), pp. S-1 and S-11.
7. Silverman, L., and G. R. Gardner. "Potassium Pallado-Sulfite Method for Carbon Monoxide Detection," *Am. Ind. Hyg. Assoc. J.* 26:2 (March–April 1965).
8. Kusnetz, H. L. "Evaluation of Chemical Detector Tubes," paper presented at Chemical Section, National Safety Congress, Chicago, (October 27, 1965).
9. Saltzman, B. E. "Direct Reading Colorimetric Indicators," in *Air Sampling Instruments*, Section S, 4th ed. (Cincinnati OH: American Conference of Governmental Industrial Hygienists, 1972).
10. Saltzman, B. E. "Basic Theory of Gas Indicator Tube Calibration," *Am. Ind. Hyg. Assoc. J.* 23:112–126 (March–April 1962).

11. Morganstern, A. S., R. M. Ash, and J. R. Lunch. "The Evaluation of Gas Detector Tube Systems: I. Carbon Monoxide," *Am. Ind. Hyg. Assoc. J.* 31:630–632 (September–October 1970).
12. Ash, R. M. and J. R. Lynch. "The Evaluation of Gas Detector Tube Systems: Benzene," *Am. Ind. Hyg. Assoc. J.* 32:410–411 (June 1971).
13. Ash, R. M., and J. R. Lynch. "The Evaluation of Gas Detector Tube Systems: Carbon Tetrachloride," *Am. Ind. Hyg. Assoc. J.* 32:552–553 (August 1971).
14. Ash, R. M. and J. R. Lynch. "The Evaluation of Gas Detector Tube Systems: Sulfur Dioxide," *Am. Ind. Hyg. Assoc. J.* 32:490–491 (July 1971).
15. Roper, C. P. "An Evaluation of Perchloroethylene Detector Tubes," *Am. Ind. Hyg. Assoc. J.* 32:847–849 (December 1971).
16. Keenan, R. C. "Direct Reading Instruments for Determining Concentrations of Aerosols, Gases, and Vapors," (Cincinnati, OH: NIOSH, 1973), chap. 16.
17. Current marketing literature for 1988 and 1989 of the following manufacturers:

    A. Photovac International Inc., 741 Park Avenue, Huntington, NY 11743 (Phototip Portable GC and Microtip).
    B. National Dräger, P.O. Box 120, Pittsburgh, PA 15230 (N.D. Detector Tubes).
    C. Sentex Sensing Technologies, Inc., 553 Broad Avenue, Ridgefield, NJ 07657 (Scentograph).
    D. Sensidyne Gas and Particulate Detection Systems, 16333 Bay Vista Drive, Clearwater, FL 34620 (Odor Monitor, Portable F.I.D., Detector Tubes, Mini Monitors).
    E. Metrosonics, Inc., General Products Division, P.O. Box 23075, Rochester, NY 14692 (db-307 Metrologger).
    F. Interscan Corporation, P.O. Box 2496, Chatsworth, CA 91313-2496 (Toxic Gas Detectors).
    G. Industrial Scientific Corp., 355 Steubenville Pike, Oakdale, PA 15071-1093 (Combustible Gas Monitors).

CHAPTER 11

Introduction to Sampling, Sampling Plans, and Chain-of-Custody

INTRODUCTION

Sampling is the physical collection of a representative portion of a universe (or environment). A representative sample must be collected and handled in such a fashion as to preserve its original physical form and chemical composition, as well as to prevent contamination or changes in concentration of the materials to be analyzed. It is imperative to ensure sample integrity and maintain quality assurance standards for a sample to be "typical" of the larger body of the material of concern.

Because hazardous incidents are so varied, sampling equipment and methods are also varied. However, if the procedures outlined in this text are used as sampling protocol, the degree of uniformity necessary for hazardous incident sampling can be obtained. The purpose of a written sampling plan is to detail a plan of action so as to meet the goal of collecting representative samples. A plan should be pre-pared prior to the actual sampling to eliminate confusion and save time. Only after the following questions have been answered should samples be collected.

1. What are the objectives of collecting samples?
2. What types of samples are needed?
3. Where are the best sampling locations?
4. How many and how often are samples required?
5. How are the samples to be collected and handled?

When answering these questions, the following factors must be taken into consideration:

1. The topographic, geologic, and hydrologic characteristics of the site. For example, surface and groundwater and soil types.
2. Meteorologic conditions, i.e., temperature, air pressure, precipitation, wind velocity, and seasonal variations.
3. The flora and fauna of the area, i.e., bioaccumulation and biotransformation in the plant and animal life, especially agricultural.
4. Geographic and demographic information, i.e., population and proximity to site and public health threats.
5. Physical properties and hazardous characteristics of materials involved.

OBJECTIVES OF SAMPLING

Samples are taken to accomplish different objectives. In many cases, they are taken to determine compliance with existing regulations. Other times, they are taken to obtain data to classify, treat, recover, recycle, or determine compatibility characteristics of the materials in question.

It is important to determine the specific purpose of collecting the sample because this will dictate many of the subsequent choices of sample methodology, sites, types, and numbers. For example, samples may be taken at an abandoned hazardous materials storage facility to determine the compatibility of materials. In this case, the samples must be obtained directly from the drums, drains, tanks, ponds, lagoons, etc. containing the materials. This will require special collection apparatus and safety equipment. However, if the impact of this facility on the local water quality is to be studied, then both off-site (upstream and downstream of the incident) and on-site samples should be obtained. The samples collected off-site may require only standard equipment.

TYPES OF SAMPLES

Environmental vs Hazardous

1. *Environmental samples.* Considered to contain no contaminants or low concentrations of contaminants. Generally taken in an area which may be contaminated.
2. *Hazardous samples.* Considered to contain high concentrations of contaminants. Usually collected from drums, tanks, lagoons, spills, and areas in the immediate vicinity of the incident.

The importance of this distinction is twofold:

1. *Personnel safety requirements.* Any sample thought to contain enough hazardous materials to pose a safety threat should be designated as hazardous and handled to ensure personnel safety.

2. *Transportation requirements.* Hazardous samples must be shipped according to DOT regulations.

Grab vs Composite

There is some debate over how to collect representative samples of hazardous materials.

Grab

A discrete sample that is collected at one specific time. Usually, this is considered representative of one specific site at a specific time. However, if the source is fairly consistent over a period of time and/or geographical area (such as a waste storage or disposal facility), then the sample can be considered representative of that time period or area. Sources that vary greatly over time or distance (for example, release of contaminants into moving water or air) sharply reduce the representative value of a grab sample. In such situations, samples should be taken more frequently.

Composite

A nondiscrete sample composed of more than one specific sample collected at various sampling sites or times and/or mixed and treated as one is called a composite sample. Composite samples may give an "average" concentration or composition and may be an alternative to analyzing numerous grab samples. Grab sample concentrations can change significantly due to collection, transport, or storage; these composites cannot be used. Composites should be utilized *only* after it has been demonstrated that there is *no significant change* under the existing sampling conditions and methodology.

When collecting hazardous samples, it is advisable to collect only grab samples. Compositing samples poses a safety risk if it involves samples of unknown hazardous content. The chemical changes that may occur during compositing also support collection of grab samples.

Location of Sample Points

Sampling sites are influenced by the objectives of the study. Exact locations should be chosen after taking into consideration the factors that can influence the concentrations of the material of concern. Choosing sampling sites requires sound judgment, and only after all logical choices are exhausted should they be selected randomly.

1. *Judgment sampling.* Used first. Relies on experience and available information to determine the spots which will provide the most representative sample.
2. *Random sampling.* Depends on the theory of random chance probabilities in order to choose the most representative sample. This process is utilized when numerous sampling locations are available but there is no satisfactory reason

Table 11.1. Table of Random Numbers and How to Use Them[a]

03	47	43	73	86	36	96	47	36	61	46	98	63	71	62
97	74	24	67	62	42	81	14	57	20	42	53	32	37	32
16	76	62	27	66	56	50	26	71	07	32	90	79	78	53
12	56	85	99	26	96	96	68	27	31	05	03	72	93	15
55	59	56	35	64	38	54	82	46	22	31	62	43	09	90
16	22	77	94	39	49	54	43	54	82	17	37	93	23	78
84	42	17	53	31	57	24	55	06	88	77	04	74	47	67
63	01	63	78	59	16	95	55	67	19	98	10	50	71	75
33	21	12	34	29	78	64	56	07	82	52	42	07	44	38
57	60	86	32	44	09	47	27	96	54	49	17	46	09	62
18	18	07	92	46	44	17	16	58	09	79	83	86	19	62
26	62	38	97	75	84	16	07	44	99	83	11	46	32	24
23	42	40	64	74	82	97	77	77	81	07	45	32	14	08
52	36	28	19	95	50	92	26	11	97	00	56	76	31	38
37	85	94	35	12	83	39	50	08	30	42	34	07	96	88
70	29	17	12	13	40	33	20	38	26	13	89	51	03	74
56	62	18	37	35	96	83	50	87	75	97	12	25	93	47
99	49	57	22	77	88	42	95	45	72	16	64	36	16	00
16	08	15	04	72	33	27	14	34	09	45	59	34	68	49
31	16	93	32	43	50	27	89	87	19	20	15	37	00	49

[a] How to use the Table of Random Numbers:

1. Based on available information, segregate the containers (i.e., drums, sacks, etc.) according to waste types.
2. Number the containers containing the same waste types consecutively , starting from 01.
3. Decide on how many samples you wish to take. This number is usually determined by the objective of the sampling. For regular surveillance sampling, the collection of one or two samples is usually adequate. In this case, random sampling is not necessary, but for regulatory or research purposes, more samples (such as one sample for every group of five containers) taken at random will generate more statistically valid data. Hence, if there were 20 drums containing the same type of waste, 5 drums have to be sampled.
4. Using the set of random numbers above, choose any number as a starting point.
5. From this number, go down the column, then to the next column to the right, or go in any predetermined direction until you have selected five numbers between 01 and 20, with no repetitions. Larger numbers are ineligible.

for choosing one over another. Tables of random numbers, which are readily available from many sources, should be used to eliminate any bias of the sample collector. Random sampling will usually be used only at dump sites involving thousands of drums of unknown content. They should be separated into groups according to content based on information provided by manifests, records, labels, etc. Drums of each group are chosen at random intervals according to random sampling techniques. Table 11.1 states how to use the table of random numbers.

Numbers of Samples

When gathering evidence for possible legal actions, multiple samples must be collected to maintain an adequate quality assurance program.

1. *Duplicate samples.* Duplicate or multiple samples are essentially identical. These samples must be collected at the same time, at exactly the same location, with the same apparatus, and placed into identical containers prepared the same way and filled to the same volume. All duplicate samples are preserved and handled in identical fashion.
2. *Reproducibility check.* A duplicate sample is analyzed using the same procedure and instrument employed in the original analysis. This verifies reproducibility of data over a period of time.
3. *Split samples.* Portions of sample, or a duplicate sample, can be analyzed by another accepted analytical technique or another qualified laboratory in order to verify the original results.
4. *Spiked samples.* Known amount of the contaminant is added to a sample at concentrations where the test method is accurate. Hazardous samples should be spiked only in the field; the amount added must be coordinated with the laboratory to check the accuracy of the field sampling procedures. Field spiking may also indicate sample changes during transport.
5. *Field blank.* Distilled, deionized, contaminant-free water is rinsed over a sampling tool and collected in a lab bottle to confirm field decontamination cleanliness.
6. *Sample (trip) blank.* A sample of distilled, deionized, contaminant-free water that is collected, containerized, treated (if used), and handled in the same manner as the samples. This can be an indicator of sample contamination throughout the entire process.

METHODOLOGY AND EQUIPMENT

Sampling Methods

At present, there are numerous standardized methods for collecting environmental samples. Many of these methods are specified primarily by governmental (e.g., EPA) or scientific organizations such as the American Society of Testing Materials (ASTM). Common publications which spell out specific sampling requirements for a particular analysis are Standard Methods for the Examination of Water and Wastewater, Methods for Chemical Analysis of Water and Wastewater, Methods for Chemical Analysis of Water and Waste, and Test Methods for Evaluating Solid Waste - Physical Chemical Methods for Evaluating Solid Waste - Physical Chemical Methods (SW246, 3rd). Sampling procedures can also be found in *Federal Register.* If there is conflicting information, discuss the most recent sampling method with the applicable government agency or use the most recent U.S. government method.

Personnel collecting hazardous samples should remember to use protective clothing and equipment to minimize exposure. The use of special collection equipment presents a problem. Limited information is available, and no universally accepted standardized methods have been devised for collection of hazardous samples.

Sampling Equipment

Equipment to collect and contain hazardous samples should be

1. Disposable or easily decontaminated. Collection device may be reused again only after thorough cleaning.
2. Inexpensive, especially for disposable items.
3. Easy to operate. Personnel may be incumbered by safety clothing and respiratory equipment.
4. Nonreactive, so that it does not contaminate samples.
5. Safe to use.

Selection of Samples and Containers

Hazardous wastes are usually complex mixtures of semisolids, liquids, or solids. The liquid and semisolid mixtures vary greatly in viscosity, corrosivity, volatility, explosivity, and flammability. Solid wastes can range from powder to granule to big lumps.

Wastes are contained in drums, barrels, sacks, bins, tanker trucks, vacuum trucks, ponds, and other containers. No single type of sampler can collect representative samples of all types of waste, so a number are commonly employed (see Figure 11.1). The sample container will also vary according to the materials being sampled. If limited information is known about a sample, use glass bottles with Teflon®* lined caps (see Figure 11.2).

Sampling Technique

Another important factor is use of the same technique for sampling. The same member of the work party should collect all the samples of a particular type (member A collects all drum samples, member B collects all soil samples, member C collects all stream samples, etc.). These practices ensure that data obtained from sample analyses are representative of the waste sampled and not a result of erratic sampling techniques.

Documentation

All information pertinent to field activities will be recorded in various forms: log books, sample tags, photographs, etc. Proper documentation and document control are crucial to both defense and prosecution of enforcement actions. Therefore, each field worker must make detailed records of inspections, investigations, photographs taken, etc. and thoroughly review all notes before leaving the site.

The purpose of document control is to assure that all documents for a specific project are accounted for when the project is completed. Accountable documents

* Registered trademark of E. I. du Pont de Nemours and Company, Inc., Wilmington, Delaware.

SAMPLE CONTAINERS AND CLOSURES*

Category	Recommended Container	Recommended Closure
Metals, inorganics, weak acids/bases	Glass or Plastic	Plastic caps with plastic or Teflon liner
Organic solvents, hydrocarbons, chlorinated, hydrocarbons	Glass	Plastic caps with Teflon liner
Photosensitive materials	Amber plastic or glass	Plastic caps with plastic or Teflon liner
Strong acids/bases	Glass	Plastic caps with Teflon liner
Hydrofluoric acid, phosphoric acid	Plastic	Plastic caps with plastic liner

*For more specific applications to Standard Methods for the Examination of Water and Wastewater of the USEPA Methods for Chemical Analysis of Water and Wastes.

SAMPLES FOR LIQUID/SOLID HAZARDOUS MATERIALS

Category	Sampler
Liquids, sludges, slurries	Open tube Pond sampler Manual hand pump Weighted bottle sampler Kemmer sampler Extended bottle sampler
Powdered or granular solids	Grain sampler Sampling trier Trowel/scoop/spoon Waste pile sampler
Soil	Soil auger

Figure 11.1. Illustration of some various sampling containers for soil, water, and air.

include items such as log books, field data records, analytical records, and photos. Each document should bear a serial number and should be listed, with the number, in a project document inventory assembled at the project's completion. Black waterproof ink must be used in recording all data bearing serial numbers documents.

CONTAINERS, CLOSURES/CLOSURE LINERS

Containers

The most important factors to consider when choosing containers for hazardous material samples are compatibility, resistance to breakage, and volume. Containers must not melt, leach, rupture, or leak as a result of chemical reactions

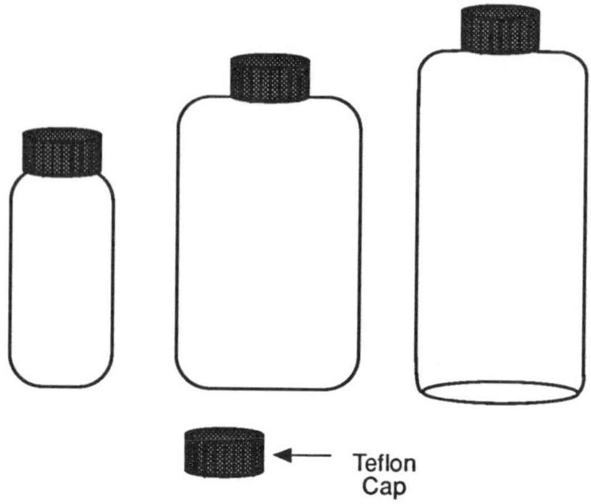

Figure 11.2. Illustration of various glass bottles with Teflon-lined caps.

with constituents of a sample. Thus, it is important to have some idea of the composition of the sample. The containers must have walls thick enough to survive sample collection and transport to the laboratory. Containers with wide mouths are easier to use. Also, the containers must be large enough to contain the required volume of the sample or the entire volume of a sampler.

Plastic and glass containers are generally used for collection and storage of hazardous material samples. Commonly available plastic containers are made of high density or linear polyethylene (LPE), conventional polyethylene, polypropylene, polycarbonate, Teflon fluorinated ethylene propylene (FEP), polyvinyl chloride (PVC), or polymethylpentene. Teflon FEP is the most inert, giving it the widest range of application. Glass containers are used only when the constituents of the material are known to react with the plastic.

Glass containers are relatively inert to most chemicals and can be used to collect and store almost all hazardous material samples. Two exceptions are strong alkali solutions and hydrofluoric acid. Glass bottles with wide mouths (to facilitate sample collection) are recommended for samples containing petroleum distillates, chlorinated hydrocarbons, pesticides, solvents, and other substances incompatible with plastic.

Several types of glass containers are available. Flint glass bottles are inexpensive and available in various shapes and sizes. Borosilicate glass is more inert, but the selection of containers is smaller.

Closures/Closure Liners

Sample containers must have tight, screw-type lids. Plastic bottles are usually provided with screw caps made of the same material as the bottles. Cap liners usually are not required. Glass containers usually come with glass or rigid plastic

Exhibit 11.1. Sample Preservation

Complete and unequivocal preservation of samples, either domestic sewage, industrial wastes, or natural waters, is a practical impossibility. Regardless of the nature of the sample, complete stability for every constituent can never be achieved. At best, preservation techniques can only retard the chemical and biological changes that inevitably continue after the sample is removed from the parent source. The changes that take place in a sample are either chemical or biological. In the former case, certain changes occur in the chemical structure of the constituents that are a function of physical conditions. Metal cations may precipitate hydroxides or form complexes with other constituents; cations or anions may change valence states under certain reducing or oxidizing conditions; other constituents may dissolve or volatilize with the passage of time. Metal cations, such as iron and lead may also absorb onto container surfaces (glass, plastic, quartz, etc.). Biological changes taking place in a sample may change the valence of an element or a radical to a different valence. Soluble constituents may be converted to organically bound materials in cell structures, or cell lysis may result in release of cellular material into solution. The well-known nitrogen and phosphorus cycles are examples of biological influence on sample composition. Therefore, as a general rule, it is best to analyze the samples as soon as possible after collection. This is especially true when the analyte concentration is expected to be in the low microgram per liter range.

Methods of preservation are relatively limited and are intended generally to

1. Retard biological action.
2. Retard hydrolysis of chemical compounds and complexes.
3. Reduce volatility of constituents.

Preservation methods are generally limited to pH control, chemical addition, refrigeration, and freezing.

screw caps. Caps often have paper liners coated with wax. Other liner materials are polyethylene, polypropylene, neoprene, and Teflon FEP.

Preservation and Storage of Samples

Samples must be analyzed immediately after collection for best results. Hazardous wastes are such complex mixtures that it is difficult to predict exactly the physical, biological, and chemical changes that occur in the samples. The pH may change significantly in a matter of minutes, sulfides and cyanides may be oxidized or evolve as gases, and hexavalent chromium may slowly be reduced to the trivalent state. Certain cations may be partly lost as they are absorbed on the walls of the sample containers. Microorganisms may grow in certain constituents. Volatile compounds may be rapidly lost.

In a number of cases, such changes may be slowed down or prevented by refrigeration at 4 to 6°C, or by adding preservatives (see Exhibit 11.1). However, these treatments may be only partially effective. Refrigeration may reduce loss of volatile components and acid gases such as hydrogen sulfide and hydrogen cyanide, but it also introduces the possibility that some salts may precipitate at lower temperatures. On warming to room temperature for analysis, the precipitates may not redissolve, thus giving inaccurate results. Preservatives may retard

Table 11.2. Table of Sample Preservation Methods

Preservative	Action	Applicable To
HgCl	Bacterial inhibitor	Nitrogen forms, phosphorus forms
Acid (HNO_3)	Metals solvent, prevents precipitation	Metals
Acid (H_2SO_4)	Bacterial inhibitor	Organic samples (COD, oil & grease, organic carbon)
	Salt formulation with organic bases	Ammonia, amines
Alkali (NaOH)	Salt formulation with volatile compounds	Cyanides, organic acids
Refrigeration	Bacterial inhibitor	Acidity-alkalinity organic materials, BOD, color, odor, organic P, organic N, carbon, etc., biological organism (coliform, etc.)

constituents to stable hydroxides, salts, or compounds, but they may also convert to other forms (such as the products of nitration, sulfonation, and oxidation, of organic compounds). Thus, subsequent analyses may not identify the original components or concentrations. Table 11.2 lists some sample preservation methods.

Safety must also be considered because a preservative may be highly reactive with the sample. Addition of a preservative may change the DOT packaging, labeling, and shipping requirements for a sample. Shipping hazardous samples packed in ice may not be permitted by DOT regulations.

SAMPLING EQUIPMENT AND PROCEDURES

Surface Waters

Surface waters include rivers, streams, lakes, ponds, ditched water, and collected/ponded runoff. Technically any water which is on the ground surface, even from a spring, is considered a surface water. Water collected from a spring, however, is generally considered indicative of groundwater conditions.

Equipment

Containers: Two 1/2-gal amber jugs, two 40-ml VOA vials, and 1 or 2 1-qt plastic bottles and standard for a full priority pollutant scan. Samplers: Kemmerer bottle, weighted bottle, dipper, bailer, buckets.

Procedure

Flowing Waters (Brooks, Rivers, Etc.). Samples should be obtained from upstream, downstream, and point source (discharge) locations. In general, the sample should be obtained from a characteristic location in the stream, not from an eddy or stagnant side pool. Ideally; it should be obtained from the middle of the main channel in the stream bed (i.e., where the stream would be flowing in a period of low water) and at a depth midway between the bottom and the surface. Tides must be accounted for when sampling estuaries; upstream and downstream will reverse with the direction of tidal flow.

Sampling becomes more complicated when there are one or more tributaries upstream of the site or flowing through or originating on site. Here samples might have to be obtained from upstream on each tributary and downstream below points of confluence.

The simplest method is to wade into the stream, face upstream, and submerse the sample bottle (except for the inorganics if prespiked).

Standing Waters (Ponds, Bays, Pools, Etc.). Sampling can vary from very simple to complex. Simple procedures involve obtaining one or more samples from standing or wading in from the shore. This might involve obtaining a sample near the point of contaminated influent and one at the opposite end of the body and/or near the point of discharge. More complex procedures include obtaining a number of peripheral samples, using transects or grid patterns to sample across the body of water and sampling at varying depths. Boats may be needed over deep waters.

VOAs should be filled so that there is a positive meniscus before capping. After capping, turn the VOA upside down and shake. There should be no air bubbles. Other containers should be filled to the level of the upper curvature, not totally full, leaving an "ullage" space for sample expansion.

Safety

These are usually low concentration, often off-site samples, and Level D is generally acceptable.

Groundwater

Groundwater includes any water beneath the land surface — saturated surface soil, shallow unconfined aquifers, and deeper confined aquifers.

Equipment

Containers: same as surface water. Samplers: Kemmerer bottle, bailer (top or bottom load), pumps. Other: M-scope or similar device, tape measure.

Procedure

Leachate seeps and springs: If there is sufficient flow, sample bottles can be filled directly; otherwise, a small pool can be dug and bottles filled from the pool.

Test Pits

Valuable information can be obtained quickly and inexpensively if the unconfined aquifer is close to the surface. Hand shoveling or a backhoe are used to excavate a pit which intersects the water table. The sample is obtained from pooled water, either by directly submersing the sample bottles if the hole is shallow (less than about 5 ft) or by lowering a bucket or using a dipper. An observation well or piezometer can be installed during the backfilling operation. The depth to groundwater and nature of sediments should be recorded. One must follow confined space entry procedures and safety precedures for trenching.

Residence Wells

Groundwater samples can be easily obtained from residence wells. Depths of home wells are very variable. Older homes usually have a dug well or shallow (30- to 50-ft depth) augured wells. Newer wells may be drilled to several hundred foot depths and penetrate bedrock. Homeowners often have little knowledge of the characteristics of their well. Information must be obtained from the driller.

If possible, samples should be obtained "upstream" of holding tanks, water treatment apparatus, and as much of the plumbing as possible. Some wells have a cock at the well head. If the well is regularly used, running the faucet for 5 to 10 min prior to sampling is sufficient. Unused wells will require a longer purging to avoid stagnant water.

Process Wells

Industrial and municipal water supply wells are useful, particularly for sampling of deep aquifers. Considerations are similar to those for sampling home wells. Most wells of this type have a sample cock at the well head. On occasions it is necessary to hand sample inactive industrial wells.

Monitoring Wells and Piezometers

Monitoring wells, observation wells, and piezometers are small-diameter (generally less than 4 in. I.D.) wells specifically installed for obtaining groundwater samples or measuring (piezometers) the depth to groundwater. Sampling consists of two phases:

1. Bailing the well.
2. Obtaining the sample.

Either phase can be performed either by hand methods or by using pumps.

Bailing

Prior to obtaining the sample, wells are bailed (purged) to remove standing or stagnant water. This is to ensure that the sample, obtained after the well has recharged (refilled), is representative of water conditions in the aquifer. It is preferable to remove 3 to 5 water volumes of the well prior to sampling. In wells with slow recharge rates or in dry weather this may not be possible; in removing one volume, the well is bailed dry and takes many hours to recharge.

The submerged casing volume of the well can be determined using the following formula:

$$V = \pi r^2 h$$

where

V	=	well volume
r	=	casing radius, inside of the well
h	=	height of water column. This is usually determined by subtracting the depth of the water surface as measured on site from the total depth of the well. The latter measurement is obtained from well logs or from on-site measurement.
π	=	3.14

Hand bailers or one of a variety of suction or submersible pumps are used to pull water from the well. The bailed water should be discarded away from and down gradient of the well. It is possible to overpump wells, inducing unrepresentative contamination into the well. If possible, bailers and pumps should not be inserted to the bottom of the well so as to avoid disturbing sediments.

Wells which are screened the full length of the water column may not require bailing. Groundwater flows through the well and does not stagnate.

If hand bailers are used, a separate length of rope should be used for each well. A wash tub or other container should be used to catch the rope during each retrieval, preventing the rope from becoming contaminated. If possible, pumps and bailers should be decontaminated between each well. (Decontamination can be done by steam pressure cleaning.)

Sampling

The sample is obtained from the recharged well. Ideally the well has recharged to the same water level as prior to bailing. Again this may not always be possible or it may be necessary to bail wells several days in advance of sampling.

Samples are usually obtained using hand methods — Kemmerer bottle or bailer — in order to avoid any contamination. The bailer is tied to a length of rope, lowered to depth in the well (avoid going to the bottom and disturbing sediments), and retrieved to fill sample bottles.

Procedure — Typical Well Sampling Operation

1. Approach well, in a more stringent level of protection, i.e., B or C, unless contaminants are known, monitor with explosimeter and vapor detector. Downgrade to an appropriate level of personnel protection (LLP).
2. Measure the casing "stick up" above ground, depth to water surface (use M-scope or similar device), if necessary well depth, and interior casing diameter. Complete submersed casing volume.
3. Bail well. Decon pump or bailer. Discard bailer rope.
4. Allow well to recharge. Record the time duration for recharge (time from completion of bailing to return to prebailing water depth). A second well can be bailed during the waiting period.
5. Sample well. Fill bottles. Discard bailer rope, decon bailer.
6. Close well.

Safety

Well heads and sample liquid should be monitored with an explosimeter and vapor detection device. Appropriate (Level B is not uncommon) level of protection should be chosen. With test pits, care must be taken to avoid cave-ins. Personnel should not descend into pits over 4 ft in depth.

Soils

Soil samples include surface and subsurface, deep unconsolidated sediments, samples from unconsolidated fill materials, and dredge spoils (unless considered as a bottom sediment(s) — see below). It is, on occasion, a judgment call as to whether a sample is considered to be a "stained soil" or a "waste." The difference is based on the observability of soil particles, ratio of soil to waste, and expected concentration.

Equipment

Containers: Two 8-oz jars (one for organic, one for inorganic) are standard for a full-priority pollutant analysis. Sampling: spoon, tongue depressor, trowel, shovel, auger, split-spoon, solid waste sampler.

Procedure

In its simplest form, soil sampling involves taking a surface scrape from an arbitrarily chosen location — one with observed staining, stressed vegetation, in a known spill location, backfilled lagoon, etc. More complex programs involve establishing transects or grids to locate sample points, use of random sampling methods, and sampling at various depths.

Chosen equipment is dependent upon the depth from which the sample is to be obtained, as well as the soil type and compaction. Wooden tongue depressors, which are cheap and disposable, are useful for transferring the sample from the sampling tool into the sample container.

Depending upon laboratory protocol, sample jars are usually filled between one half and completely full.

Surface Scrapes. Twigs, stones, vegetation, and other refuse are removed. Surface soil, from 0 to 2 in. in depth is transferred to the sample jar.

Shallow Depths. Depending upon soil type and compaction, soil samples can be obtained by hand methods from depths to about 4 ft. Unless soil conditions are ideal (soft, no stones, adequate cohesion to maintain an open hole), obtaining sample(s) from greater depths is impracticle by hand. Trowels, shovels, soil augers, and US-SCS bucket augers can be used to obtain the sample from depth.

Deep Samples. Hand-held power auger (post hole digger), truck-mounted augers, well drilling equipment, and split spoon samplers are used to sample deeper soil. Representative samples from each sediment layer are commonly obtained during well installations. With the exception of hand-held augers, deep soil sampling requires large-scale, specialized equipment, trained personnel, and is best undertaken by a qualified contractor.

Safety

Hazards include contact with contaminated soil or buried waste, rupturing buried drums, inhalation of vapors, electrical power lines, and underground gas piping. Use of protective gloves, boots, and coveralls is generally required. Respiratory protection may be necessary to protect against particulates, ambient vapor, and sudden releases of high concentration vapor. Care is needed when augering or drilling in areas suspected of having buried drums.

Sediments and Sludges

Although the terms are somewhat interchangeable, sediments will be considered to be solid and semisolid (ooze) materials from the bottoms of surface waters. Sludges are viscous to semisolid materials commonly found on the bottom of lagoons, tanks, storage ponds, and drums. Sampling methods are similar for sludges and sediments, but the former usually requires a much higher level of safety. Sampling of sludges from lagoons and drums is covered in following sections.

Equipment

Containers: Two 8-oz jars (one inorganic and one organic) are standard for a full-priority pollutant analysis. Samplers: sediment sampler (clam shell), corers, dippers, buckets, shovels, and triers.

Procedures

Flowing Surface Waters (Streams, Rivers). Samples may be obtained upstream, downstream, and at point discharges. Theoretically, samples should be obtained from the middle of the channel. This is often unproductive in high-energy streams, as the bottom is swept clear of all but the coarsest sediments. In this instance, samples can be obtained from quiet pools and eddies where finer materials settle out. Long-handled dippers are useful for obtaining the sample. Large rivers may require several samples along a transect across the river. Samples can be obtained from a boat or a bridge using a sediment sampler or long-handled dipper.

Stationary Waters (Ponds, Lakes). Samples can be obtained peripherally, along a transect, or at grid points. Boats, long-handled dippers, and/or sediment sampler may be necessary. The easiest method on small ponds is to stretch a rope across the pond, along the desired transect. Personnel in the boat can then pull the boat along, using the rope, and use the rope to station the boat over the desired sample point. Personnel on shore can also be used to pull the boat along.

Safety

Sampling of sediments on the bottoms of surface waters can usually be conducted at Level D personnel protection, pending background information and air monitoring. Hazards are primarily contact with contaminated sediments which can be avoided by wearing gloves, boots, and coveralls. Boats require special precautions. Life jackets should be available. Supervisors of the project should take the U.S. Coast Guard's Course on Boating Safety when overseeing projects involving use of boats.

REORD KEEPING AND CHAIN OF CUSTODY

Air Monitoring

Site cleanup activities, leaking or open containers, and air-blown dusts are all sources of fugitive emissions at hazardous waste site. These emissions adversely affect the health and safety of workers in the area or dictate the levels of protection needed by site workers, the boundaries of site control zones, and the extent of air monitoring. An effective ambient air monitoring strategy, tailored to a particular site, generates the data needed to adequately evaluate the precautions necessary regarding control of personnel exposures to hazardous materials. These data can also be useful when determining the extent of any remedial actions that may be necessary.

Objectives

The objectives of an area monitoring program during site work activities are threefold:

1. To indicate the quantity of airborne contaminants at the site and, thus, the levels of protection needed to control personnel exposures.
2. To delineate the zones in which various levels of protection are required.
3. To generate information over the lifetime of the site work activities that may provide an indication of necessary changes in protective measures for on-site personnel or for affected populations downwind from the site.

With these objectives in mind, the monitoring plan development should be of prime importance during initial site survey activities. During this phase, a team using hand-held instruments and making visual observations gains a preliminary indication of the presence of ionizable gases or vapors as well as the location of "hot spots," which may require a higher level of protection than other areas. Based upon this information, site control boundaries are marked, establishing an exclusion zone (area of potential chemical exposure), a support zone (clean area where the command post and support equipment will be located), and a contamination reduction zone (buffer area between exclusion zone and support zone).

Since the ambient air monitoring network will be used to supply information regarding the appropriateness of the specified site control boundaries, location of sampling stations would give the greatest resolution of air contaminant transport. However, the complexity of area monitoring must be kept in perspective according to its role within the overall cleanup operation. Its role is to act as a verification

Exhibit 11.2. Records

Adequate records should be maintained on each sample that is taken. Record information should include

1. Sample description — type (groundwater, surface water) and volume.
2. Sample source — well number and location.
3. Sampler's identity — chain of evidence should be maintained; each time transfer of a sample occurs, a record including signatures of parties involved in transfer should be made (this procedure can have legal significance).
4. Time and date of sampling.
5. Significant weather conditions.
6. Sample laboratory number.
7. Pertinent well data — depth, depth to water surface, pumping schedule, and method.
8. Sampling method — vacuum, bailer, pressure.
9. Preservations (if any) — type and number (e.g., NaOH for cyanide, H_3PO and $CuSO_4$ for phenols, etc.).
10. Sample containers — type, size, and number (e.g., 3-L glass stoppered bottles, 1-gal screw-cap bottle, etc.).
11. Reason for sampling — initial sampling of new landfill, annual sampling, quarterly sampling, special problem sampling in conjunction with contaminant discovered in nearby domestic well, etc.
12. Appearance of sample — color, turbidity, sediment, oil on surface, etc.
13. Any other information which appears to be significant (e.g., sampled in conjunction with state, county, local regulatory authorities; samples for specific conductance value only; sampled for key indicator analysis; sampled for extended analysis; resampled following engineering corrective action, etc.).
14. Name and location of laboratory performing analysis.
15. Sample temperature upon sampling.
16. Thermal preservative (e.g., transportation in ice chest).
17. Analytical determinations (if any) performed in the field at the time of sampling and results obtained (e.g., pH , temperature, dissolved oxygen, and specific conductance, etc.).
18. Analyst's identity and affiliation.

Sample Containers

For most samples and analytical parameters, either glass or plastic containers are satisfactory. Some exceptions are encountered such as the use of plastic for silica determinations and glass for phenols or oil and grease determination. Containers should be kept full until samples are analyzed to maintain an aerobic condition.

As a general guide in choosing a sample container, the ideal material of construction should be nonreactive with the sample and especially the particular analytical parameter to be tested.

Cleanliness of containers is of utmost importance. An effective procedure for cleaning containers is to wash with detergent, sequentially followed by tap water rinse, nitric acid rinse, tap water rinse, hydrochloric acid rinse, tap water rinse, and finally a rinse with deionized or distilled water. In addition, the containers should be rinsed at least once with the sample at the time of sampling.

---

tool in conjunction with site surveys with hand-held instruments and as a source of additional data not generated by these instruments. As such, a limited array of properly positioned samplers could provide adequate data while minimizing the labor and expense associated with a complex area monitoring program. To achieve this, the following sampler array is proposed.

Exhibit 11.3. Recommendation for Sampling and Preservation of Samples According to Measurement[a]

Measurement	Vol. Req. (ml)	Container[b]	Preservative	Holding Time[c]
		100 Physical Properties		
Color	50	P,G	Cool, 4°C	24 hr
Conductance	100	P,G	Cool, 4°C	24 hr[d]
Hardness	100	P,G	Cool, 4°C HNO_3 to pH 2	6 months[e]
Odor	200	G only	Cool, 4°C	24 hr
pH	25	P,G	Det. on site	6 hr
Residue				
Filterable	100	P,G	Cool, 4°C	7 days
Nonfilterable	100	P,G	Cool, 4°C	7 days
Total	100	P,G	Cool, 4°C	7 days
Volatile	100	P,G	Cool, 4°C	7 days
Settleable matter	1000	P,G	Non Req.	24 hr
Temperature	1000	P,G	Det. on site	No holding
Turbidity	100	P,G	Cool, 4°C	7 days
		200 Metals		
Dissolved	200	P,G	Filter on site HNO_3 to pH 2	6 months[e]
Suspended	200		Filter on site	6 months
Total	100	P,G	HNO_3 to pH 2	6 months[e]
Mercury, dissolved	100	P,G	Filter on site HNO_3 to pH 2	38 days (glass) 13 days (hard plastic)
Total	100	P,G	HNO_3 to pH 2	38 days (glass) 13 days (hard plastic)
		300 Inorganics, Nonmetallics		
Acidity	100	P,G	Non req.	24 hr
Alkalinity	100	P,G	Cool, 4°C	24 hr
Bromide	100	P,G	Cool, 4°C	24 hr
Chloride	50	P,G	None req.	7 days
Chlorine	200	P,G	Det. on site	No holding
Cyanides	500	P,G	Cool, 4°C NaOH to pH 12	25 hrs
Fluoride	300	P,G	None req.	7 days
Iodine	100	P,G	Cool, 4°C	24 hr
Nitrogen				
Ammonia	400	P,G	Cool, 4°C H_2SO_4 to pH 2	24 hr
Kjeldahl, total	500	P,G	Cool, 4°C H_2SO_4 to pH 2	24 hr[f]
Nitrate plus nitrite	100	P,G	Cool, 4°C H_2SO_4 to pH 2	24 hr[f]
Nitrate	100	P,G	Cool, 4°C	24 hr
Nitrite	50	P,G	Cool, 4°C	48 hr

Exhibit 11.3. (Continued)

Measurement	Vol. Req. (ml)	Container[b]	Preservative	Holding Time[c]
		100 Physical Properties		
		300 Inorganics, Nonmetallics		
Dissolved oxygen probe	300	G only	Det. on site	No holding
Winkler	300	G only	Fix on site	4–8 hr
Phosphorus				
Orthophosphate	50	P,G	Filter on site	24 hr[f]
Dissolved	50	P,G	Cool, 4°C	
Hydrolyzable	50	P,G	Cool, 4°C H_2SO_4 to pH 2	24 hr[f]
Total	50	P,G	Cool, 4°C H_2SO_4 to pH 2	24 hr[f]
Silica	50	P only	Cool, 4°C	7 days
Sulfate	50	P,G	Cool, 4°C	7 days
Sulfide	500	P,G	2 ml zinc acetate	24 hr
Sulfite	50	P,G	Det. on site	No holding
		400 Organics		
BOD	1000	P,G	Cool, 4°C	24 hr
COD	50	P,G	H_2SO_4 to pH 2	7 days[f]
Oil & grease	1000	G only	Cool, 4°C H_2SO_4 or HCl to pH 2	24 hr
Organic carbon	25	P,G	Cool, 40°C H_2SO_4 or HCL to pH 2	24 hr
Phenolics	500	G only	Cool, 4°C H_3PO_4 to pH 4 1.0 g $CuSO_4$/L	24 hr
MSAS	250	P,G	Cool, 4°C	24 hr
NTA	50	P,G	Cool, 4°C	24 hr

[a] More specific instructions for preservation and sampling are found with each procedure as detailed in this manual. A general discussion on sampling water and industrial wastewater may be found in ASTM, part 31, pp. 72–78 (1976) Method D-3370.

[b] Plastic (P) or Glass (G). For metals, polyethylene with a polypropylene cap (no liner) is preferred.

[c] It should be pointed out that holding times listed above are recommended for properly preserved samples based on currently available data. It is recognized that for some sample types, extension of these times may be possible, while for other types, these times

Background Station

Because many hazardous waste sites are located near industries which may be generating emissions impacting the work area, background samples must be taken upwind of the waste site to establish a baseline quantity of air contaminants not generated by the wastes themselves.

Support Zone Station

Monitoring should be conducted near the command post or other support facilities to ensure that they are in fact located in a "clean" area and that the area remains "clean" throughout the operation period.

Contamination Reduction Zone Station

Operations within this zone center around the decontamination corridor. Monitoring should be performed along the decontamination line to ensure that decontamination workers are properly protected and that on-site workers are not removing their protective gear in a contaminated area.

Monitoring Plan Implementation

Hazardous waste sites may contain hundreds of potentially dangerous materials which could become airborne during cleanup operations. However, most of these materials can be classified according to three general categories:

1. Organic vapors
2. Inorganic gases
3. Aerosols

Organic vapors will most likely constitute the greatest proportion of air contaminants at hazardous waste sites and should receive the most attention. Activated carbon has long been known as an excellent sorbent media for many organic vapors. However, solid sorbents, other than carbon, are routinely employed to sample specific organic compounds that do not absorb or desorb well with activated carbon systems. A wide variety of solid sorbent media is available and new sorbents are continuously being introduced onto the market, making it possible to use what may be considered the best sorbent for each contaminant anticipated at a waste site. Therefore, a small number of sorbents, chosen for collection of the widest range of materials or for specifically identified hazardous substances which may be generated from a particular waste site will have to suffice for area monitoring purposes. An air sampling pump (such as ones used by industrial hygienists) equipped with a suitable manifold is capable of simultaneously collecting a number of tubes or cassettes. An activated carbon tube is used to collect vapors of materials with a boiling point above 0°C; these include most odorous organic and organic substances, such as solvent vapors. A tube containing a porous polymer such as Tenax could be used to collect substances that adsorb poorly onto activated carbon, such as high-molecular-weight hydrocarbons, organophosphorus compounds, and the vapors of certain pesticides. Tenax also adsorbs organic materials more efficiently than carbon at low temperatures. Organic vapors that exhibit a relatively high dipole moment such as the aromatic amines could be collected on a suitable polar sorbent such as silica gel. The fourth position on the sampling manifold could hold a tube packed with another specialty sorbent selected for the specific site. For example, a Florisil tube could be used if PCBs are expected at a particular location. Long-term colorimetric detector tubes could also be used here when monitoring for specified compounds.

Exhibit 11.4. User's Guide to the EPA

Contract Laboratory Program

Part 1

Introduction

I. EPA Contract Laboratory Program

The EPA Contract Laboratory Program (CLP) is an example of one type of sample documentation, analysis, and quality assurance program for analysis of samples. It is designed to provide chemical analysis through a national contract laboratory network. These services are prefunded under Superfund as support for the cleanup and investigation of targeted hazardous waste sites. Laboratories in the program operate under fixed-price contracts and in accordance with specified methodologies.

The analytical methods and protocols used are in accordance with EPA Quality Assurance procedures. The performance of contract laboratories is continually monitored through quality assurance evaluations conducted by the Environmental Monitoring & System Laboratory/Las Vegas (EMSL/LV). EMSL/LV also supports the Program by developing and/or approving all methods, standards, and protocols used by contract laboratories. In order to ensure that analytical data are satiable for use in enforcement and litigation, the National Enforcement Investigation Center (NEIC) conducts quarterly evaluations of laboratory sample control, documentation, security of evidence, and document control and inventory. The Contract Laboratory Program thus provides a wide range of analytical services of known quality on a national scale. The CLP ensures that all samples are analyzed according to a uniform and consistent protocol and reduces analytical costs through the economics of scale.

II. CLP Sample Classification System

Of particular interest is the CLP Sample Classification System. A summary of sample categories is found in Exhibits 11.5A through D. The classification of samples as low, medium, or high concentration is for defining proper sample volume, sample container, and analytical procedure. In addition, this classification determined the proper packaging and shipping requirements. Under CLP guidelines, a low-concentration sample is packaged and shipped as an "environmental" sample, and a medium- or high-concentration sample is packaged and shipped as a "hazardous" sample.

Inorganic gases that may be present at hazardous waste sites would primarily be polar compounds, such as the halacid gases, which could be adsorbed on the silica gel tube described earlier and analyzed by ion chromatography.

Aerosols are solid or liquid particulates which have been dispersed by wind or mechanical entrainment. At a hazardous waste site, the aerosols that may be encountered include contaminated and noncontaminated soil particles, heavy metal particulates, pesticide dusts, and organic or inorganic liquid droplets. The most effective method for monitoring these materials is by filtration using a separate sampling pump.

Dust Acquisition

With the proposed monitoring network capable of sampling a wide range of organic and inorganic compounds, one major consideration remains; exactly how should the samples be collected and what should they be analyzed for? These are interrelated problems and will have to be solved, to a great extent, on a site-specific basis. However, some general guidance and recommendations as to methodology in solving the problem is appropriate here.

In general, it will be desirable to determine some relationship between actual contaminant concentrations at the site and published worker exposure guidelines. It is therefore recommended that all samples be taken over an 8-hr period whenever possible to allow direct comparison to 8-hr time-weighted average threshold limit values.

Aerosols

Samples of toxic and nontoxic particulates should be taken at a flow rate of 2.0 L/min using a standard industrial hygiene pump and filter assembly with 37-mm cellulose ester fabric filters. For total particulate collection, a filter having a 0.8-mm pore size is recommended. The sample can be analyzed gravimetrically for total particulate loading and then analyzed destructively or nondestructively for metals, inorganics, and organics. The filter is sectioned, optical particle sizing performed, and the appropriate analysis conducted.

Sorbent Samples

The choice of sorbent material, sorbent mass, and sample volume will vary according to the types and concentrations of substances anticipated at a particular waste site. However, 100-mg sorbent tubes with a 50-mg backup section, collecting a sample of 5 to 10 L, should serve as an appropriate starting point.

Polar sorbent material such as silica gel will collect polar substances which are not adsorbed well onto activated carbon and some of the porous polymers. The silica gel sample can be split and analyzed for the inorganic haloacid gases and aromatic amines.

Activated carbon and porous polymers will collect a wide range of organic compounds. Exhaustive analysis to identify and quantify all the collected species is prohibitively expensive and technically infeasible for the field laboratory. On the other hand, a "total organic" analysis is almost completely meaningless when determining personnel protective measures. Therefore, it is recommended to analyze the samples for site-specific principle hazardous constituents (PHCs) based upon the concentration and relative toxicity of the various airborne materials found at a particular waste site.

The selection of PHCs should be based upon the types of materials anticipated at a given waste site, from generator's records, and from initial site survey information. A valuable tool to aid in the selection of PHCs would be an activated carbon sample, taken during the initial site survey activities, and exhaustively analyzed of off-site by GC/MS to identify the major peaks within selection categories. This one exhaustive analysis, along with what is already known about a particular site, could provide enough information for the selection of PHCs.

Standards of these compounds could then be prepared for field analysis of the continuing area monitoring network samples. Special adsorbents and sampling conditions can be used for specific PHCs if desired, while continued multimedia sampling will provide a base for analysis of additional PHCs that may be identified during the course of cleanup operations.

ORBO-80 is a special sorbent material developed by Surpelco, Inc. of Bellefonte, Pennsylvania. It is unique because it is among the first type of media developed for 8-HR air monitoring of diisocyanate compounds. Analysis for diisocyanates and isocyanate compounds is performed on high performance liquid chromatography (HPLC). See Appendix for Surpelco Inc.'s Field Air Monitoring Data Sheet.

Biological Sampling

Biological sampling is regularly performed to obtain information on surface water quality. There are three basic methods — population survey, tissue analysis, bioassay.

Population Survey

Invertebrate stream surveys are routinely undertaken to determine water quality, particularly below sewage outfalls. A kick net or other method is used to collect a survey sample of invertebrates (worms, larvae, nymphs) at selected locations along the stream. The number of each different species in the sample is recorded. Certain species are more tolerant of different pollutants or levels of pollution. Presence, absence, or shifts in population size are indicative of water quality.

Tissue Analysis

This is most commonly performed with fish, benthic organisms (worms, mollusks), and occasionally with some plants. It is most useful for metals and heavy organics such as PCBs and pesticides. The organism is caught and the tissue analyzed. Tissue concentrations are related to pollutant concentrations — over a given time span, generally the life of the organism. Benthic organisms are stationary, reflecting conditions in a given area. Fish are mobile, providing a wider ranging, more general indication of conditions. Little work has been done with plants, although it appears that some rooted aquatics, such as cattails, absorb heavy metals.

Bioassay

A test population of small fish, such as minnows, is submerged in the test environment for a given time period. Mortality rates are indicative of water quality and suitability for the test species.

Each of these methods require specialized equipment and expertise. Information can be obtained from local or state Fish and Game Department, Water Quality Departments, or biological laboratories.

Drum Sampling

Characterization of wastes on-site often includes sampling of drums. This is generally regarded as the most dangerous task in hazardous waste assessment and should be undertaken in a controlled, organized fashion. There is no substitute for experience.

Drums are constructed of metal (steel and occasionally stainless steel) or fiberboard and may have plastic liners. They are either "ring-top" — the entire top is removable and is held in place by a steel ring — or unified construction in which the only access is through the bung or small vent hole. Ring-tops may also have bungs.

The great danger in drum handling and sampling is that the drum may rupture, resulting in spillage, explosion, or a sudden release of vapor or pressurized liquid.

Usually, during investigative actions, it is not necessary to sample all drums on site. A few are selected for representative sampling. The majority will be sampled by the cleanup contractor. Avoid unusual, bulging (contents under pressure), heavily rusted, or structurally weak drums. Where possible, sample open drums — those with open or loose bungs. Obtaining a sample from spillage or leakage is a good alternative.

Equipment

Containers: Two 8-oz glass jars for each sample. Sampling: hollow glass tubes (11 mm I.D.), COLIWASA, triers, nonsparking (NS) bung wrench, large adjustable wrenches, bolt cutters, large screw driver, deheader, remote drum opener, backhoe with spike.

Procedure

1. Obtain as much background information as possible on the possible contents of the drums. Check the DOT drum classification numbers, markings on drums, condition of drum, and facility records.
2. Choose drums for sampling. Stencil and I.D. number on the drum.
3. Decide whether or not to stage or segregate drums for sampling. On some sites the drums will be closely packed or jumbled and may have to be moved. On other sites, drums may already be segregated and little or no staging will be necessary.

   Advantage of staging drums:
   a. Provides a larger, safer work space.
   b. Drum is less likely to react with other drums.
   c. Provides access to the drum.
   d. Drum can be placed upright on a level, firm surface.

   Disadvantage of staging drums:
   a. Additional movement of drum provides more chance for something to go wrong.
   b. Drums may spill, rupture, or explode.
   c. Can be difficult and strenuous.
   d. May require special equipment.
   e. Is time consuming.

4. Only one drum should be opened at a time if there is a chance of a reaction between drums. Fit the bung wrench to the bung, lean away from the drum and slowly crack open the bung. Leave immediately if the drum fumes, burps liquids, or moves. If there is no reaction continue opening the bung. Ring-tops can be sampled through the bung or by removing the ring top. Use a wrench to loosen the tightening bolts or bolt cutter if rusted and pry off the ring with a screw driver. Remote openers can also be used to spin out the bung or pierce a hole in the drum. A backhoe with a spike on the bucket is handy and is commonly used during cleanup operations when large numbers of drums must be opened for sampling prior to disposal.
5. Monitor the drum for explosive vapor.
6. Liquid and semiliquid sludges are easily sampled using a hollow glass tube (3 to 4 ft in length, 11 mm I.D.) and "soda straw" technique. Carefully insert the tube into the drum to the bottom, place thumb over top and remove. This method retrieves a sample column. The sample can be released into one jar or with judicious thumbing, individual layers can be released into each of several jars.
7. Reseal the drum.
8. Check the sample for radioactivity. If the meter reads greater than 10 mr/hr, leave the sample drum and evacuate the area.

Safety

1. Full Level B is usually mandatory for drum sampling. Level A may occasionally be needed, but is usually avoided due to the cumbersomeness of the apparatus.
2. Consider the surrounding environment and demographics. Downwind populations may be effected by a vapor release or explosion.
3. All care must be used to avoid shocking or sparking of the drums.
4. Drum sampling is often very messy and sorbent and other cleanup and decon materials should be handy.
5. A fire extinguisher and water supply (buckets, sprayers, "indian pumps") should be at the hotline or readily accessible.

Other Closed Containers

Carboys

Glass or plastic carboys and plastic 55-gal drums are occasionally encountered. These are usually used to contain strong acids or bases and other chemicals which would react with steel drums. The can usually can be opened by unscrewing the cap. Methods and safety are the same as for the sampling of drums.

Labpacks

Labpacks are drums, usually fiberboard but also of steel, and other similar containers used to receive packets of laboratory wastes. Wastes include bottles, jars, and vials of chemicals (explosion potential), poisons, radioactives, refuse from pathological and tissue culture studies, and hospital wastes. Wastes are supposed to be placed in color-coded plastic bags and sealed before placing in the drum, as required by federal regulation RCRA if a biowaste.

Labpacks are extremely dangerous. Any drum containing miscellaneous small chemical bottles, vials, discarded petri dishes, or small colored plastic bags should be noted as a labpack and not tampered with.

Tanks

250-gal fuel tanks, above- and below-ground storage tanks up to 30,000 gal capacity, and larger railroad tank cars and truck tankers waste is stored in. Tanks can be viewed as large drums and may require a variety of sampling procedures. Some tanks have bottom valves for which a sample can be withdrawn. Problems are that bottom sludges often plug the valve or, in a tank with stratified contents, only the bottom layer is sampled. Tanks are usually sampled from the top.

Procedure

1. As with drums, obtain as much background information as possible on tank contents. Choose and identify tanks to be sampled.
2. Sampling with the bottom valve: Place the sample container under the tap and slowly open the valve. Use a grounding strap if the sample container is metal.
3. Sampling from the top: Using nonsparking wrenches, slowly open the top hatch. Do not allow the hatch cover to bang against the tank or slam shut. Monitor the tank for explosive vapors.
4. With small tanks a long, hollow glass rod can be used to pull a sample column; or a long handled dipper can be used to sample bottom sludges. Larger tanks will require the use of a sediment sampler, bomb sampler, kemmer, or weighted bottle to remove samples. If the tank contents are stratified, samples should be obtained from each layer and placed in separate containers.

Safety

1. Level B personnel protection is mandatory for tank sampling.
2. This can be messy work and sorbent and decon materials should be handy.
3. Care should be taken when working on tank tops. Footing can be slippery and protective equipment cumbersome.
4. Check the integrity of the tank, also of the footings, stations, supports, and ladders.
5. Explosions are a possibility. Continual monitoring with an explosimeter is necessary. Disturbances of the contents of the tank during sampling may cause a vapor buildup.

Lagoons

Sampling of lagoons and holding ponds can be complicated; obtaining samples may be difficult, and it is occasionally necessary to obtain samples from more than one liquid level as well as bottom sludges.

Equipment

Containers: Two 8-oz glass jars for each sample (one organic, one inorganic). If liquids are suspected of being low concentration, then full glassware (two 1/2-gal amber jugs, two 40-ml VOAs, one 1-qt plastic bottle) may be necessary. Samplers: dipper, shovel, sediment sampler, trier, boat, crane.

Procedure

Sampling is a combination of the procedure used in stationary surface water sampling and sediment sampling. Liquid and sludge samples can be obtained peripherally, along transects, or using a grid pattern. If there is more than one layer of liquids (i.e., oil over water), samples should be obtained at respective depths. If the lagoon is not too deep, a hollow glass rod can be used to obtain a vertical column for observation (see Chapter 9, Drum Sampling) or sampling.

Large lagoons, particularly those which have been used for indiscriminate disposal, may not be homogeneous, either vertically or horizontally, and will require a more complex sampling plan in order to assure representative samples.

Samples can be obtained by reaching from the bank with a long-handled dipper or hollow glass rod, from a boat, or by using a crane to suspend a weighted sample bottle or sediment sample. Small lagoons can be sampled by stretching a tape for use as a trolley line; a sampler is pulled out to the desired location and lowered to the desired depth.

Safety

Lagoon sampling requires a high level of safety. Level C or high personnel protection is usually mandatory. Disturbing the lagoon may release vapors. Lagoon banks are steep, and a safety line is used to steady sampling personnel. Boats are particularly dangerous and are not highly recommended. Personnel must wear life preservers.

Solids and Waste Piles

Solid materials, including dried and solidified sludges, can be surface dumped or landfilled. Sampling procedures are similar to those used for soil.

Equipment

Containers: Two 8-oz glass jars (one organic, one inorganic) are standard. Jars are filled one half to full, depending upon laboratory protocol. Sampling: wooden tongue depressors, spoons, scoops, trowels, shovels, soil auger, grain sampler, trier, waste pile sampler, backhoe.

Procedure

Sampling procedures will vary according to the size and composition of the waste deposit pile, landfill, sludge field.

A sample can be grabbed from each different waste type observed. A composite sample can be obtained by combining small amounts from several waste types.

Large, nonhomogeneous deposits may require use of a grid pattern to obtain representative samples.

Waste deposits may also be layered, requiring sampling at several depths. Triers and grain and waste pile samplers can be used to obtain cross-sectional samples from the deposits. A soil auger can be used on thicker deposits. In some situations, it may be necessary to dig trenches and test pits using shovels or a backhoe.

Safety

Air monitoring will help determine the appropriate level of respiratory protection. Concentrated powdery or very dry wastes may necessitate protection to prevent inhalation of particulates. Protective clothing is necessary to prevent contact contamination. Test pits and trenches over 4 ft in depth require special OSHA safety considerations.

SAMPLING EQUIPMENT AND METHODS

Introduction

A variety of equipment and techniques are used for sampling hazardous materials. Correct use of liquid and solid material samplers, selecting sample containers and closures, and preserving samples are important to obtaining analytical data that give meaningful results.

The various samplers currently available can be applied in certain situations. It is important, however, to be familiar with limitations associated with their use (see Exhibit 11.5a, b, and c).

Liquids

Open-Tube/Drum Sampler

An open glass tube is the most versatile of hazardous liquid samplers. This is true because of its wide range of applications, relatively low cost, and ease of operation.

Description. The open-tube sampler is made of hollow glass (or plastic tubing), usually 4 ft long. The inside diameter (I.D.) is generally 1/4 to 1/2 in. depending on viscosity of sample. Open-tubes made specifically for sampling (called drum samplers) can be purchased. The only difference is that drum samplers have a constricted orifice at the top to facilitate plugging the opening to maintain a vacuum in the tube.

Procedures for Use

1. Slowly lower tube into the liquid to the desired depth. It may be desirable not to mix the contents of the container so that a more representative sample can be collected.

Exhibit 11.5A. Summary of EPA CLP Categories for Analysis of Water for Organics

Category	Concentration (Any Single Contaminant)	Sample Container	Analysis
Low concentration (environmental)	<10 ppm	Extractables 1-gal glass jar volatile organic analysis (VOA) 40-ml glass VOA vial	Initial GC screen GC/MS identification
Medium concentration (hazardous)	10–150,000 ppm (10 ppm–15%)	Low range 1- to 2-L glass jar High range 8-oz glass jar	Initial CG screen CG/MS identification
High concentration (hazardous)	15–100%	8-oz glass jar	pH Acidity/alkalinity Conductivity Cyanide Sulfide Oxidants CG screen GC/MS identification

Exhibit 11.5B. Summary of EPA CLP Categories for Analysis of Soil/Sediments for Organics

Category	Concentration (Any Single Contaminant)	Sample Container	Analysis
Low concentration (environmental)	<10 ppm	8-oz glass vial	Initial GC screen GC/MS identification
Medium concentration (hazardous)	10–150,000 ppm (10 ppm–15%)	4-oz glass vial	Initial CG screen CG/MS identification
High concentration (hazardous)	15–100%	8-oz glass jar	ph Acidity/alkalinity Conductivity Cyanide Sulfide CG screen GC/MS identification

2. At desired depth, stopper tube with a rubber stopper or thumb or crease a piece of flexible tubing attached to the open end of the sample tube. Stoppering establishes a vacuum in the tube, enabling withdrawal of the sample from the container.

Exhibit 11.5C. Summary of EPA CLP Categories for Analysis of Water for Inorganics

Category	Concentration (Any Single Contaminant)	Sample Container	Analysis
Low concentration (environmental)	<10 ppm	1-L polyethylene bottle	Elements Priority pollutants Specified inorganics
Medium concentration (hazardous)	10–150,000 ppm (10 ppm–15%)	8-oz glass vial	Elements Priority pollutants Specified inorganics
High concentration (hazardous)	15–100%	8-oz glass vial	pH Acidity/alkalinity Conductivity Cyanide Sulfide Oxidants Additional requirements to be determined

Exhibit 11.5D. Summary of EPA CLP Categories for Analysis of Soil/Sediments for Inorganics

Category	Concentration (Any Single Contaminant)	Sample Container	Analysis
Low concentration (environmental)	<10 ppm	4-oz glass vial	Elements Priority pollutants Specified inorganics
Medium concentration (hazardous)	10–150,000 ppm (10 ppm–15%)	4-oz glass vial	Elements Priority pollutants Specified inorganics
High concentration (hazardous)	15–100%	8-oz glass vial	pH Acidity/alkalinity Conductivity Cyanide Sulfide Oxidants Additional requirements to be determined

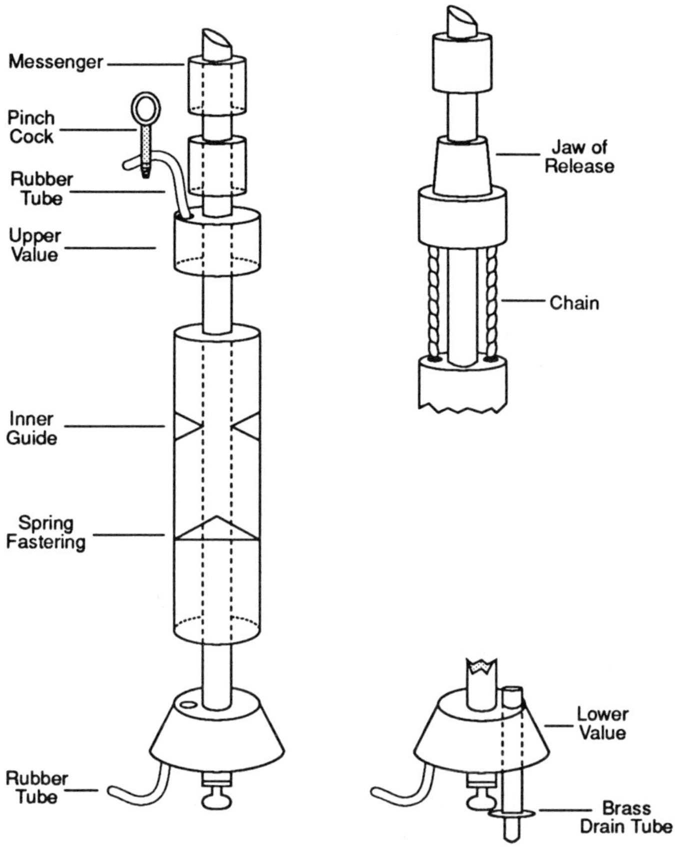

Figure 11.3. Illustration of a Kemmerer bottle.

3. Slowly withdraw tube from vessel and expel sample into appropriate sample container. Avoid handling the portion of sampler that was in the liquid.
4. Dispose of tube on site by placing it back into the container that was sampled.

Discussion. The Kemmerer bottle is a messenger-activated water sampling device (see Figure 11.3). In the open position water flows easily through the device. Once lowered to the desired depth, a messenger is dropped down the sample line, tripping the release mechanism and closing the bottle. In the closed position the bottle is sealed, both on top and bottom, from any additional contact with the water column and can be retrieved.

Most commercially available Kemmerer bottles are of brass or plastic construction. Modification of existing systems with nonreactive materials such as Teflon, glass, or stainless steel would be only partially successful due to the complicated machining necessary for the release mechanism. Other modifications such as a stoppered bottom drain are simpler and useful in minimizing sample disturbance during transfer to the appropriate containers.

Uses. The Kemmerer bottle is currently the most practical method of collecting discrete, at-depth samples from surface waters or vessels where the collection depth exceeds the lift capacity of pumps. The application is limited, however, with the incompatibility of various construction materials with some analytical techniques. Proper selection, i.e., all metal assemblies for organic analysis or all plastic assemblies for trace element analysis, will overcome this deficiency.

Procedures for Use.

1. Inspect Kemmerer bottle for thorough cleaning and insure that sample drain valve is closed (if bottle is so equipped).
2. Measure and then mark sample line at desired sampling depth.
3. Open bottle by lifting top stopper-trip head assembly.
4. Gradually lower bottle until desired level is reached (predesignated mark from Step 2).
5. Place messenger on sample line and release.
6. Retrieve sampler; hold sampler by center stem to prevent accidental opening of bottom stopper.
7. Rinse or wipe off exterior of sample body (wear proper gloves and protective clothing).
8. Recover sample by grasping lower stopper and sampler body with one hand (gloved), and transfer sample by either (a) lifting top stopper with other hand and carefully pouring contents into sample bottles, or (b) holding drain valve (if present) over sample bottle and opening valve.
9. Allow sample to flow slowly down side of sample bottle with minimal disturbance.
10. Preserve the sample if necessary as per guidelines.
11. Check that a Teflon liner is present in the cap if required. Secure the cap tightly.
12. Label the sample bottle with an appropriate tag. Be sure to complete the tag with all necessary information. Record the information in the field logbook and complete all chain-of-custody records.
13. Place the properly labeled sample bottle in an appropriate carrying container maintained at 4°C throughout the sampling and transportation period.
14. Decontaminate sampler and messenger or place in plastic bag for return to lab.

Discussion. The weighted bottle sampler consists of a glass bottle, a weight sinker, a bottle stopper, and a line for opening the bottle and lowering and raising the sampler during sampling (Figure 11.4). There are variations of this sampler, as illustrated in the ASTM methods D 270 and E 300. This sampler can be either fabricated or purchased commercially.

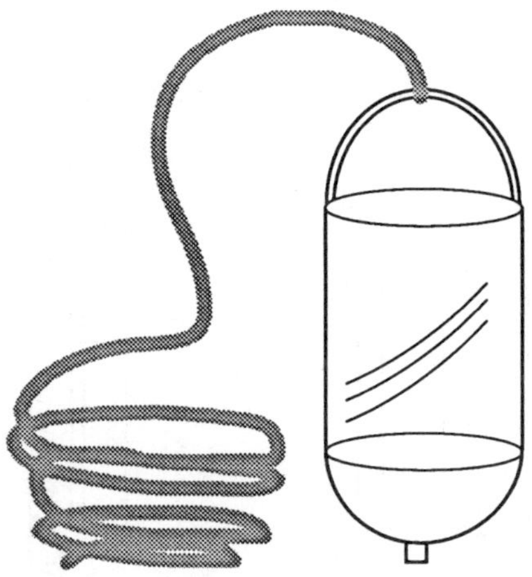

Figure 11.4. Illustration of weighted bottle sampler.

Uses. Weighted bottle samplers are used to sample liquids at depth. These samplers are difficult to use in very viscous liquids. In addition, the outside of the bottle is exposed to the waste. This is undesirable if the bottle is used as the sample container. Alternatives to the weighted bottle sampler are the Kemmerer and the extended bottle sampler.

Procedures for Use.

1. Gently lower the sampler into the liquid to the desired depth so as not to remove the stopper prematurely.
2. Pull out the stopper with a sharp jerk of the sampler line.
3. Allow the bottle to fill completely as evidenced by the cessation of air bottles.
4. Raise the sampler and cap the bottle.
5. The bottle can be used as the sample container, but it must be thoroughly decontaminated.

Collection of Water Samples from Depth with an Extended Bottle Sampler

Discussion. The extended bottle sampler consists of a 1.8-m (6-ft)-long aluminum tube (Figure 11.5A through D). A stainless steel clamp is attached to the end of the tube and can be adjusted to hold a sample jar of desired size. The sample jar cap can be remotely removed and replaced while the bottle is submerged by turning a 1.8-m (6-ft) handle grip rod, which attaches to the cap by means of a screw clamp or a suction cup.

Uses. The extended bottle sampler is a grab sampler designed to sample subsurface liquids to a maximum depth of 1.5 m. Since the outside of the sample bottle is exposed to the waste, it must be decontaminated prior to shipment or reuse.

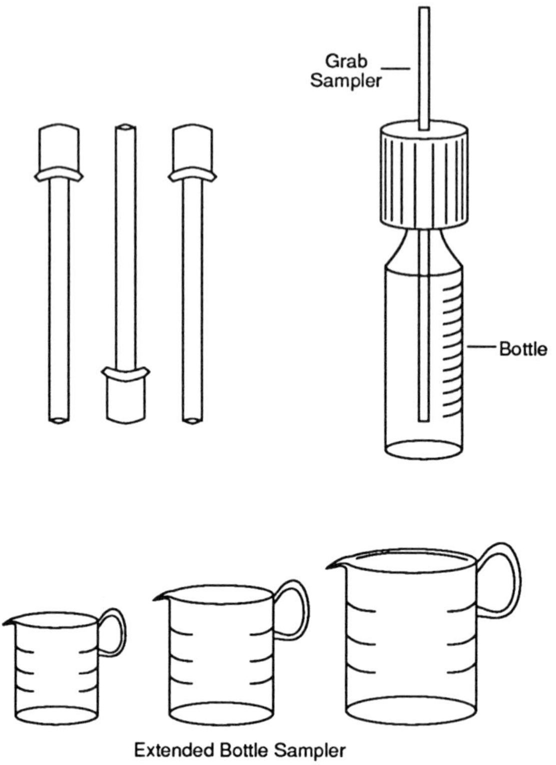

Figure 11.5. Illustration of extended bottle sampler.

Procedures for Use.

1. Place an uncontaminated, capped bottle in the stainless steel clamp.
2. Attach the rod to the cap by means of the screw clamp or suction cup.
3. Lower the sampler into the liquid to the desired depth.
4. Turn the handle grip rod to remove the cap.
5. Allow the bottle to fill and then replace the cap.
6. Raise the sampler and thoroughly decontaminate it prior to its next use.

Composite Liquid Waste Sampler (COLIWASA)

The composite liquid wations. After sampling hazardous liquids, however, it is difficult to decontaminate and may cause cross-contamination.

Description. The main parts of the COLIWASA are sampling tube, closure-locking mechanism, and closure system. The sampling tube consists of 5 ft by 1 5/8-in.-I.D. translucent pipe, usuaste sampler (COLIWASA) (Figure 11.6) is a valuable tool in some applically polyvinyl chloride (PVC) or borosilicate glass plumbing tube. The closure-locking mechanism consists of a short-length, channeled aluminum bar attached to the sampler's stopper rod by an adjustable swivel.

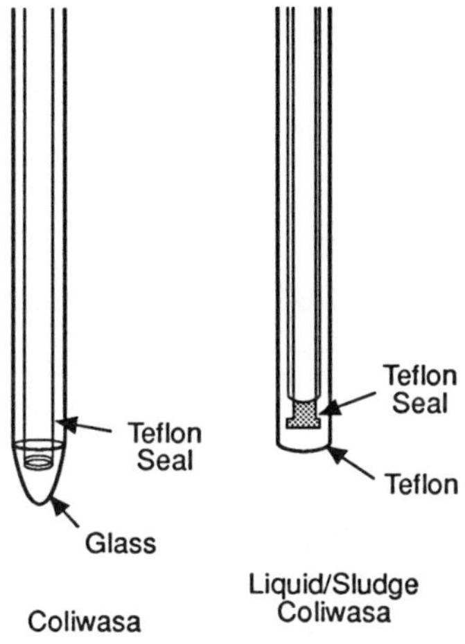

Figure 11.6. Illustration of composite liquid waste sampler (COLIWASA).

An aluminum bar serves both as a Y-handle and lock for the sampler's closure system.

Procedures for Use.

1. Put sampler in open position by placing stopper rod handle in T-position and pushing rod handle until it sits against sampler's locking block.
2. Slowly lower sampler into liquid so that liquid level inside and outside the tube are about the same. If the level inside the sampler tube is lower than outside sampler, sampling rate is too fast, resulting in a nonrepresentative sample.
3. When sampler stopper hits bottom of waste container, push the tube downward against stopper to close sampler. Lock sampler in close position by turning T-handle until it is upright and one end rests tightly on locking block.
4. Slowly withdraw the sampler from container.
5. Carefully discharge sample into a suitable sampler container by opening sampler. This is done by slowly pulling lower end of T-handle away from locking block, while lower end of sampler is positioned in a sample container.

Pond (Dip) Sampler

Pond (dip) sampler can be used to collect liquids or sludges from ponds, pits, lagoons, or open vessels, but only as far as its limited reach.

Description. The pond (dip) sampler consists of a container in an adjustable clamp attached to the end of a telescoping pole 8 to 15 ft long. The pole can be made of wood, plastic, or metal because the sample is collected in a jar or beaker which is secured in the claim (Figure 11.7).

Procedures for Use. Ladle liquids into sample containers.

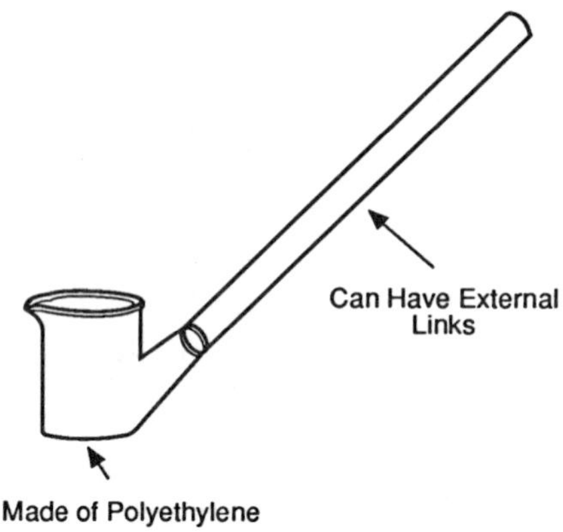

Figure 11.7. Illustration of pond (dip) sampler.

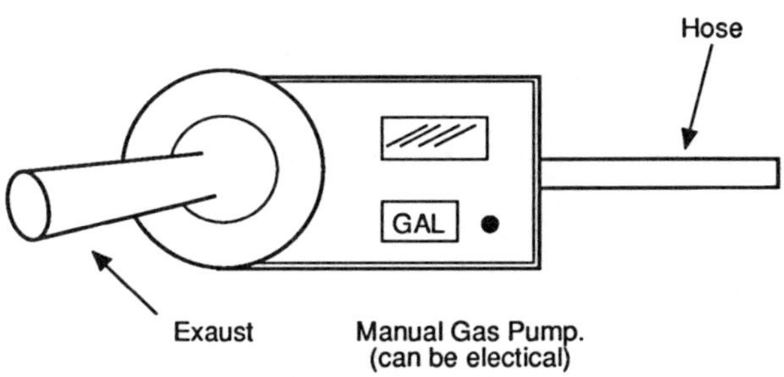

Figure 11.8. Illustration of manual pump.

Manual Pumps

The principle drawback of manual pumps is that they require large amounts of disposable tubing, which must be compatible with the waste being sampled (Figure 11.8).

Description. Manual pumps most commonly operate by peristalsis, bellows, diaphragms, or siphon. They are available in various sizes and configurations.

Procedures for Use. Operate according to manufacturer's instructions. In most cases this involves placing sample inlet hose into liquid, then manually activating a crank or inlet hose into liquid, then manually activating a crank or bellows. To avoid contamination of the pump, a liquid trap is inserted in the sample inlet hose. As with the sample, care must be taken to avoid sample contamination.

Solids

Grain Sample

Grain samplers are best used for collecting granular or loose solids. Moist, compressed, and large-particle solids are difficult to collect.

Mechanical

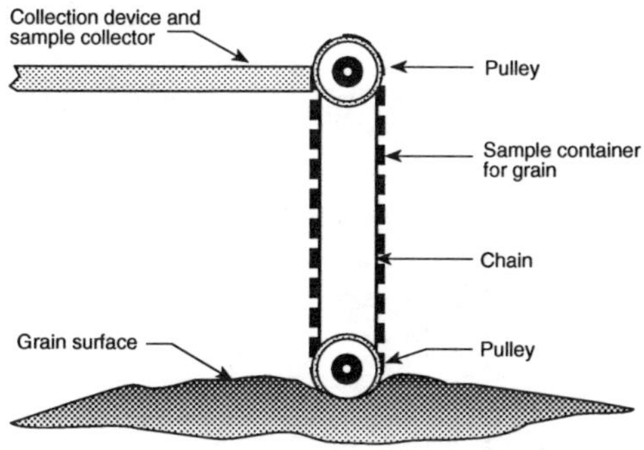

Manual

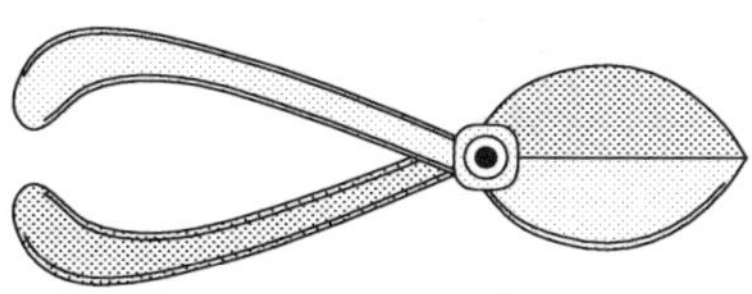

Figure 11.9. Illustration of grain sampler.

Description. The grain sample consists of two slotted telescoping tubes, usually made of brass or stainless steel (Figure 11.9). The outer tube has a conical, pointed tip on one end that permits the sampler to penetrate the material being sampled. The sampler is opened and closed by rotating the inner tube. Grain samplers are generally 24 to 40 in. long by 1/2 to 1 1/2 in. in diameter.

Procedures for Use.

1. Insert sampler (in the close position) into material being sampled from a point near a tip edge or corner, through center, and to point diagonally opposite point of entry.
2. Rotate inner tube into open position.
3. Wiggle sampler a few times to allow material to enter open slits.
4. Place sampler in close position and withdraw.
5. Place sampler in a horizontal position, if possible, with slots facing upward.
6. Rotate and slide out outer tube from the inner tube.
7. Transfer sample in inner tube into a suitable sample container. If inner tube is not removable, turn sampler upside down and pour contents out of an opening in end of the tube.
8. Store sampler in a plastic bag until decontamination.

Sampling Trier

Sampling triers are used to sample compressed solids. One difficulty, however, is the removal of the core sample cut with the trier.

Description. A typical sampling trier is a long tube about 24 to 40 in. long and 1/2 to 1 in. in diameter, with a slot that extends almost its entire length (Figure

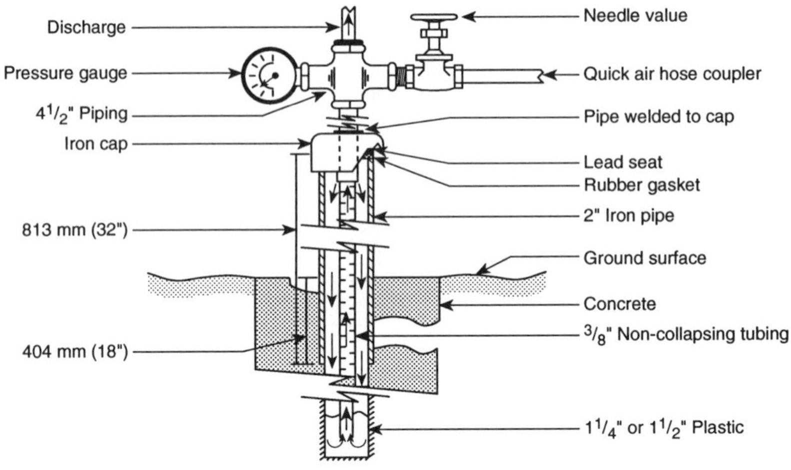

Figure 11.10. Illustration of a typical sampling tier.

11.10). The tip ant edges of the tube slot are sharpened to allow the trier to cut a core when rotated in a solid material. Sampling triers are usually made of stainless steel with wooden handles.

Procedures for Use.

1. Insert trier into solid material at 0 to 45 angle from horizontal. This will minimize spillage from sample. Tilt containers if necessary.
2. Rotate trier once or twice to cut a core of material.
3. Slowly withdraw trier, making sure that slot is facing upward.
4. Transfer sample into suitable container with aid of a spatula and/or brush.
5. Store sample in a plastic bag until decontamination.

Waste Pile Sampler

The waste pile sampler is used to sample wet and large-diameter solids.

Description. A waste pile sampler is essentially a large sampling trier. It is commercially available, but it is easy to fabricate from sheet metal or plastic pipe. Polyvinyl chloride plumbing pipe 5 ft long by 1 1/4 in. I.D. with 1/8 in. wall thickness is adequate. The pipe is sawed lengthwise (about 60/40 split) except for the last 4 in. The narrower piece is sawed off, leaving a slot in the pipe. The edges of the slot and the tip of the pipe are sharpened to permit the sampler to cut into the material being sampled. The unsplit 4 in. of the pipe serves as the handle.

Procedures for Use.

1. Insert sampler into material being sampled at 0 to 45° from horizontal, tilting container if necessary.
2. Rotate sampler two or three times in order to cut a core of material.
3. Slowly withdraw sampler making sure that the slot is facing upward.

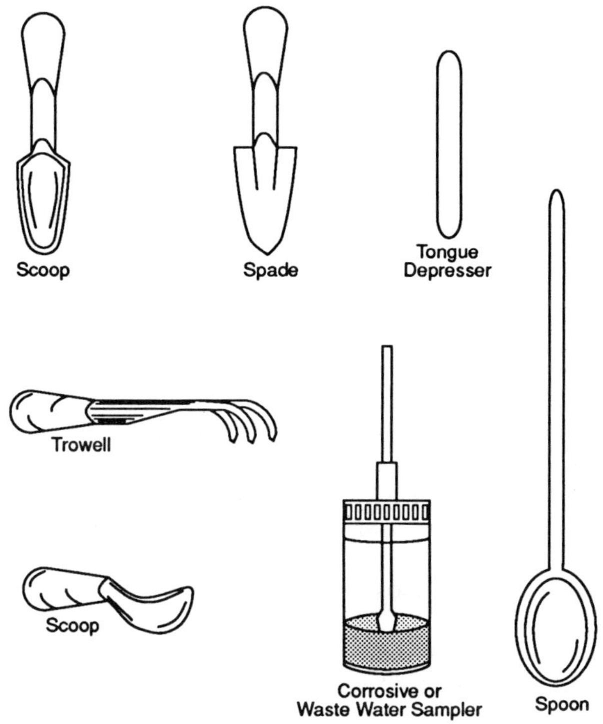

Figure 11.11. Illustration of a waste pile sampler, trowel, scoop, and spoons used in solid sampling.

4. Transfer sample into a suitable container with the aid of a spatula and/or brush.
5. Store sampler in a plastic bag until decontamination.

Trowel

Trowel, scoops, and spoons are inexpensive, but they can sample only to a depth of 4 to 5 in. (Figure 11.11).

Description. A garden-variety trowel looks like a small shovel. The blade is usually about 3 by 5 in. with a sharp tip. A laboratory scoop is similar, but the blade is usually more curved and has a closed upper end to contain material. Scoops come in different sizes and materials. Stainless steel or polypropylene scoops with 2 3/4 by 6-in. blades are preferred. Another alternative for small samples is a stainless steel table spoon.

Procedures for Use.

1. Collect small, equal portions of sample from surface or near the surface of material to be sampled.
2. Deposit samples in a suitable container.

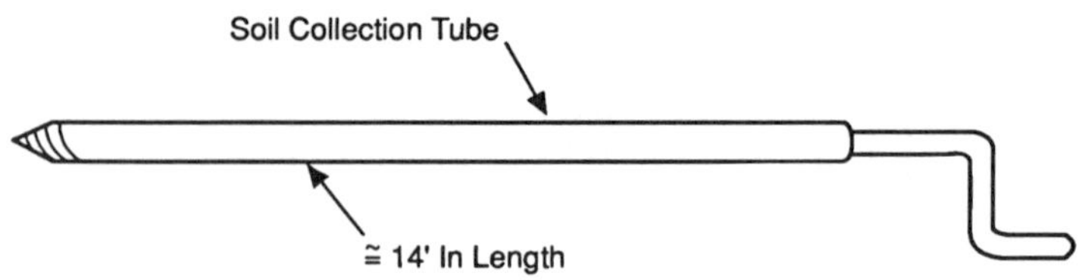

Figure 11.12. Illustration of a manual soil auger.

3. Dispose of sample or place in a plastic bag until decontaminated.

Soil Auger (Manual)

A manual soil auger can sample deeper and more compacted soils and solids than other samplers, but it is difficult to get the sample into a container (Figure 11.12).

Description. The soil auger consists of a hard metal central shaft and sharpened spiral blades. When the tool is rotated clockwise by its handle, it cuts the soil as it moves downward and deposits most of the loose soil upward. Augers are available in various sizes and configurations.

Procedure for Use. Follow manufacturer's instructions.

Cleaning and Storage Procedures. All samples must be clean before use. After use, they must be washed with warm detergent solution (for example, Liquinox or Alconox), rinsed several times with tap water, rinsed with distilled water, drained of excess water, and air dried, dried with a stream of warm, dry air, or wiped dry. Samplers used on petroleum products and oil residues may first have to be wiped with absorbent cloth to eliminate the residues. The equipment is then rinsed with an organic solvent, followed by washing with detergent solution and rinsing with water. A necessary piece of equipment for cleaning the tube of a COLIWASA is a bottle brush that fits tightly in the tube. The brush is connected to a rod long enough to reach the entire length of the sampler tube. This ramrod and fiber-reinforced paper towels clean the COLIWASA tube quickly.

Improper cleaning of sampling equipment will cause cross-contamination of samples. Such contamination is particularly important in samples taken for legal or regulatory purposes. Also, contamination becomes important when sampling wastes from different sources at the same time. If samples are to be taken for legal or regulatory purposes, or if they are expected to contain low concentrations of hazardous substances, a clean, unused sampler is needed.

SAMPLE WITHDRAWAL, STORAGE, AND PRESERVATION

Introduction

The sampling of ground and surface waters associated with the monitoring of land disposal sites is a critically important operation. The analytical results obtained from the samples and the subsequent decisions based upon the analytical data are vitally dependent upon the validity of the samples obtained.

Every effort must be made to assure that the sample is representative of the particular body of water being sampled. A detailed sampling plan, acceptable to all interested parties, should be developed prior to any sampling operations.

The physical, chemical, and bacteriological integrity of the sample must be maintained from the time of sampling to the time of testing in order to keep any changes at a minimum. The time between sampling and testing should be kept at the absolute minimum.

Sample Collection

Sample Withdrawal Methods

Sample withdrawal methods include the use of pumps, compressed air, bailings, and samplers. The primary consideration is to obtain a representative sample of the groundwater body by guarding against mixing the sample with stagnant (sanding) water in the well casing. In a nonpumping well, there will be little or no vertical mixing of the water, and stratification will occur. The well water in the screened section will mix with the groundwater due to normal flow patterns, but the well water above the screened section will remain isolated and become stagnant. Persons sampling should realize that stagnant water may contain foreign material inadvertently or deliberately introduced from the surface, resulting in unrepresentative data and misleading interpretation of the same.

To safeguard against collecting nonrepresentative stagnant water in a sample, the following guidelines and techniques should be adhered to during sample withdrawal.

As a general rule, all monitoring wells should be pumped or bailed prior to withdrawing a sample. Evacuation of a minimum of 1 volume of water in the well casing and preferably 3 to 5 volumes is recommended for a representative sample. In a high-yielding groundwater formation and where there is no stagnant water in the well above the screened section, evacuation prior to sample withdrawal is not as critical. However, in all cases where the monitoring data is to be used for enforcement actions, evacuation is recommended.

For wells that can be pumped or bailed to dryness with the sampling equipment being used, the well should be evacuated and allowed to recover prior to sample withdrawal. If the recovery rate is fairly rapid and time allows, evacuation of more than one volume of water is preferred.

For high-yielding monitoring wells which cannot be evacuated to dryness, bailing without prepumping the well is not recommended; there is no absolute safeguard against contaminating the sample with stagnant water. The following procedures should be used:

1. The inlet line of the sampling pump should be placed just below the surface of the well water and 3 to 5 volumes of water pumped at a rate equal to the well's recovery rate. This provides reasonable assurance that all stagnant water has

been evacuated and that the sample will be representative of the groundwater body at the time. The sample can then be collected directly from the pump discharge line.

2. The inlet line of the sampling pump (or the submersible pump itself) should be placed near the bottom of the screen section, pumped approximately one well volume of water at the well's recovery rate, and the sample collected directly from the discharge line.

A nonrepresentative sample can also result from excessive prepumping of the monitoring well. Stratification of the leachate concentrations in the groundwater formation may occur, and excessive pumping can dilute or increase the contaminant concentrations from what is representative of the sampling point of interest.

In light of this discussion, bailing by hand is not a recommended well-sampling method unless adequate precautions are taken. Bailing is accomplished in small-diameter wells by lowering and raising a weighted bottle or capped length of pipe on a length of rope. Rarely can a sufficient quantity of water be removed to adequately eliminate stagnant water from the sample unless many time-consuming trips in and out with the bailer are made. Persons sampling a well often will use the first bailer full of water as the sample because of the ease with which the sample can be collected. The reliability of this sample is nul, and this fact must be impressed upon the sample collector. Of course, in situations where the well can be bailed dry or there is only about a meter of water in the bottom of a well, representative samples can be obtained with a bailer as the casing can readily be emptied.

A Kemmerer water bottle sample is a commonly used bailer. In transferring the sample from the sample to the sample bottle, contact with air and agitation of the sample should be minimized; slow and careful transfer, placing the tip of the sampler's exit tube to the side of the sample bottle, is recommended. To minimize cross-contamination, the bailers should be thoroughly flushed out with tap water and the first sample from the next well to be sampled prior to collecting the sample for analysis.

Where the water table is within suction lift, small-diameter wells can be sampled with peristaltic, centrifugal, or pitcher pumps. Peristaltic pumps have rather low pumping capacities. The sample is conducted through inert silicone rubber tubing, thereby reducing the possibility of sample contamination by constituents from the sampling apparatus. For this reason, the procedure is attractive. Small highly portable centrifugal pumps are available with pumping rates from 0.3 to 2.5 L/sec (5 to 40 gpm). Removing stagnant water and flushing the discharge setup clean will pose little difficulty, allowing collection of representative samples. If concentrations of less than 1 ppm are being investigated, extra care must be taken in sampling; in the extreme parts per billion range, the peristaltic pump would have to be used.

Pitcher pumps can be easily carried to a site, screwed onto a well, and used to pump a sample. No power source is required other than the investigator, and equipment costs are low. A typical vacuum sampling technique would involve the use of a vacuum or suction pump and a portable generator as a power source.

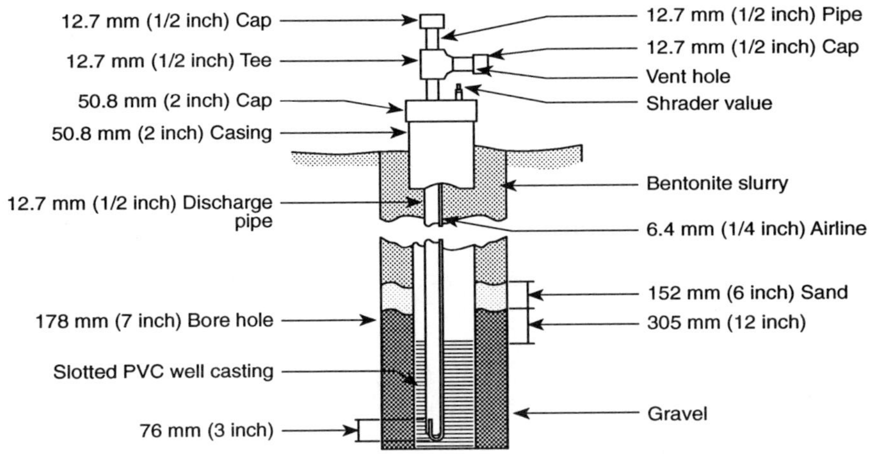

Figure 11.13. Illustration of air lift sampler.

Vacuum can also be supplied from an automobile or truck engine manifold or from a hand pump, replacing the vacuum pump. In Orange County, Florida, such a vacuum pump was used to collect groundwater samples under anaerobic conditions. At that location, a 1.27-cm (1/2-in.) tube is permanently installed in the monitoring well, thereby eliminating the possibility of cross-contamination between wells.

As an alternative to these pumps, an inexpensive bailer pump can be constructed from readily available materials. This pump consists of a length of garden hose with a foot valve at its bottom end and fittings at its top end that allows a vacuum to be applied to the base. A water sample is collected by moving the hose up and down and activating the foot valve, causing the partial vacuum to assist in bringing water to the surface. Vacuum is obtained from an automobile engine. It should be remembered, however, that this sampler should be used where the well contains only a small volume of water in order that clearing stagnant water from the casing does not become an inordinately time-consuming process.

An inexpensive air lift sampler can be constructed from polyethylene or any reasonably flexible tubing, as shown in Figure 11.13. Since the tubing is flexible, it can be readily coiled and moved conveniently from well to well. Primary limitations on the sample are the amount of air pressure that can be safely applied to the tubing and finding a suitable source of compressed air. A high-pressure hand pump would serve nicely for a shallow water table, but a small air compressor may be required for lift greater than 9 m (30 ft). The advantage of this sample is that the apparatus can be designed for permanent installation in the monitoring well. The smaller the tubing used, the less that can be pumped; this can be compensated for by pumping longer to obtain the minimum sample volume required for analysis.

With a pressure-type sampling method, the sample is obtained by connecting the sample bottle directly to the 1.27-cm (1/2-in.) water discharge outlet. To prevent contact with oxygen, the sample bottle could be flushed out with an inert gas prior to collecting a sample. The built-in feature of this method used to both prepump and sample the well can affect considerable savings on labor and also eliminate the possibility of cross-contamination between wells as can occur with portable pumping and sampling devices. This method is also of notable value in obtaining bacteriological samples where external sources of contamination must be avoided.

Somewhat more elaborate pumping equipment is required in small-diameter wells where the water level is below suction lift. The easiest, but not the driest, way to collect a sample is to install an airline in the well and blow the water out. However, trying to adjust airflow so that water flows smoothly over the top of the casing instead of blowing violently into the air is a difficult task. The addition of some simple, relatively inexpensive hardware to cap the well can make sampling a straightforward and easy process. Sommmerfelt and Campbell (1975) have described such an installation and Trescott and Pinder (1970) have pumped water from as deep as 58 m (190 ft) with this type of installation. Air pressure can be supplied from a gasoline-powered air compressor, an engine air pump, or a compressed air cylinder. The source of air pressure selected will depend upon well site accessibility and budgetary constraints. Where a well is not accessible to vehicle transportation, a good-quality hand pump (available in stores handling racing bicycles) can be used as an air supply.

The best, but most expensive, method of collecting water samples is with a small submersible pump. McMillion and Keeley (1968) have designed a truck-portable submersible pump capable of being set at depths up to 91.4 m (300 ft) inside a casing as small as 114 mm (4 1/2 in.) I.D. with a capacity of 0.44 to 0.88 L/sec (7 to 14 gpm). With this equipment, a pump can be set as desired in the well and operated until a representative sample can be obtained. Cross-contamination between wells from the sampling equipment will be virtually eliminated by the flushing action during prepumping. Provided there are no air leaks in the pumping system, samples could also be collected for dissolved gas analyses.

Samples for bacteriological examination must be collected in sterile containers. Detailed sampling procedures for bacteriological samples are given in *Standard Methods*, 13th ed., pp. 657–660, and *Biological Analysis of Water and Wastewater*, AM 302, Millipore Corp. (1975), pp. 4–6.

Samples can be taken directly from wells with a sterile bottle in a weighted frame which can be lowered below the water surface and opened below surface. Samples can also be obtained by means of various pumping devices, as previously described. However, great care must be taken to minimize sample contamination by the sampling equipment. Sample volumes of approximately 250 ml are usually satisfactory and bacteriological testing.

Sampling and preservation of samples are addressed in 1973 Annual Book of ASTM Standards, Part 23, Water and Atmospheric Analysis — pp. 72–75: "Standard Methods of Sampling Homogeneous Industrial Wastewater;" pp. 76–91: "Standard Methods of Sampling Water."

Exhibit 11.6. Sample Receipt

On ______________________________,
(date) (company name)
representative ____ received permission
from ____ to remove material from his/her
property, contained in __________________________ 2-qt glass organic sample bottle(s),
____________________ 40-ml glass volatile organic s ample bottle(s), ____________ 8-
oz glass hazardous sample jar(s), and ____________________ inorganic 1-quart polyethyl-
ene sample bottle(s).

____
Property Owner, Signature & Date

____
Company Name

____
Representative Signature & Date

REFERENCES

1. American Public Health Association, American Water Works Association, Water Pollution Federation, *Standard Methods for the Examination of Water and Wastewater*, 15 ed. (American Public Health Association, 1981).
2. American Society for Testing and Materials, *ASTM Standards*, Standards D1452, D1586, D3550, D270, E300, (Philadelphia, PA:The Society).
3. Environmental Monitoring and Support Laboratory, "Handbook for Analytical Quality Control in Water and Wastewater Laboratories," U.S. EPA, 600/4-79-019 (March 1979).
4. Environmental Monitoring and Support Laboratory, "Methods for Chemical Analysis of Water and Wastes," U.S. EPA, 600/4-79-020 (March 1979).
5. Environmental Research Information Center, Office of Research and Development, "Sampling of Water and Wastewater," U.S. EPA, 600/4-77-039 (August 1977).
6. Kittrell, F. W. "A Practical Guide to Water Quality Studies of Streams," U.S. Government Printing Office, Washington, DC (1969).
7. Lind, Owen T. "Handbook of Common Methods in Limnology," (St. Louis, MO: C.V. Mosby Co. 1974).
8. Municipal Environmental Research Laboratory, "Samples and Sampling Procedures for Hazardous Waste Streams," U.S. EPA, 600/2-80-018 (January 1980).
9. Office of Technology Transfer, "Handbook for Monitoring Industrial Wastewater," U.S. EPA (August 1973).

Exhibit 11.11. Samplers for Hazardous Materials

Sampler	Applications	Limitations	Suppliers
Open-tube/ COLIWASA	Liquids, slurries	Not for containers over 1.5 m (5 ft) deep	Open-tube: laboratory supply houses
Plastic		Not for materials containing ketones, nitrobenzene, dimethylformalide, mesityl oxide, tetrahydrofuran, or many common solvents such as acetone	COLIWASA: Nasco, 901 Janesville Ave.,
Glass		Not for materials containing hydrofluoric acid and concentrated alkali solutions	
		COLIWASA may be difficult to decontaminate and may cause cross-contamination of samples	
Pond (dip) sampler	Liquid, sludges	Not for sampling beyond 3.5 m (11.5 ft)	Fabricate; clamps available Cole-Parmer Instrument Co. 7425 N. Oak Park Ave. Chicago, IL 60648
Manual pump	Liquids	Requires large amounts of disposable tubing that must be compatible with material	Local hardware stores and/or laboratory supply houses
Weighted bottle sampler	Liquids	Difficult to use with very viscous liquids; exterior of sample bottle exposed to hazardous materials	Fabricate
Extended bottle sampler	Liquids	Difficult to use with very viscous liquids; exterior of sample bottle exposed to hazardous materials	Ace Glass Company Vineland, NJ 08360

Kemmerer sampler	Liquids	Must be compatible with sample	Laboratory supply houses
Grain sampler	Granular solids	Limited application for sampling moist and sticky solids with a diameter of 0.6 cm (1/4 in.)	
Sampling trier	Solids	Possibly difficult in retaining core sample of very dry granular materials during sampling	Laboratory supply houses
Trowel/scoop/ spoon	Solids, soil surface	Not for sampling deeper than 8 cm (3 in.); difficult to obtain reproducible mass of samples	Laboratory supply house
Waste pile	Loose solids	Not for sampling solid wastes with dimensions greater than half the diameter of the sampling tube	Fabricate from pipe
Soil auger (manual) (3–4 in.)	Soil deeper than 102–126 cm	Does not collect undisturbed core sample	Weyco Distribution, Sacramento, CA

CHAPTER 12

Site Operations at a Hazardous Waste Site

INTRODUCTION

Team size and organization will depend upon the degree of difficulty and the requirements of individual investigations. The only requirements are that each team member receive the training necessary to successfully complete those tasks on the investigation and that each person have sufficient training so as to be able to complete any emergency tasks that may occur on the site. That means that every person on the site who is part of the basic team must be able to respond to an emergency by using all safety equipment and, if necessary, going downrange.

TEAM SIZE

A "buddy system" is required for the investigation of hazardous waste areas or a response to an emergency. The minimum personnel requirements for the different levels of operation are discussed in the following sections.

Minimum Requirements for Level C or D

Two persons are required for a Level C or D operation (see Exhibit 12.1).

The two-person team is the minimum for a hazardous substance site investigation, and its capabilities are very limited. The two-person team is best suited for off-site surveys and inspections or obtaining environmental (nonhazardous, off-site) samples. Ground truthing of aerial photographic surveys, inspection of files, or interviews can all be accomplished by the two-person team.

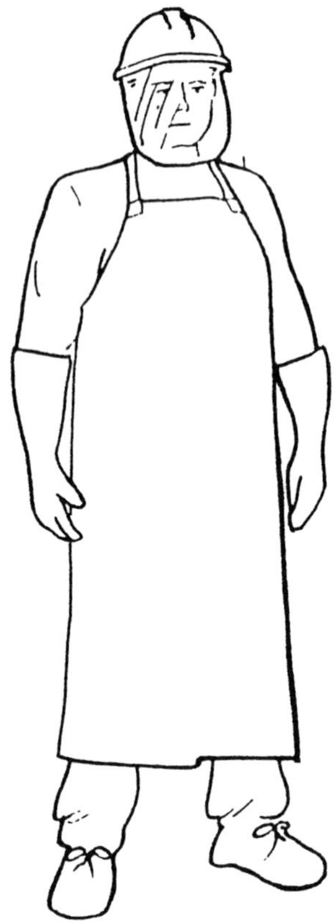

Exhibit 12.1. Sampler in Level D.

Two-Person Team

Person I ("Clean" Person)

Project manager
Log book
Photography
Monitoring equipment
Decon of Person II
Tape recorder (for interviews or field survey reports)
Walky-talky

Person II ("Dirty" Person)

Safety officer
Sampling
Decon of Person I

Shared:

Sample management
Mobilization
Demobilization

Advantages

1. Less personnel-related costs.
2. Less personnel exposed to hazards (medical or risk management).

Disadvantages

1. Busy field team — may get hectic.
2. Awkward decon procedures.
3. No emergency or rescue people available.
4. Field work takes longer — would not work on larger sites.
5. QC difficult to maintain.
6. Limits some types of work where manpower is needed.

Minimum Requirements for Level B

Three persons are required for a Level B operation, including two persons downrange and a third person, dressed at the same level of protection as the downrange people. These three persons perform the functions of safety person, decontamination person, and emergency rescue person. It is preferable to have a fourth person dressed at Level D in the clean area to act as the safety person, while the number three person serves as the decontamination/emergency rescue person. A three-person team is sufficient only for the simplest of Level B operations. Communication between these individuals is done by hand signals or two-way radio.

The three-person team is used where the likelihood of needing emergency rescue is low. This field investigation team is best used in non-IDLH (immediately dangerous to life and health) atmospheres where the primary objective is to map, photograph, or inventory. It is assumed that at no time will the work party be exposed to hazardous situations. In the event of an emergency, the third member does not enter the site until outside assistance has been summoned; even then, entry should only be made when non-life-threatening situations exist.

Three-Person Team

Method 1

Person I ("Clean")	Person II ("Dirty")	Person III ("Dirty")
Project manager	Assist sampler	Sampler
Log book	Decon	Decon
Photography	Safety officer	Demobilization
Monitoring equip.	Assist with sample management	Radio
Decon	Mobilization	
Sample management	Radio	
Radio		

Method 2

Person I (Offsite "Clean")	Person II (Onsite "Clean")	Person III ("Dirty")
Safety officer rescue	Project manager	Sampler
Sample management	Photography	Decon
Mobilization	Monitoring equipment	Demobilization
Community contact	Log book	Mobilization
Radio	Assists sampler	Radio
Tape recorder	Decon	Tape recorder
	Assists with sampler	
	Management	
	Mobilization	
	Radio	
	Tape recorder	

Advantages

1. More manpower available.
2. Easier to mobilize and demobilize.
3. Better safety and rescue system.
4. Better decon procedures.

Disadvantages

1. Still too difficult to maintain sample QC (too many hands).
2. Project manager too busy — may contaminate (equipment).
3. No "line of site" on large site.
4. Safety officer too busy to watch field procedures.
5. More likely to miss documentation.

Optional

Use of a video camera would assist for documentation and site overview.

Minimum Requirements for Level A

Most Level B and some restricted Level A operations can be conducted with a four-person team. These operations would include work on active sites where facility personnel are present or on inactive sites with potentially IDLH atmospheres. The objectives of a four-person team at a site requiring Level B or Level A protection might include sampling of ponds, soils, or open containers and inspections at sites known for poor housekeeping (i.e., sites with a history of spills, leaks, or other accidents) (see Exhibit 12.2).

The team consists of the standard two-person work party, a combination site safety officer and personnel decontamination station (PDS) operator/equipment specialist, and a common post supervisor who may assist in the PDS operation. Because life-threatening hazards are assumed or known to be present at Level B

Exhibit 12.2. Level A Dermal and Respiratory Protection

or Level A sites, it is essential that all personnel be fully acquainted with their duties. During an emergency, the command post supervisor stays in the support area to maintain communication, while the site safety officer/PDS operator/equipment specialist enters the exclusion area to aid the work party. Once the work party is in the contamination reduction area, the command post supervisor can then offer assistance on the PDS or provide fresh equipment from the support area.

Four-Person Team (Minimum 5 People)

Person I Offsite ("Clean")

Sample management
Mobilization
Call for help

Person II Onsite ("Clean")

Safety officer
Decon
Rescue
Monitoring instruments
Demobilization

Person III Onsite ("Clean")
Project manager
Log Book
Photography
Community contact

Person IV ("Dirty")
Sampler
Mobilization
Demobilization

Advantages

1. Best use of personnel.
2. Clean zones isolated from contaminated areas.
3. Effective sample management.
4. Safety officer can focus on health and safety.
5. Personnel available on site in case of difficulties.
6. Line of site easier to maintain.
7. Continuous air monitoring is possible.
8. Most routine work can be completed safely.

The five-person team is the minimum size for most Level A and Level B operations or when known percutaneous hazards exist or there is an absence of historical information. The site hazards that require Level A protection, combined with the limitations and stresses placed on personnel by wearing Level A protection, necessitate a full-time PDS operator/equipment specialist who can also serve in emergency response. In the event of a serious emergency such as a fire, explosion, or acutely toxic release, both the site safety officer and the PDS operator/equipment specialist may need to enter the exclusion area dressed in Level A gear. The command post supervisor remains in the support area to direct outside help to the site and then assume the functions of PDS operator/equipment specialist.

Teams of Seven or More

Certain hazardous substance sites require sampling operations that necessitate larger or alternating work parties and additional support personnel in the contamination reduction area. The seven-person team employs the basic structure of the five-person team plus an additional work party for alternating work loads. The eight-person team includes an additional PDS operator/equipment specialist to assist in the continuous decontamination tasks involved with alternating work parties and to decontaminate and pack samples as they are received.

It is not unusual to employ teams of 12 where such tasks as drum opening may require three work parties working concurrently. This operation may involve teams to move the drums, open the drums, and sample and reseal them under rigorous safety procedures. Larger teams can be designed with additional work parties and support personnel to safely, gather the site data and ensure communication and site control.

Emergency Response Team

Extensive assignments requiring long hours and large work parties (more than five persons) necessitate the use of a standby emergency response team. The protective gear used by the team must provide at least as much protection as that used by the downrange field team. Members of the emergency response team are half-dressed in protective gear so that they can quickly enter the exclusion area in an emergency. This team is particularly valuable at dangerous sites where protective equipment produces stress and heat loads on the work party.

TEAM ORGANIZATION

Introduction

Once it has been determined that a comprehensive field investigation is necessary to fully characterize a potentially hazardous site, a detailed work plan must be prepared and a field investigation team selected. The primary function of the field investigation team is to gather information according to the approved work plan. These guidelines describe the components, duties, and sizes of the field team that are necessary in order to safely meet the stated goals of the investigation. These guidelines are applicable to any field investigation work involving hazardous waste disposal sites.

Responsibilities

There are at least eight roles which may be required for a field investigation team:

1. Project team leader
2. Field team leader
3. Site safety officer/industrial hygienist
4. Personnel decontamination station operator/equipment specialist
5. Command post supervisor
6. Initial entry party
7. Work party
8. Emergency response team

The number of roles needed at each site is dictated by the potential hazards and specific site needs. Dual role assignments may be acceptable when hazardous and physical conditions at a site are well documented and the nature of the work is limited.

Project Team Leader

The project team leader is primarily an administrator when not participating in the field investigation as field team leader or command post supervisor. The project team leader is responsible for the following:

1. All the team does or fails to do.
2. Preparing and organizing all project work.
3. Selecting team personnel and briefing them on specific assignment.
4. Obtaining the owner's permission to enter the site.
5. Coordinating with the field team leader to complete the work plan.
6. Completing final reports and preparing the evidentiary file.
7. Ensuring that safety and equipment requirements are met.
8. Coordinating with the lead agency.

Some of this responsibility may be delegated to the field team leader and site safety officer.

Field Team Leader

The field team leader is responsible for the overall operation and safety of the field team. As mentioned earlier, this role can be filled by the project team leader or the designated representative.

The field team leader may join the work party. The field team leader is responsible for the following:

1. Safety procedure enforcement through delegation to the site safety officer.
2. Field operations management.
3. Public relations and state and federal liaison.
4. Site control.
5. Compliance of field documentation and sampling methods with evidence collection procedures.
6. Execution of the site work plan.

Site Safety Officer

The site safety officer has primary responsibility for all safety procedures and operations at the site. The site safety officer will report to the person responsible for safety in the organization rather than to the field team leader or project team leader. This provides for two separate lines of authority, thereby allowing decisions based on safety to be represented on an equal basis with decisions based on the pressures for accomplishing the investigation according to schedule. If a situation is an imminent danger, safety would have the authority to shut down operations until corrective actions were taken.

The site safety officer either remains on the clean side of the exclusion area while monitoring the work plan and site activities or may accompany the downrange team in order to supervise hazardous work. The site safety officer is also responsible for the following:

1. Determining the level of personnel protection required.
2. Updating equipment or procedures based on new information gathered during the site inspection.
3. Changing the levels of protection based on site observations and/or on additional site data as it becomes available.
4. Enforcing the safety requirements.
5. Stopping work as required to ensure personal safety and protection of property or where noncompliance with safety requirements are found.
6. Determining and posting routes to capable medical facilities and emergency telephone numbers (including poison control centers), and arranging for emergency transportation to medical facilities.
7. Notifying local public emergency officers (i.e., police and fire department) of the nature of the team's operations and posting their telephone numbers.
8. Entering the exclusion area in emergencies when at least one other member of the field team is available to stay behind and notify emergency services (or after the emergency services have been notified).
9. Examining work party members for symptoms of exposure or stress.
10. Ensuring that each team member has been given the proper medical clearance by a qualified medical consultant. Monitor all team members to determine compliance with the applicable physical requirements as stipulated in the health and safety program.
11. Maintaining communications and line-of-site contact with the work party.
12. Providing emergency medical care and first aid as necessary at the site.
13. Reporting employee accidents or loss time injuries/exposures on site.
14. Reporting #13 on OSHA 200 form.

The site safety officer is a general term for description of a professional individual who may be an industrial hygienist (I.H.) or safety engineer. Each specialized in their fields. An I.H. can recognize, monitor, evaluate, and make recommendations of the risk, while the safety engineer evaluates the hazard and potential for accidents.

It should also be brought out that in addition to the above-cited specialties, a person with general safety construction background would assist the site safety officer. This is very important when dragging pits/trenches for site sampling and guarding against cave-ins.

PDS Operator/Equipment Specialist

PDS operator/equipment specialist functions are two roles that do not require concurrent attention. As the equipment specialist, he is charged with

1. Ensuring that all equipment is properly maintained and operating.
2. Inspecting all equipment before and after use.

3. Ensuring that all required equipment is available prior to arrival at work site.
4. Decontaminating all personnel, samples, and equipment returning from the exclusion area.

The PDS operator/equipment specialist is responsible for designing and setting up the PDS and for preparing the necessary decontamination solutions to ensure that chemical contamination is not transported into the clean area by equipment, samples, protective clothing, or personnel. The PDS operator/equipment specialist also manages the mechanics of removing contaminated clothing from the work party and the proper disposal of discarded contaminated clothing and decontamination solutions.

Command Post Supervisor

The command post supervisor functions is the clearinghouse for communications. He does not enter the exclusion area to assist the work party. Should an emergency arise, the command post supervisor notifies emergency support personnel by phone, air horn, radio, or some other communication device to respond to the situation. Depending on the team size and the nature of the emergency, the command post supervisor may be needed to assist the site safety officer in effecting a rescue. Usually, the command post supervisor may be called upon to assist the PDS operator/equipment specialist in operating the PDS during an emergency and to assist the site safety officer in emergency medical measures. The field team leader usually assumes the responsibilities of command post supervisor.

The command post supervisor is also responsible for the following:

1. Maintaining a log of communications and site activities, such as duration of work periods with respirators.
2. Assisting the site safety officer in sustaining communication and line-of-sight contact with the work party.
3. Maintaining good public relations in the absence of the field team leader.

WORK ZONES

Introduction

The activities required during responses to environmental incidents involving hazardous substances may contribute to the movement of materials (contaminants) from the site to unaffected areas. Response personnel working and equipment used around the substances may become contaminated and carry the material into clean areas. Material may become airborne because of its volatility or due to disturbances of contaminated soil. To minimize the transfer of hazardous substance(s) from the site due to site activities, contamination control procedures

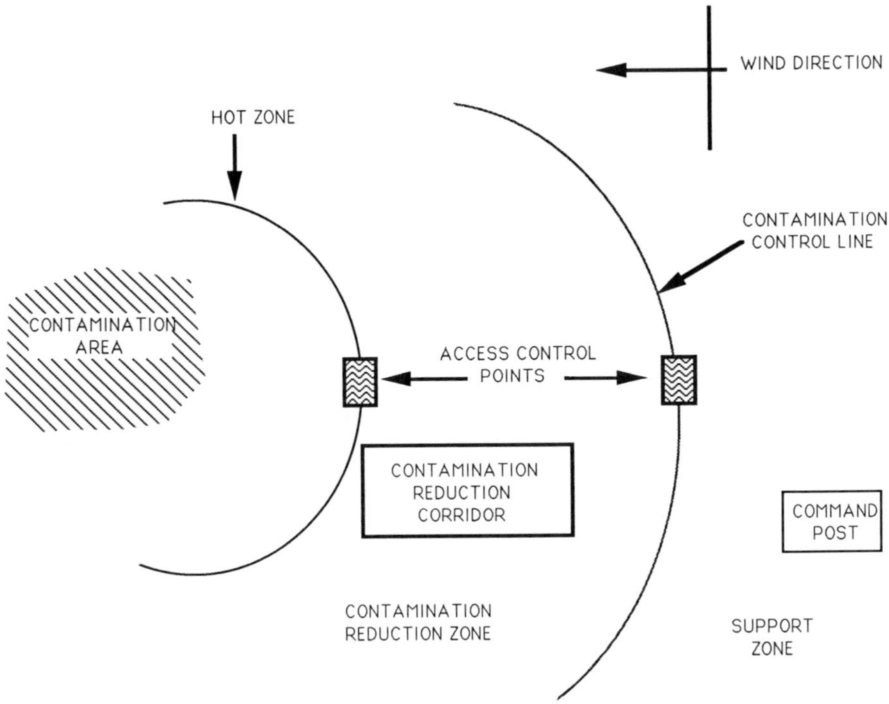

Figure 12.1. Diagram of site work zones.

are needed. Two general methods are used: establishing site work zones and removing contaminates from people and equipment (see Figure 12.1).

Control at the Site

A site must be controlled to reduce the possibility of (1) exposure to any contaminants present, and (2) their transport by personnel or equipment from the site. The possibility of exposure or transfer of substances can be reduced or eliminated in a number of ways, including:

1. Setting up security and physical barriers to exclude unnecessary personnel from the general area.
2. Minimizing the number of personnel and equipment at the site consistent with effective operations.
3. Establishing work zones within the site.
4. Establishing control points to regulate access to, and egress from, work zones.
5. Conducting operations in a manner to reduce the exposure of personnel and equipment and to eliminate the potential for airborne dispersion.
6. Implementing appropriate decontamination procedures.

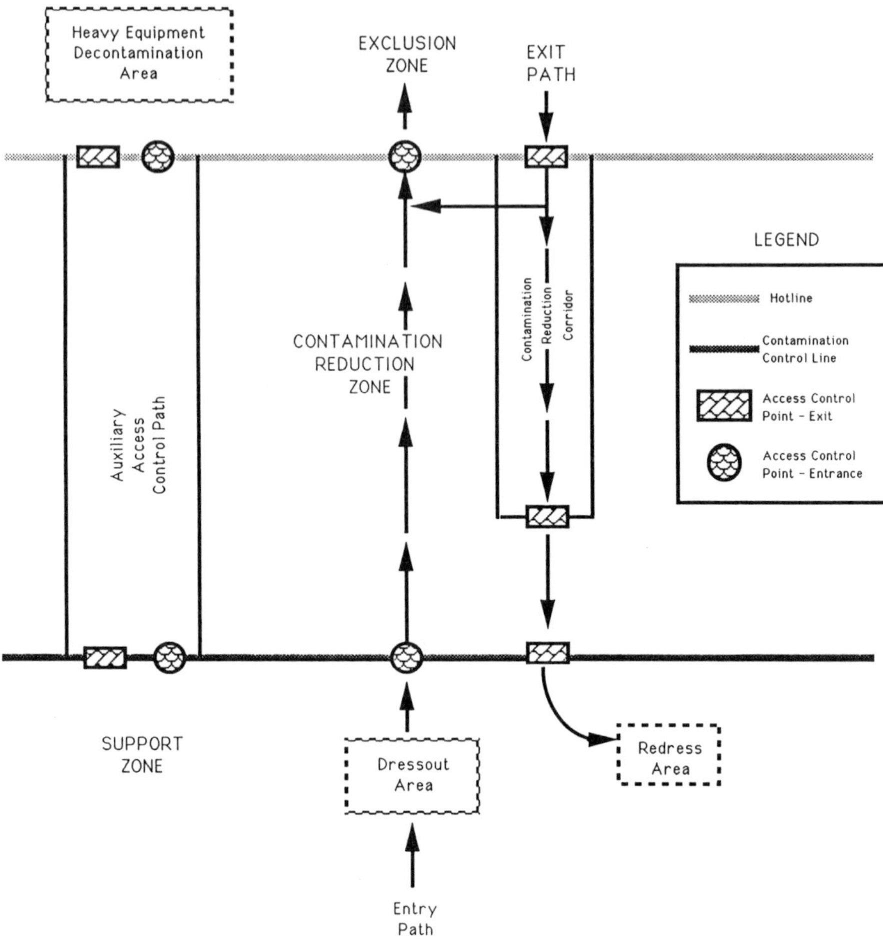

Figure 12.2. Diagram of three continuous zones.

Delineation of Work Zones

One method of preventing or reducing the migration of contamination is to delineate zones on the site where prescribed operations occur. Movement of personnel and equipment between zones and onto the site itself would be limited by access control points. By these means, contamination would be expected to be contained within certain relatively small areas on the site and its potential for transfer minimized. Three contiguous zones (see Figure 12.2) are recommended:

- Zone 1: exclusion zone
- Zone 2: contamination reduction zone
- Zone 3: support zone

Zone 1: Exclusion Zone

The exclusion zone, the innermost of three concentric areas, is the zone where contamination does or could exist. All people entering the exclusion zone must wear the prescribed levels of protection. An entry and exit checkpoint must be established at the periphery of the exclusion zone to regulate the flow of personnel and equipment into and out of the zone to verify that the procedures established to enter and exit are followed.

The outer boundary of Zone 1, the hotline, is initially established by visually surveying the immediate environs of the incident and determining where the hazardous substances involved are located; where any drainage, leachate, or spilled material is; and whether any discolorations are visible. Guidance in determining the boundaries is also provided by data from the initial site survey indicating the presence of organic or inorganic vapors/gases or particulates in the air, combustible gases, radiation, and/or the results of water and soil sampling.

Additional factors that should be considered include the distances needed to prevent fire or an explosion from affecting personnel outside the zone; the physical area necessary to conduct site operations; and the potential for contaminants to be blown from the area. Once the hotline has been determined, it should be physically secured, fenced, or well-defined by landmarks. During subsequent site operations, this boundary may be modified and adjusted as more information becomes available.

All personnel within the exclusion zone must wear the required level of protective gear. Personal protective equipment is designated on the basis of site-specific conditions, including the type of work to be done and the hazards that might be encountered. Occasionally, within the exclusion zone, different levels of protection are justified. Subareas are specified and conspicuously marked as to which level of protection is required. The level of protection is determined by the measured concentration of substances in the air, the potential for contamination, and the known or suspected presence of highly toxic substances.

Different levels of protection in the exclusion zone might also be designated by job assignment. For example, collecting samples from open containers might require Level B protection, while a walk-through to monitor the ambient air would require Level C protection. The assignment, when appropriate, of different levels of protection within the exclusion zone generally makes for a more flexible, effective, and less costly operation, while still maintaining a high degree of safety.

Zone 2: Contamination Reduction Zone

Between the exclusion zone and the support zone is the contamination reduction zone, which provides a transition between contaminated and clean zones. Zone 2 serves as a buffer to further reduce the probability of the clean zone becoming contaminated or being affected by other existing hazards. It provides additional assurance that the physical transfer of contaminating substances on people, equipment, or in the air, is limited through a combination of decontamination, distance between the exclusion and support zones, air dilution, zone restrictions, and work practices.

Initially, the contamination reduction zone is considered to be a noncontaminated area. At the boundary between the exclusion and contamination reduction zones, decontamination stations are established, one for personnel, one for heavy equipment, and if prevailing winds are from different directions. Depending on the size of the operation, more than two stations may be necessary. Exit from the exclusion zone is through a decontamination station.

As operations proceed, the area around the decontamination station may become contaminated, but to a much lesser degree than the exclusion zone. On a relative basis, the amount of contaminants should decrease from the hotline to the support zone due to the distance involved and the decontamination procedures used.

The boundary between the support zone and the contamination reduction, zone, called the contamination control line, separates the possibly low contamination area from the clean support zone. Access to the contamination reduction zone from the support zone is through a control point. Personnel entering there would wear the prescribed PPE, if required, for working in the contamination reduction zone. Entering the support zone requires removal of any protective equipment worn in the contamination reduction zone.

Zone 3: Support Zone

The support zone, the outermost part of the site, is considered to be a noncontaminated or clean area. Support equipment (command post, equipment trailer, etc.) is located in the zone; traffic is restricted to authorize response personnel. Since normal work clothes are appropriate within this zone, potentially contaminated personnel clothing, equipment, and samples are not permitted, but are left in the contamination reduction zone until they are decontaminated.

The location of the command post and other support facilities in the support zone depends on a number of factors, including

1. *Accessibility.* The topography, available open space, and the locations of highways and railroad tracks are major considerations.
2. *Wind direction.* The support facilities should be located upwind of the exclusion zone. However, shifts in wind direction and other conditions may be such that an ideal location determined on the basis of wind direction alone does not exist.
3. *Resources.* There should be adequate roads, power lines, water, and shelter.

OTHER CONSIDERATIONS

Modifications

The use of a three-zone system, access control points, and exacting decontamination procedures provides a reasonable assurance against the translocation of contaminating substances. This site control system is based on a "worst-case" situation. Less stringent site control and decontamination procedures may be used if more definitive information is available on the types of substances involved and the hazards they present. This information can be obtained through air monitoring, instrument survey and sampling, and technical data concerning the characteristics and behavior of the material present.

Area Dimensions

The distance between the hotline, contamination control line, and command post and the size and shape of each zone have to be based on conditions specific to each site. Considerable judgment is needed to ensure that the distances between zone boundaries are large enough to allow room for the necessary operations, provide adequate distances to prevent the spread of contaminants, and eliminate the possibility of injury due to explosion or fire. Long-term operations would involve developing reasonable methods to determine if material is being transferred between zones to assist in modifying site boundaries.

The following criteria should be considered in establishing area dimensions and boundary distances:

1. Physical and topographical features of the site.
2. Weather conditions.
3. Field/laboratory measurements of air contaminants and environmental samples.
4. Air dispersion calculations.
5. Potential for explosion and flying debris.
6. Physical, chemical, toxicological, and other characteristics of the substances present.
7. Cleanup activities required.
8. Potential for fire.
9. Area needed to conduct operations.
10. Decontamination procedures.
11. Dimensions of contaminated area.
12. Potential for exposure.

Communications

Internal (Onsite) Communications. The internal communications network is used among the designated work zones. It is to alert the team to emergencies, to pass along safety information (air time left), to communicate changes in work scope, or to maintain site control. Team members key in internal communications are the command post supervisor, site safety officer, and work party members.

Means of communication which can be used are limited only by resources available and the imagination of the team. Regardless of what is used, there should be a primary and back-up system. An important fact to remember is that any device used downrange for communication must be intrinsically safe so as to avoid creating a spark in a potentially explosive atmosphere. Common devices used on site are

1. Radios — either citizen's band (CB) or FM (high and low band).
2. Noise-making devices — bell, compressed-air horn whistle, siren, megaphone, or vehicle horn.
3. Visual signals — smoke or flares (only from support zone), flags, signal boards, lights, or arm and hand signals.

Emergency communication systems should be established and only used when an emergency exists. Common emergency signals used include hand, audible, or radio signals. Examples of emergency hand and audible signals are:

1. *Clutching throat* — personal distress.
2. *Arm waved in a circle over the head* — if given downrange: need assistance; if given from command post: evacuate area.
3. *Three short blasts followed by a pause and repeated* — from downrange: need assistance; from CP: evacuate area.

With radio communications, code words should be established. All systems must be checked each day prior to site entry to insure they work properly.

External Communications (Site to Off Site and Emergency Response). External communications are used for emergency response or to report to management results, changes, or problems associated with the site work. They are used to maintain contact with those off-site personnel who are not necessarly at the operation site. Primary external communications are the telephone and radio. If phone lines are not temporarily installed, the location of the nearest phone must be known by all team members and the correct change available. In the case of an emergency, this is usually the command post supervisor. The necessary phone numbers should be available with the change and phone location stated.

Site Security

During Work Hours. Site security during the work hours is controlled first at the support zone and second at the access control points of the contamination reduction zone. If the site is not fenced or surrounded by a barrier or prohibitive terrain, site security presents a more difficult problem. On small area sites, if personnel are available, patrolling the site boundary may be feasible as long as line-of-sight is maintained. On larger area sites, posting signs around the perimeter may have some effect. In this case, the team should be alert for unauthorized personnel who may approach from a direction away from the support zone. When working on an active facility or property which is inhabited, it is not always feasible to prevent unauthorized personnel from entering the exclusion zone. Evacuation of the area by team members may be necessary if the unauthorized personnel create a danger to the health and safety of the work parties. An explanation of the work being performed and a request to avoid the area will hopefully invite the cooperation of property owners, facility employees, or local citizens. If the site presents a significant risk to the health and safety of the local populace, local police support should be requested for additional site security.

During Off Hours. Site security during off hours can be handled in two primary ways. A fence or barrier around the site is required to prevent unauthorized personnel from entering the site. In addition, equipment must be locked within the fence or trailer or removed from the site each day. The other alternative is to employ a security guard to watch the area during off hours or have the local police patrol. With this alternative, the equipment must still be locked up or removed from the site to prevent theft or tampering.

Evacuation Routes

Evacuation routes both from the exclusion zone and from the support zone to emergency medical care should be well delineated and kept clear. The primary evacuation route from the exclusion zone is through the contamination reduction zone access control points. Even if there is not time to process team members through the PDS, it serves as a means of accounting for personnel. This exit route will also be upwind, insuring that the team is leaving the hazard rather than heading into it. Alternative evacuation routes should be made available from the exclusion zone. Consideration should be given to wind direction and access to the support zone when alternative routes are marked. Also, a place to assemble must be designated in order to account for all personnel.

Evacuation routes from the support zone to emergency medical care should be known by at least the command post supervisor and site safety officer. Explicit directions and a map should be in the emergency vehicle on site. Keys to the vehicle should be readily accessible or left in the ignition. The access road from the site must be kept clear of other vehicles. This is especially important to remember if press crews are at the site creating additional traffic.

Organization of the Entry Party

Before entering a site, you need to know

1. The toxic hazards associated with the site.
2. The specific mission or tasks to be performed at the site.

This information then is used to develop the level of protection to be used. In general, before entering a site

1. Team entry and exit routes should be planned.
2. A detailed work plan should be developed and practiced if necessary.
3. An emergency evacuation signal should be established, such as portable air horns, beating a wrench on a container, etc.

In addition, care must be taken during self-contained breathing apparatus checkout to prevent damage to the units.

Downrange Operations

The following should be considered as a general protocol for site entry:

1. Entry
   a. Team members hook up air masks.
   b. Team captain conducts check of team members and arranges members in proper entry sequence.
   c. Team crosses hotline and proceeds toward sample area.
   d. Appropriate team members operate the oxygen and explosive gas monitor ing equipment.
2. Soil sampling
   a. Sample carrier uncaps bottle and holds it out to receive sample.
   b. Sampler uses spatula to dig sample and put in bottle.
   c. Photographer takes two pictures of process.
   d. Sample carrier recaps bottle.
   e. Team captain records time, type of sample, and sample point.
   f. Sampler does not ever
      i. Kneel on the ground.
      ii. Squat in such a manner as to transfer contaminant from boots to clothing.
      iii. Lean or sit on trees, drums, or equipment.
   g. Sampler uses ground cloth for kneeling or placement of equipment.
3. Water sampling
   a. Sample carrier uncaps bottle and holds it out to receive sample.
   b. Sampler takes the water sample and puts it in bottle. (Dexterity problems usually occur, causing the water sample to flow on the outside of the bottle and over the gloves of the sample carrier.)
   c. Photographer takes picture of process.
   d. Sample carrier recaps bottle.
   e. Team captain records time, type of sample, and sample point.
   f. Sample holder changes outer gloves in order to
      i. Avoid contaminating other samples that might be taken.
      ii. Prevent contamination of equipment.

   Gloves could be changed to either a heavier or lighter weight to perform a different operation.
4. Exit
   a. The team, with the team captain in the rear, moves from the sample point to the hotline.
   b. Samples are deposited for processing by hotline operators.

Special Considerations

The condition of team members should be given special consideration at all times. Team members may become

1. Fatigued from the wearing of the mask and equipment.
2. Thirsty because of minor dehydration.
3. Hot from a temperature buildup caused by the protective clothing.

During hot weather operations, there are the additional dangers of heat stroke, heat prostration, and dehydration. To prevent these problems, it may be necessary to

1. Provide fluid replenishment, frequent rest periods, and work crew changes.
2. Consider starting work earlier in the morning to avoid the hottest part of the day.
3. Establish temperature limits at which operations will be terminated.

During cold weather operations, there is a danger of frostbite and malfunction of equipment. To prevent these problems, it may be necessary to provide

1. Frequent rest periods.
2. Warming tents or areas.
3. Frequent checks of protection equipment.
4. Work crew changes.

During prolonged operations, the following problems may arise:

1. Accumulated work crew fatigue.
2. Increased danger of accidents.
3. Logistics problems associated with air tanks, protective clothing changes, food, and equipment.

To prevent these problems, consideration should be given to providing:

1. Larger crews.
2. Larger support element.
3. More supervisors.
4. Multiple work crews.

As one can see, decontamination, the process of removing or neutralizing contaminations that have accumulated on personnel and equipment, is critical to health and safety at hazardous waste sites. Decontamination protects workers from hazardous substances that may contaminate and eventually permeate the protective clothing, respiratory equipment, tools, vehicles, and other equipment used on site; it protects all site personnel by minimizing the transfer of harmful materials into clean areas; it helps prevent mixing of incompatible chemicals; and it protects the community by preventing uncontrolled transportation of contaminants from the site. This chapter does not cover decontamination of radioactivity-contaminated personnel or equipment. A health physicist should be consulted if this situation arises.

SURVEY AND RECONNAISSANCE

Introduction

The team(s) initially entering the site must accomplish one or more of the following objectives:

1. Characterize the hazards that exist or potentially exist affecting the public health, the environment, and response personnel.
2. Verify existing information and/or obtain data about the incident.
3. Evaluate the need for prompt mitigation actions.
4. Collect supplemental information to determine the safety requirements for personnel initially and subsequently entering the site.

Depending on the time available, the risks that may exist before the team enters the site, as much information as possible needs to be collected concerning the types of hazards, degree of hazards, and risks that may exist. Using available information (shipping manifests, transportation placards, existing records, container labels, etc.) or offsite studies, the team assesses the hazards, determines the need to go to the site, and identifies initial safety requirements.

Preliminary On-Site Evaluation

The initial on-site survey is to determine, on a preliminary basis, hazardous or potentially hazardous conditions. The main effort is to rapidly identify the immediate hazards that may affect the public, response personnel, and the environment. Of major concern are the real potential dangers — fire, explosion, oxygen-deficient atmospheres, radiation, airborne contaminants, containerized or pooled hazardous substances — that could affect workers during subsequent operations. All efforts preceding and involving initial site entries must be conducted in accordance with OSHA Standard 1910.120, paragraph (c).

Organic Vapors and Gases

If the type(s) of organic substances(s) involved in an incident is (are) known and the material is volatile or can become airborne, air measurements for organics should be made with one or more appropriate, properly calibrated survey instruments.

When the presence or types of organic vapors/gases are unknown, instruments such as a photoionizer (HNU Systems or AID/OVM-580) and/or a portable gas chromatograph (Century Systems OVA), operated in the total readout mode, should be used to detect organic vapors. Until specific constituents can be identified, the readout indicates total airborne substances to which the instrument is responding. Identification of the individual vapor/gas constituents may permit the instruments to be calibrated and used for more specific analysis.

Sufficient data should be obtained during the initial entry to map or screen the site for various levels of organic vapors. These gross measurements can be used on a preliminary basis to (1) determine levels of personnel protection, (2) establish site work zones, and (3) select candidate areas for more thorough qualitative and quantitative studies.

Higher-than-background readings on the HNU Systems photoionizer or the Century Systems OVA gas chromatograph may also indicate the displacement of oxygen or the presence of combustible vapors.

Inorganic Vapors and Gases

The ability to detect and quantify nonspecific inorganic vapors and gases is extremely limited. Presently, the photoionizers have minimal detection capability, while the Century Systems OVA gas chromatograph has none. If specific inorganics are known or suspected to be present, measurements should be made with appropriate instruments, if available. Colorimetric tubes can be used if substances present are known (or can be narrowed to a few) and appropriate tubes are available.

Radiation

Although radiation monitoring is not necessary for all responses, it should be incorporated in the initial survey where radioactive materials may be present, i.e., fires at warehouses or storage facilities for hazardous materials, transportation incidents involving unknown materials, or abandoned waste sites.

Normal γ-radiation background is approximately 0.01 to 0.02 mrad/hr on a gamma survey instrument. Work can continue with elevated radiation exposure rates; however, if the exposure rate increases to 3 to 5 times above gamma background, a qualified health physicist should be consulted. At no time should work continue with an exposure rate of 10 mrad/hr or above without the advice of a health physicist. The EPA's Office of Air, Noise, and Radiation has radiation specialists in each region, as well as at headquarters, Montgomery, Alabama, and Las Vegas, Nevada, to assist.

The absence of gamma readings above background should not be interpreted as the complete absence of radioactivity. Radioactive materials emitting low-energy γ–, α–, or β-radiation may be present, but for a number of reasons may not cause a response on the instrument. Unless airborne, these radioactive materials should present minimal hazard, but more thorough survey should be conducted as site operations continue in order to completely eliminate the presence of any radioactive material.

Oxygen Deficiency

At sea level, ambient air typically contains approximately 21% oxygen. Atmospheres consisting of less than 19.5% oxygen require air-supplied respiratory protection equipment. Oxygen measurements are of particular importance for work in enclosed spaces, low-lying areas, or in the vicinity of accidents that have produced heavier-than-air vapors which could displace ambient air. These oxygen-deficient areas are also prime locations for taking further organic vapor and combustible gas measurements, since the air has been displaced by other substances. Oxygen-enriched atmospheres increase the potential for fires.

Combustible Gases

The presence or absence combustible vapors or gases must be determined. If readings approach or exceed 10% of the lower explosive limit (LEL), extreme caution should be exercised in continuing the investigation. If readings approach or exceed 20% of the LEL, personnel should be withdrawn immediately.

Before resuming any on-site activities, project personnel, in consultation with experts in fire and explosion prevention, must develop procedures for continuing operations.

Visual Observations

While at the site, the initial entry team should make visual observations that would help in evaluating site hazards, i.e., dead fish or other animals; land features; wind direction; labels on containers indicating explosive, flammable, toxic, or corrosive materials; conditions conductive to splash or contact with unconfined liquids, sludges, or solids; and other general conditions.

Direct-Reading Instruments

A variety of toxic air pollutants, including organic and inorganic vapors, gases, or particulates, can become airborne on a hazardous substance site. Direct-reading field instruments will not detect or measure all of these substances. Thus, negative readings should not be interpreted as the complete absence of airborne toxic substances. Verification of negative results can only be done by collecting air samples and analyzing them in a laboratory.

Initial Surveys

In general, the initial survey is considered a relatively rapid screening process for collecting preliminary data on site hazards. The time needed to conduct the initial survey depends on the urgency of the situation, the type of incident, the information needed, the size of site, the availability of resources, the level of protection required for initial entry personnel, etc. Consequently, initial surveys may take hours or days to complete and may consist of more than one entry onto the site.

Priority for Initial Entry Monitoring

Of immediate concern to initial entry personnel are atmospheric conditions that could affect the immediate safety of these personnel (Exhibit 12.3). These condi-

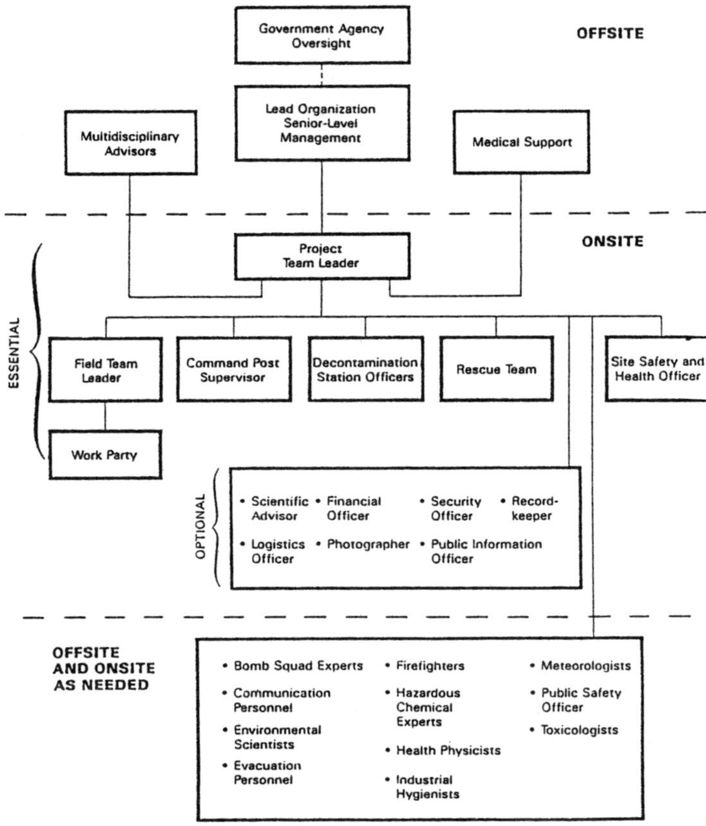

tions are airborne toxic substances, combustible gases or vapors, lack of oxygen, and to a lesser extent, ionizing radiation. Priorities for monitoring these potential hazards should be established after a careful evaluation of conditions.

When the type(s) of material(s) involved in the investigation is (are) identified and its (their) release into the environment suspected or known, the chemical/physical properties of the material(s) and the prevailing weather conditions may help determine the order of monitoring. An unknown substance(s) or situation(s) presents a more difficult monitoring problem.

In general, when poorly ventilated spaces, i.e., buildings, ship holds, boxcars, or bulk tanks, are entered, combustible vapors/gases and oxygen-deficient atmospheres should be monitored first by team members wearing, as a minimum, Level B protective equipment. Measurement of toxic gases/vapors and radiation, unless known not to be present, should be the next priority.

For open, well-ventilated areas, flammable gases and oxygen deficiency are lesser hazards and require lower priority. However, areas of lower elevation at the site (such as ditches and gulleys) and downwind areas may have flammable gas mixtures in addition to toxic vapors or gases and may lack sufficient oxygen to sustain life. Entry teams should approach and monitor whenever possible from the upwind area.

Periodic Monitoring

The monitoring surveys made during the initial site entry phase are for a preliminary evaluation of atmospheric hazards. In some situations, the information obtained may be sufficient to preclude additional monitoring, i.e., a chlorine tank determined to be releasing no chlorine. Materials detected during the initial site survey call for a more comprehensive evaluation of hazards and analyses for specific components. A program must be established for monitoring, sampling, and evaluating hazards for the duration of site operations. Since site activities and weather conditions change, a continuous program to monitor atmospheric changes must be implemented using a combination of stationary sampling equipment, personnel monitoring devices, and periodic area monitoring with direct-reading instruments.

In accordance with OSHA Standard 1910.120 paragraph (h), the minimum requirements constituting periodic monitoring include when

1. Work begins on a different portion of the site.
2. Contaminants other than those previously identified are being handled.
3. A different type of operation is initiated (e.g., drum opening as opposed to exploratory well drilling).
4. Employees are handling leaking drums or containers or working in areas with obvious liquid contamination (e.g., a spill or lagoon).

Peripheral Monitoring

Whenever possible, atmospheric hazards in the areas adjacent to the on-site zone should be monitored with direct-reading instruments, and air samples should be taken before the initial entry for on-site monitoring. Negative instrument readings away from the site should not be construed as definite indications of on-site conditions, but only as another piece of information to assist in the preliminary evaluation.

Monitoring Instruments

It is imperative that personnel using monitoring instruments be thoroughly familiar with their use, limitations, and operating characteristics. All instruments have inherent constraints in their ability to detect and/or quantify the hazards for which they were designed. Unless trained personnel use instruments and assess data readout, air hazards can be grossly misinterpreted, endangering the health and safety of response personnel. In addition, only intrinsically safe instruments should be used until the absence of combustible gases or vapors can be confirmed.

SITE EMERGENCIES

The nature of work at hazardous waste sites makes emergencies a continual possibility, no matter how infrequently they may actually occur. Emergencies happen quickly and unexpectedly and require immediate response. At a hazardous waste site an emergency may be as limited as a worker experiencing heat stress, or as vast as an explosion that spreads toxic fumes throughout a community. Any hazard on site can precipitate an emergency; chemicals, biological agents, radiation, or physical hazards may act alone or in concert to create explosions, fires, spills, toxic atmospheres, or other dangerous and harmful situations. Table 12.1 lists common causes of site emergencies.

Site emergencies are characterized by their potential for complexity: controlled toxic chemicals may be numerous and unidentified; their effects may be synergistic. Hazards may potentiate one another — for example, a flammable spill feeding a fire. Rescue personnel attempting to remove injured workers may themselves become victims. This variability means that advance planning, including anticipation of different emergency scenarios and thorough preparation for contingencies, is essential to protect worker and community health safety.

Planning

When an emergency occurs, decisive action is required. Rapidly made choices may have far-reaching, long-term consequences. Delays of minutes can create life-threatening situations. Personnel must be ready to immediately rescue or respond; equipment must be on hand and in good working order. In order to handle emergencies effectively, planning is essential. For this purpose, a contingency plan should be developed.

A contingency plan is a written document that sets forth policies and procedures for responding to site emergencies. It should incorporate the following:

- Personnel:
  Roles
  Lines of authority
  Training
  Communication
- Site:
  Mapping
  Security and control
  Refuges
  Evacuation routes
  Decontamination stations
- Medical/first aid
- Equipment
- Emergency procedures
- Documentation
- Reporting

Overall, a contingency plan should:

- Be designed as a discrete section of the Site Safety Plan.

Table 12.1. Causes of Emergencies at Hazardous Waste Sites

Work-Related
Minor accidents (slips, trips, falls)
Chemical exposures
Medical problems (heat stress, heat stroke, aggravations of pre-existing conditions)
PPE failure (air source failure, tearing or permeations of protective clothing, face piece fogging)
Physical injury (injuries for hot or flying objects, loose clothing entangling in machinery, serious falls, vehicle accident)
Electrical (burns, shock, electrocution)
Waste-Related
Fire
Explosion
Leak
Release of toxic vapors
Reaction of incompatible chemicals
Collapse of containers
Discovery of radioactive materials

- Be compatible and integrated with the pollution response, disaster, fire, and emergency plans of local, state, and federal agencies.
- Be rehearsed regularly using drills and mock situations.
- Be reviewed periodically in response to new or changing site conditions or information.

Federal Response Organizations

Site emergencies involving significant chemical releases should be coordinated with federal response organizations. The federal government has established a National Contingency Plan (NCP) to promote the coordination and direction of federal and state response systems and to encourage the development of local government and private capabilities to handle chemical emergencies involving chemical releases.

To implement the NCP, a national organization was established, including a National Response Team (NRT), a network of Regional Response Teams (RRTs), a cadre of On-Scene Coordinators (OSCs), and a National Response Center (NRC). The NRC is the national terminal point for receipt of notification of

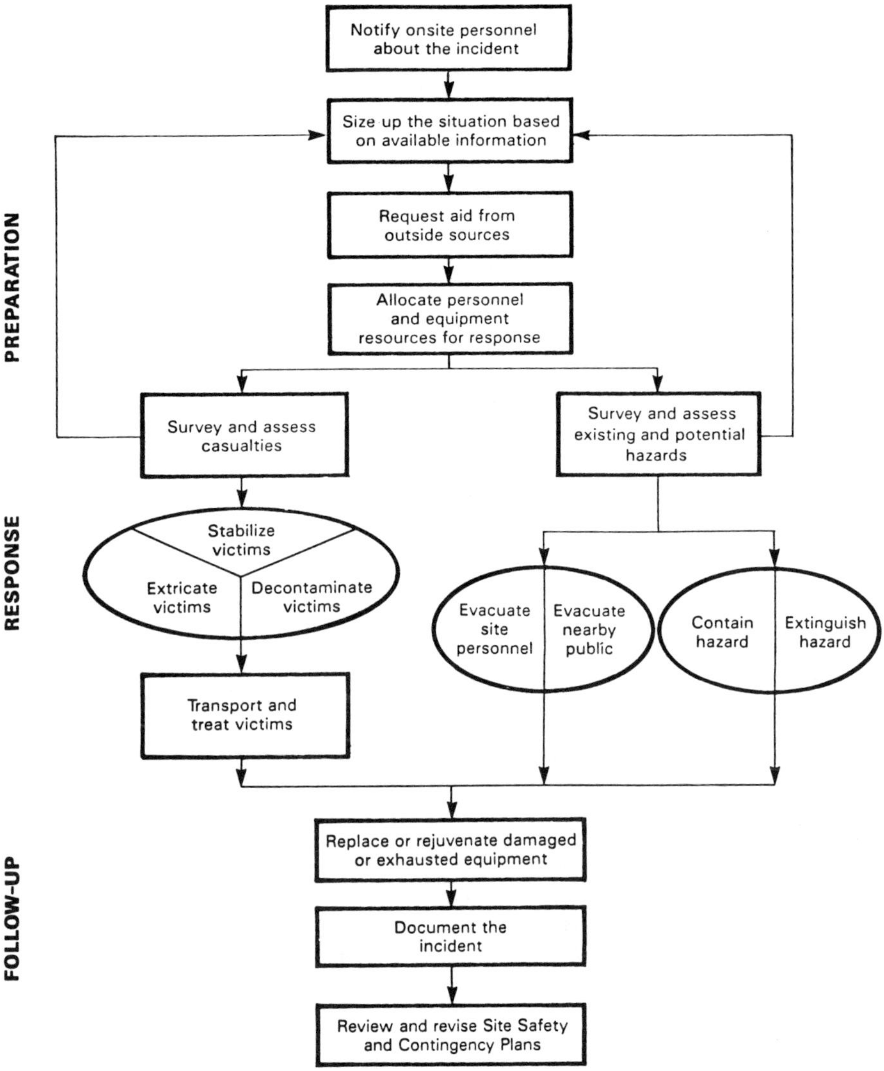

Figure 12.3. Emergency response operations.

significant chemical release, and the OSCs are the interface between the on-site personnel and the federal official responsible for ensuring that necessary response actions are taken to protect the public and the environment from the effects of a chemical release. Many federal agencies have specific technical expertise which is available to assist the OSC.

If a significant chemical release occurs at a hazardous waste site, the NRC in Washington, DC, should be contacted (Telephone: 800-424-8802). The NRC will activate federal response under the NCP.

Emergency Recognition and Prevention

On a day-to-day basis, individual personnel should be constantly alert for indicators of potentially hazardous situations and for signs and symptoms in themselves and others that warn of hazardous conditions and exposures. Rapid recognition of dangerous situations can avert an emergency. Before daily work assignments, regular meetings should be held. Discussion should include

- Tasks to be performed.
- Time constraints (e.g., rest breaks, air tank changes, etc.).
- Hazards that may be encountered, including their effects, how to recognize symptoms or monitor them, concentration limits, or other danger signals.
- Emergency procedures.

After daily work assignments, a debriefing session should be held to review work accomplished and problems observed. Figure 12.3 illustrates emergency response operations.

Follow-Up

Before normal site activities are resumed at a hazardous waste site that has been under emergency response procedures, personnel must be fully prepared and equipped to handle another emergency.

- Notify appropriate government agencies as required. For example, OSHA must be notified if there have been any fatalities or five or more hospitalizations.
- Restock all equipment and supplies. Replace or repair damaged equipment. Clean and refuel equipment for future use.
- Review and revise all aspects of the contingency plan according to new site conditions and lessons learned from the emergency response. When reviewing the information, consider typical questions such as

Cause:	What caused the emergency?
Prevention:	Was it preventable? If so, how?
Procedures:	Were inadequate or incorrect orders given or actions taken? Were these the result of bad judgment, wrong or insufficient information, or poor procedures? Can procedures or training be improved?
Site profile:	How does the incident affect the site profile? How are other site cleanup activities affected?
Community:	How is community safety affected?
Liability:	Who is liable for damage payments?

In summation, all emergency responses at a hazardous waste site for personnel, medical, equipment, etc., should be well documented. The individual in charge, usually the project team leader, should initiate the investigation and documentation of the incident. The documentation of these responses and the investigation findings may then be used to help avert recurrences, as evidence in future legal action, for assessment of liability by insurance companies, and for review by government agencies.

REFERENCES

1. NIOSH, OSHA, USCG, EPA. "Occupational Safety and Health Guideance Manual for Hazardous Waste Site Activities". DHHA (NIOSH) Publication No. 85-115. U.S. Government Printing Office, Washington, D.C. (1985).
2. Ehrhardt, Robert F., "How Clean is Clean?" Clean-Up Standards for Groundwater and Soil. Domes and Moore. Bethesda, Maryland, (July 1986).
3. Levine, S. P. and W. F. Martin, Eds., *Protecting Personnel at Hazardous Waste Sites*. (Butterworth Publishers, Boston, Mass. 1985).
4. Envrironmental Montiroing Stystems Laboratory, (1983). Characaterization of Hazardous Waste Sites - A Methods Manual. Volume II, Available Sampling Methods, U.S. EPA, reproduced by the U.S. Dept. of Commerce National Technical Information Services (NTIS), Springfield, VA.
5. U.S. EPA. Standard Operating Safety Guidelines, U.S. EPA, Washington, D.C. (1984)
6. Keith, Lawrence H., *Environmental Sampling and Analysis: A Practical Guide*. (Lewis Publishers, Chelsea, MI 1991).

CHAPTER 13

Decontamination

INTRODUCTION

Personnel working on hazardous waste sites responding to emergency spills or reducing toxic exposure of a chemical(s) may become contaminated in a number of ways:

1. Contacting vapors, gases, mists, or particulates in the air.
2. Being splashed by materials while sampling or opening containers.
3. Walking through puddles of liquids or on contaminated soil.
4. Using contaminated instruments or equipment.

Protective clothing and respirators help prevent the wearer from becoming contaminated or inhaling contaminants, while good work practices help reduce the contamination of protective clothing, instruments, and equipment.

Even with these safeguards, contamination may occur. Harmful materials can be transferred into clean areas, exposing unprotected personnel. In removing contaminated clothing, personnel may come into contact with and/or inhale the contaminants. To prevent such occurrences, contamination reduction and decontamination procedures must be developed and implemented. Such procedures must be in place before anyone enters a site and they must continue (modified as necessary) throughout the period of site operation.

Decontamination involves physically removing contaminants and/or converting them chemically into innocuous substances. How extensive decontamination must be depends on a number of factors, the most important being the types of contaminants involved. The more harmful the contaminant, the more extensive and thorough decontamination must be. Less harmful contaminants may require less decontamination. Combining decontamination, the correct doffing of protective equipment, and the zoning of site work areas minimizes cross-contamination from protective clothing to wearer, equipment to personnel, and one area to another. Only general guidance can be given on methods and techniques for decontamination. The exact procedure is determined by evaluation of a number of factors specific to the incident.

PRELIMINARY CONCERNS

Initial Planning

The initial decontamination plan is based on the assumption that all personnel and equipment leaving the exclusion zone (area of potential contamination) are grossly contaminated.

The plan includes a system for washing and rinsing, at least once, all the protective equipment worn. The washing and rinsing are done in combination with a sequential doffing of equipment, starting at the first station with the most heavily contaminated article. Each piece of clothing or operation requires a separate station.

The spread of contaminants during the washing and doffing process is further reduced by separating each decontamination station by a minimum of 3 ft. Ideally, contamination should decrease as a person moves from one station to another along the line.

While planning site operations, methods should be developed to prevent the contamination of people and equipment. For example, using remote sampling techniques, opening containers by nonmanual means, bagging monitoring instruments, using drum grapplers, watering down dusty areas, and avoiding areas of obvious contamination would reduce the probability of contamination and would preclude a more elaborate decontamination procedure.

The initial decontamination plan is based on a worst-case situation (if no information is available about the incident). Specific conditions at the site are then evaluated, including

1. Type of contaminant.
2. Amount of contamination.
3. Levels of protection required.
4. Type of protective clothing worn.

The initial plan is modified by eliminating unnecessary stations or it is otherwise adapted to site conditions. For instance, the plan might require a complete wash and rinse of protective garments. If disposable garments are worn, the wash and rinse step could be omitted. Wearing disposable boot covers and gloves could eliminate the washing and rinsing of gloves and reduce the number of stations needed.

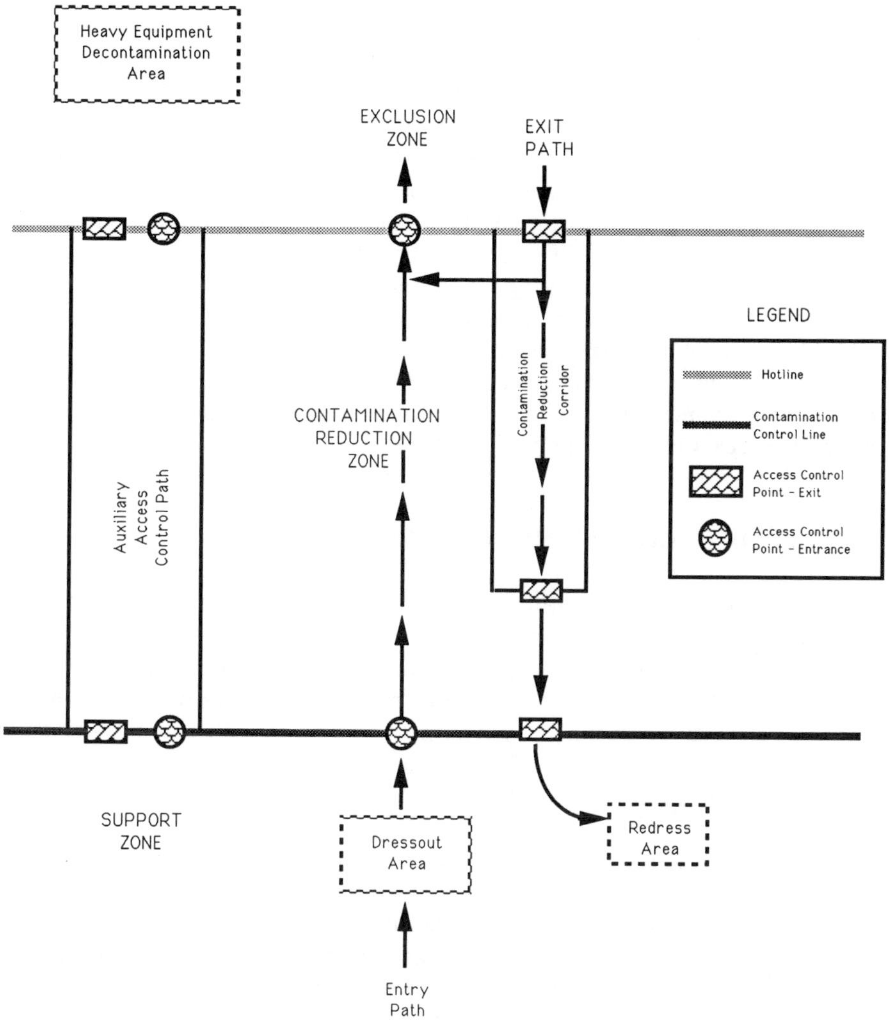

Figure 13.1. Contamination reduction zone layout.

CONTAMINATION REDUCTION CORRIDOR

An area within the contamination reduction zone (Figure 13.1) is designated the contamination reduction corridor (CRC). The CRC controls access into and out of the exclusion zone and confines personnel decontamination activities to a limited area. The size of the corridor depends on the number of stations in the decontamination procedure, the overall dimensions of work control zones, and the amount of space available at the site. A corridor of 75 by 15 ft should be adequate for full decontamination. Whenever possible, it should be a straight path.

The CRC boundaries should be conspicuously marked, with entry and exit restricted. The far end is the hotline — the boundary between the exclusion zone and the contamination reduction zone. Personnel exiting the exclusion zone must go through the CRC. Anyone in the CRC should be wearing the level of protection designated for the decontamination crew. Another corridor may be required for the entrance and exit of heavy equipment needing decontamination. Within the CRC, distinct areas are set aside for the decontamination of personnel, portable field equipment, removed clothing, etc. These areas should be marked and restricted to those personnel wearing the appropriate level of protection. All activities within the corridor are confined to decontamination.

Protective clothing, respirators, monitoring equipment, sampling supplies, and other equipment are all maintained outside of the CRC. Personnel don their protective equipment away from the CRC and enter the exclusion zone through a separate access control point at the hotline.

EXTENT OF DECONTAMINATION REQUIRED

Modifications of Initial Plan

The original decontamination plan must be adapted to conditions specific to each incident. These conditions may require more or less personnel decontamination then planned, depending on a number of factors.

Type of Contaminant

The extent of personnel decontamination depends on the effects the contaminants have on the body. Contaminants do not exhibit the same degree of toxicity (or other hazard). The more toxic a substance is, the more extensive or thorough decontamination must be. Whenever it is known or suspected that personnel can become contaminated with highly toxic or skin-destructive substances, a full decontamination procedure should be followed. If less hazardous materials are involved, the procedure can be downgraded.

Amount of Contamination

The amount of contamination on protective clothing is usually determined visually. If the clothing is badly contaminated, a thorough decontamination is generally required. Gross material remaining on the protective clothing for any extended period of time may degrade or permeate it. This likelihood increases with higher air concentrations and greater amounts of liquid contamination. Gross contamination also increases the probability of personnel contact. Swab or wipe tests may help determine the type and quantity of surface contaminants, and in some cases, even a bulk sample of the clothing in addition to skin swab/wipe tests.

Levels of Protection

The level of protection and specific pieces of clothing worn determine on a preliminary basis the layout of the decontamination line. Each level of protection presents different problems with respect to the decontamination and doffing of the equipment. For example, the decontamination of the harness straps and backpack assembly of the self-contained breathing apparatus (SCBA) is difficult. A butyl rubber apron worn over the harness makes decontamination easier. Clothing variations and different levels of protection may require the addition or deletion of stations used in the original decontamination procedure.

Work Function

The work that each person does determines the potential for contact with hazardous materials. In turn, this dictates the layout of the decontamination line. Observers, photographers, operators of air samplers, or others in the exclusion zone performing tasks that will not bring them in contact with contaminants may not need, for example, to have their garments washed and rinsed. Others in the exclusion zone with a potential for direct contact with the hazardous material will require more thorough decontamination. Different decontamination lines could be set up for different job functions, or certain stations in a line could be omitted for personnel performing certain tasks.

Location of Contamination

Contamination on the upper areas of protective clothing poses a greater risk to the worker because volatile compounds could make breathing hazardous both for the worker and for the decontamination personnel. There is also an increased probability of contaminant contact with the skin when the worker is doffing the upper part of the clothing.

Reason for Leaving the Site

The reason for leaving the exclusion zone also determines the need for and extent of decontamination. A worker leaving the exclusion zone to pick up or drop off tools or instruments and immediately returning may not require decontamination. A worker leaving to get a new air cylinder or change a respirator or canister, however, may require some degree of decontamination. Individuals departing the CRC at break time, lunch time, or the end of the day must be thoroughly decontaminated.

Effectiveness of Decontamination

There is no method of determining immediately how effective decontamination is in removing contaminants. Discolorations, stains, corrosions, and residues on objects may indicate that contaminants have not been removed. However, observable effects only indicate surface contamination and not permeation (absorption) into clothing. Many contaminants are not easily observed. One method for determining the effectiveness of surface decontamination is swipe testing. Cloth or paper patches — swipes — are wiped over predetermined surfaces of the suspect object and later analyzed in a laboratory. Both the inner and outer surfaces of protective clothing should be swipe tested. Positive results for both sets of swipe tests would indicate that surface contamination has not been removed and substances have penetrated or permeated the garment. Swipe tests can also be performed on skin or inside clothing. Laboratory analysis is required to determine if protective garments have been permeated. Both swipe and permeation testing provide after-the-fact information. Along with visual observations, results of these tests can help in ascertaining the effectiveness of decontamination.

Equipment

Decontamination equipment, materials, and supplies are generally selected on the basis of availability. The ease of equipment decontamination and disposability are also considered. Most equipment and supplies can be easily procured. Soft-bristle scrub brushes or long-handle brushes are used to remove contaminants. Buckets of water or garden sprays are used for rinsing. Large galvanized wash tubs, stock tanks, or children's wading pools can also be used as containers for wash and rinse solutions. Large plastic garbage cans or similar containers lined with plastic bags are useful for the storage of contaminated clothing and equipment, and metal or plastic cans or drums are convenient for the temporary storage of contaminated liquids. Other gear include paper or cloth towels for drying protective clothing and equipment.

Decontamination Solution

Protective equipment, sampling tools, and other equipment are usually decontaminated by scrubbing with detergent water using a soft-bristle brush, followed by rinsing with copious amounts of water. While this process may not be fully effective in removing some contaminants (in some cases, the contaminants may react with water), it is a relatively safe option compared to the use of a decontamination solution. The contaminant must be identified before a decon chemical is used, and reactions of such a chemical with unidentified substances or mixtures could be especially troublesome. A decontamination solution must be selected in consultation with an experienced chemist.

Establishment of Procedures

Once decontamination procedures have been established, all personnel requiring decontamination must be given precise instructions (and practice, if necessary). Compliance with the procedures must be frequently checked. The time it takes for decontamination must be ascertained. Personnel wearing SCBAs must leave their work areas with sufficient air to walk to the CRC and go through decontamination.

DECONTAMINATION DURING MEDICAL EMERGENCIES

Basic Considerations

Part of the overall planning for incident response is managing medical emergencies. The plan should provide the following:

1. Full training of some response team members in first aid and cardiopulmonary resuscitation.
2. Arrangements with the nearest medical facility for transportation and treatment of the injured and for the treatment of personnel suffering from exposure to chemicals.
3. Consultation with a toxicologist.
4. Emergency eye washes, showers, and wash stations.
5. First aid kits, blankets, a stretcher, and a resuscitator.

In addition, the plan should have established methods for decontaminating personnel with medical problems and injuries. There is the possibility that decontamination may aggravate a health problem or cause more serious problems. If prompt life-saving first aid and/or medical treatment is required, decontamination procedures should be omitted. Whenever possible, response personnel should accompany contaminated victims to the medical facility to advise on matters involving decontamination.

Physical Injury

Physical injuries can range from a sprained ankle to a compound fracture, or from a minor cut to massive bleeding. Depending on the seriousness of the injury, treatment may be given on the site by trained response personnel. For more serious injuries, additional assistance may be required at the site or the victim may have to be treated at a medical facility.

Life-saving care should be instituted immediately without considering decontamination. The outside garments can be removed (depending on the weather) if this does not cause delays, interfere with treatment, or aggravate the problem.

Respiratory masks and backpack assemblies must always be removed. Fully encapsulating suits or chemical-resistant clothing can be cut away. If the outer contaminated garments cannot be safely removed, the individual should be wrapped in plastic, rubber, or blankets to help prevent contaminating medical personnel and/or the inside of ambulances. Outside garments are then removed at the medical facility. No attempt should be made to wash or rinse the victim unless it is known that he has been contaminated with an extremely toxic or corrosive material that could also cause severe injury or loss of life. For minor medical problems or injuries, the normal decontamination procedure should be followed.

Heat Stress

Heat-related illnesses range from heat fatigue to heat stroke, the latter being the most serious. Heat stroke requires prompt treatment to prevent irreversible damage to health or death. Protective clothing may have to be cut off. Less serious forms of heat stress require prompt attention or they may lead to a heat stroke. Unless the victim is obviously contaminated, decontamination should be omitted or minimized and treatment begun immediately.

Chemical Exposure

Exposure to chemicals can be divided into two categories:

1. Direct contact through either touch (e.g., acid burns) or inhalation.
2. Indirect contact through gross contamination of clothing or equipment.

Injuries from contaminant inhalation can only be treated by qualified physicians. If the contaminant is on the skin or in the eyes, immediate measures must be taken to counteract its effect. First aid treatment usually involves flooding the affected area with water; however, for a few chemicals, water may cause more severe problems.

When protective clothing is grossly contaminated, contaminants may be transferred to the wearer or to treatment personnel and cause injuries. Unless severe medical problems could be created by splashing chemical, the protective clothing should be washed off as rapidly as possible and carefully removed.

PROTECTION FOR DECONTAMINATION WORKERS

Determination of Level of Protection

The level of protection worn by decontamination workers is determined by

1. Expected or visible contamination on workers.
2. Type of contaminant and associated respiratory and skin hazards.
3. Total vapor/gas concentrations in the CRC.
4. Particulates and specific inorganic or organic vapors in the CRC.
5. Results of swipe tests.

6. The presence (or suspected presence) of highly toxic or skin-destructive materials.

Level C Protection

Level C protection includes a full-face, canister-type air-purifying respirator, a hard hat with face shield (if splashing is a problem), chemical-resistant boots and gloves, and protective clothing. The body covering recommended is chemical-resistant overalls with either an apron or a jacket.

A face shield is recommended to protect against splashes, because respirators alone may not provide this protection. The respirator should have a canister approved for filtering known contaminants such as ammonia, organic vapors, acid gases, and particulates.

Level B Protection

Decontamination workers should wear Level B protection wherever and whenever they may be contaminated with highly volatile liquids, highly toxic materials, or other hazardous substances.

Level B protection includes an SCBA, a hard hat with face shield, chemical-resistant gloves, and protective clothing. The clothing suggested is chemical-resistant overalls, a jacket, and a rubber apron. The rubber apron protects the SCBA harness assembly and regulator from contaminants.

DECONTAMINATION OF EQUIPMENT

Basic Considerations

All possible measures should be taken to prevent the contamination of sampling and monitoring equipment. Sampling devices become contaminated, but monitoring instruments, unless they are splashed, usually do not. Once contaminated, instruments are difficult to clean without damaging them. Any delicate instrument that cannot be decontaminated easily should be protected while it is being used. It should be bagged, and the bag should be taped and secured around the instrument. Openings are made in the bag for sample intake.

DECONTAMINATION PROCEDURES

Sampling Devices

Sampling devices require special cleaning. The EPA regional laboratories can provide information on proper decontamination methods.

Tools

Wooden tools are difficult to decontaminate because they absorb chemicals. They should be kept on the site and handled only by protected workers. At the end of the response, such tools should be discarded. For decontaminating other tools, the EPA regional laboratories should be consulted.

Respirators

Certain parts of contaminated respirators, such as the harness assembly and leather or cloth components, are difficult to decontaminate. If grossly contaminated, they may have to be discarded. Rubber components can be soaked in soap and water and scrubbed with a brush. Regulators must be maintained according to the manufacturer's recommendations. Persons responsible for decontaminating respirators should be thoroughly trained in respirator maintenance.

Heavy Equipment

Bulldozers, trucks, backhoes, bulking chambers, and other heavy equipment are difficult to decontaminate. The method generally used is to wash them with water under high pressure and/or to scrub accessible parts with a detergent/water solution under pressure. In some cases, shovels, scoops, and lifts have been sandblasted or steam-cleaned. Particular care must be given to tires, scoops, and other components in direct contact with contaminants. Swab/wipe tests should be used to determine effectiveness of the decontamination process.

Sanitizing of PPE

In addition to being decontaminated, all respirators, protective clothing, and other personal articles must be sanitized before they can be used again. The insides of masks and clothing become soiled from exhalation, body oils, perspiration, bacteria, and viruses. The manufacturer's instructions should be followed in sanitizing the respirator mask. If practicable, protective clothing should be machine washed after a thorough decontamination; otherwise, it should be cleaned by hand.

Persistent Contamination

In some instances, clothing and equipment will become contaminated with substances that cannot be removed by normal decontamination procedures. A solvent may be used to remove such contamination from equipment if it does not destroy or degrade the protective material. If persistent contamination is expected, disposable garments should be used. Testing for persistent contamination of protective clothing and appropriate decontamination must be done by qualified laboratory personnel.

Disposal of Contaminated Materials

All materials and equipment used for decontamination must be disposed of properly. Clothing, tools, buckets, brushes, and all other equipment that is contaminated must be secured in drums or other containers and labeled. Clothing not completely decontaminated on the site should be secured in plastic bags before it is removed form the site.

Contaminated wash and rinse solutions can be kept temporarily in a step-in container (for example, child's wading pool) or in a plastic-lined trench about 4 in. deep. Such solutions are ultimately transferred to labeled drums and disposed of with other substances on the site.

DECONTAMINATION PROCEDURES FOR LEVELS OF PROTECTION A, B, AND C

Level A Decontamination

Equipment Worn

The full decontamination procedure outlined is for workers wearing Level A protection (with taped joints between gloves, boots, and suit). Such protection consists of

1. Fully encapsulating chemical-resistant suit with integral boots and gloves.
2. SCBA.
3. Hard hat (optional).
4. Chemical-resistant boots with steel toe and shank.
5. Boot covers.
6. Inner and outer gloves.

See Figure 13.2 for layout of Level A decontamination.

Procedure for Full Decontamination

Station 1: Segregated Equipment Drop

Deposit equipment used on the site (tools, sampling devices and containers, monitoring instruments, radios, clipboards, etc.) on plastic drop cloths or in different containers with plastic liners. Each will be contaminated to a different degree. Segregation at the drop reduces the potential for contamination. Equipment necessary:

1. Containers of various sizes.
2. Plastic liners.
3. Plastic drop cloths.

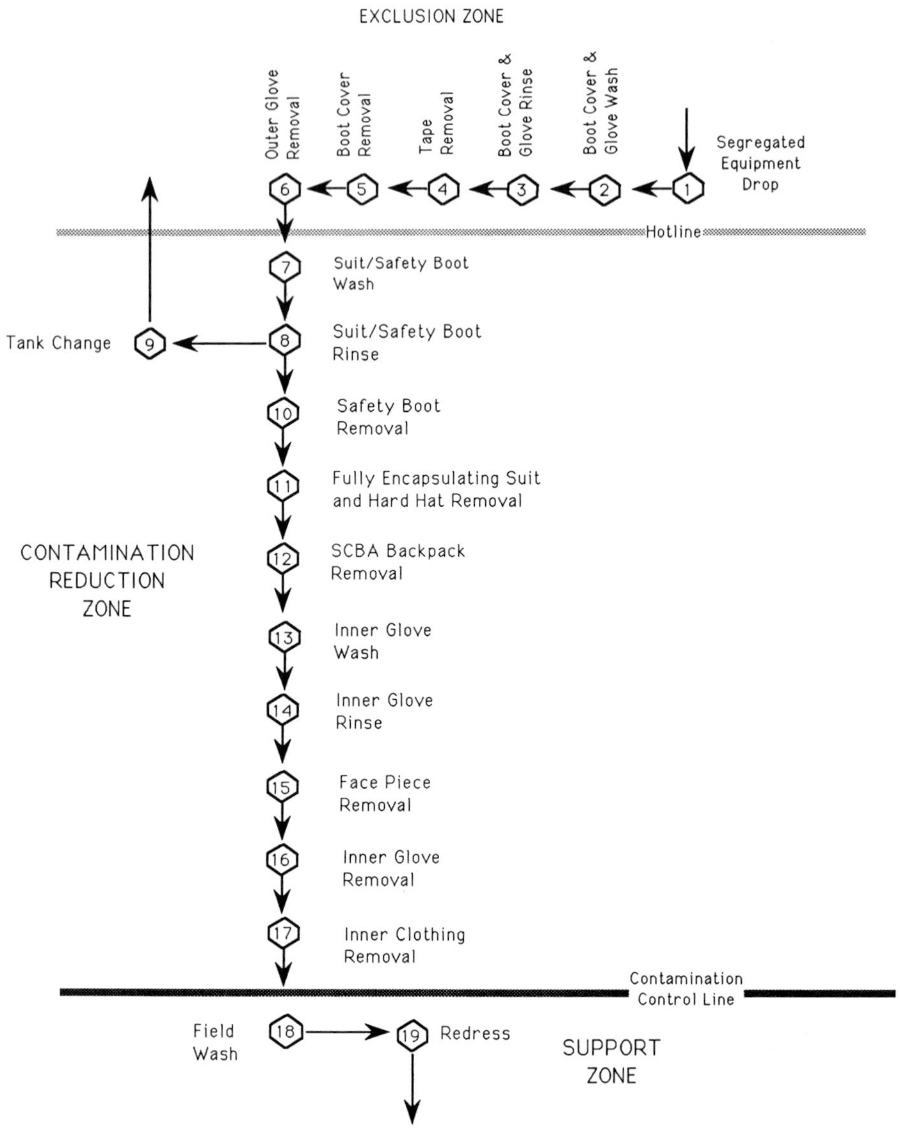

Figure 13.2. Layout for Level A decontamination.

Station 2: Boot Cover and Glove Wash

Scrub outer boot covers and gloves with decon solution or detergent/water solution. Equipment necessary:

1. Container (20 to 30 gal).
2. Decon solution.
3. Detergent/water solution.
4. Two or three long-handle, soft-bristle scrub brushes.

Station 3: Boot Cover and Glove Rinse

Rinse off decon solution from Station 2 using copious amounts of water. Repeat as many times as necessary. Equipment necessary:

1. Container (20 to 30 gal).
2. High-pressure spray unit.
3. Water.
4. Two or three long-handled, soft-bristle scrub brushes.

Station 4: Tape Removal

Remove tape around boots and gloves and deposit in container with plastic liner. Equipment necessary:

1. Container (20 to 30 gal).
2. Plastic liners.

Station 5: Boot Cover Removal

Remove boot covers and deposit in container with plastic liner. Equipment necessary:

1. Container (30 to 50 gal).
2. Plastic liner.
3. Bench or stool.

Station 6: Outer Glove Removal

Remove outer gloves and deposit in container with plastic liner. Equipment necessary:

1. Container (20 to 30 gal).
2. Plastic liners.

Station 7: Suit/Safety Boot Wash

Thoroughly wash fully encapsulating suit and boots. Scrub suit and boots with long-handle, soft-bristle scrub brush and copious amounts of decon solution or detergent/water solution. Repeat as many times as necessary. Equipment necessary:

1. Container (30 to 50 gal).
2. Decon solution.
3. Detergent/water solution.
4. Two or three long-handle, soft-bristle scrub brushes.

Station 8: Suit/Safety Boot Rinse

Rinse off decon solution or detergent/water solution using copious amounts of water. Repeat as many times as necessary. Equipment necessary:

1. Container (30 to 50 gal).

2. High-pressure spray unit.
3. Water.
4. Two or three long-handle, soft-bristle scrub brushes.

Station 9: Tank Change

If worker leaves exclusion zone to change air tank, this is the last step in the decontamination procedure. Worker's air tank is exchanged, new outer gloves and boot covers donned, and joints taped. Worker then returns to duty. Equipment necessary:

1. Air tanks.
2. Tape.
3. Boot covers.
4. Gloves.

Station 10: Safety Boot Removal

Remove safety boots and deposit in container with plastic liner. Equipment necessary:

1. Container (30 to 50 gal).
2. Plastic liners.
3. Bench or stool.
4. Boot jack.

Station 11: Fully Encapsulating Suit and Hard Hat Removal

With assistance of helper, remove fully encapsulating suit and hard hat. Hand suits on rack or lay out on drop cloths. Equipment necessary:

1. Rack.
2. Drop cloths.
3. Bench or stool.

Station 12: SCBA Backpack Removal

While still wearing face piece, remove backpack and place on table. Disconnect hose from regulator valve and proceed to next station. Equipment necessary:

1. Table.

Station 13: Inner Glove Wash

Wash with decon solution or detergent/water solution that will not harm skin. Repeat as many times as necessary. Equipment necessary:

1. Basin or bucket.
2. Decon solution.

3. Detergent/water solution.
4. Small table.

Station 14: Inner Glove Rinse

Rinse with water. Repeat as many times as necessary. Equipment necessary:

1. Water.
2. Basin or bucket.
3. Small table.

Station 15: Face Piece Removal

Remove face piece. Deposit in container with plastic liner. Avoid touching face with fingers. Equipment necessary:

1. Container (30 to 50 gal).
2. Plastic liners.

Station 16: Inner Glove Removal

Remove inner gloves and deposit in container with plastic liner. Equipment necessary:

1. Container (20 to 30 gal).
2. Plastic liners.

Station 17: Inner Clothing Removal

Remove clothing soaked with perspiration. Place in container with plastic liner. Do not wear inner clothing off the site, since there is a possibility that small amounts of contaminants have been transferred in removing fully encapsulating suit. Equipment necessary:

1. Container (30 to 50 gal).
2. Plastic liners.

Station 18: Field Wash

Shower if highly toxic, skin-corrosive, or skin-absorbable materials are known or suspected to be present. Wash hands and face if shower is not available. Equipment necessary:

1. Water.
2. Soap.
3. Small table.
4. Basin or bucket.
5. Field showers.
6. Towels.

Station 19: Redress

Put on clean clothes. A dressing trailer is needed in inclement weather. Equipment necessary:

1. Tables.
2. Chairs.
3. Lockers.
4. Clothes.

Full Decontamination (Situation 1) and Three Modifications

The preceding description outlines each station that is included in a complete worst-case decontamination protocol. It is obvious that different sites will present different hazard levels and, thus, that site-specific modifications of this protocol will be required. The following table illustrates the modification that can be made in response to a variety of conditions.

Situation 1

The individual entering the CRC is observed to be grossly contaminated, or extremely toxic substances are known or suspected to be present.

Situation 2

Same as Situation 1, except that individual needs new air tank and will return to exclusion zone.

Situation 3

Individual entering the CRC is expected to be minimally contaminated. Extremely toxic or skin-corrosive materials are not presented. No outer gloves or boot covers are on. Inner gloves are not contaminated.

Situation 4

Same as Situation 3, except that individual needs new air tank and will return to exclusion zone.

Level B Decontamination (Respiratory Protection)

Equipment Worn

The full decontamination procedure outlined is for workers wearing *Level B Protection* (with taped joints between gloves, boots, and suit). Such protection consists of

1. One-piece, hooded, chemical-resistant splash suit.
2. SCBA.
3. Hard hat.
4. Chemical-resistant boots with steel toe and shank.
5. Boot covers.
6. Inner and outer gloves.

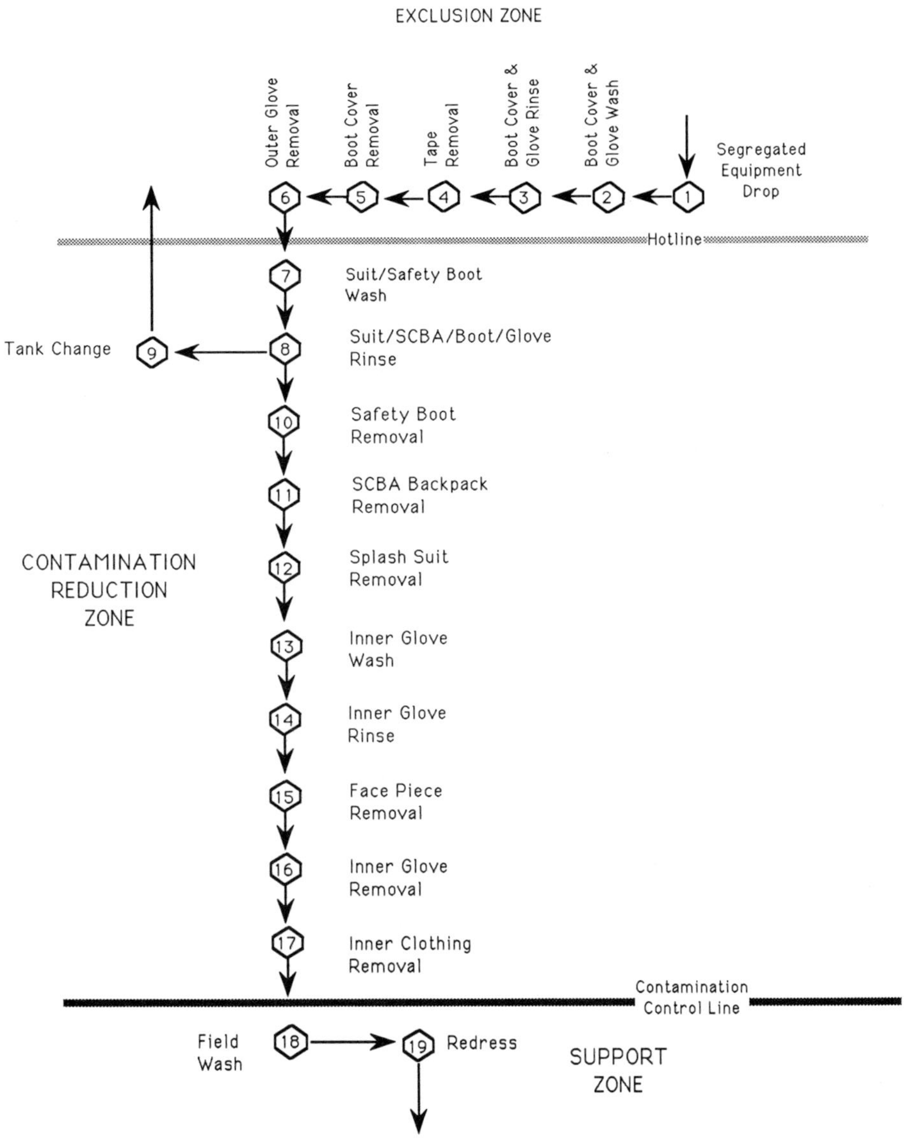

Figure 13.3. Layout for Level B decontamination.

Figure 13.3 depicts the layout for Level B decontamination.

Procedure for Full Decontamination

Station 1: Segregated Equipment Drop

Deposit equipment used on the site (tools, sampling devices and containers, monitoring instruments, radios, clipboards, etc.) on plastic drop cloths or in different containers with plastic liners. Each will be contaminated to a different degree. Segregation at the drop reduces the probability of cross-contamination. Equipment necessary:

1. Containers of various sizes.
2. Plastic liners.
3. Plastic drop cloths.

Station 2: Boot Cover and Glove Wash

Scrub outer boot covers and gloves with decon solution or detergent/water solution. Equipment necessary:

1. Container (20 to 30 gal).
2. Decon solution.
3. Detergent/water solution.
4. Two or three long-handle, soft-bristle scrub brushes.

Station 3: Boot Cover and Glove Rinse

Rinse off decon solution from Station 2 using copious amounts of water. Repeat as many times as necessary. Equipment necessary:

1. Container (20 to 30 gal).
2. High-pressure spray unit.
3. Water.
4. Two or three long-handled, soft-bristle scrub brushes.

Station 4: Tape Removal

Remove tape around boots and gloves and deposit in container with plastic liner. Equipment necessary:

1. Container (20 to 30 gal).
2. Plastic liners.

Station 5: Boot Cover Removal

Remove boot covers and deposit in container with plastic liner. Equipment necessary:

1. Container (30 to 50 gal).
2. Plastic liner.
3. Bench or stool.

Station 6: Outer Glove Removal

Remove outer gloves and deposit in container with plastic liner. Equipment necessary:

1. Container (20 to 30 gal).

2. Plastic liners.

Station 7: Suit/Safety Boot Wash

Thoroughly wash chemical-resistant splash suit, SCBA, gloves, and safety boots. Scrub suit and boots with long-handle, soft-bristle scrub brush and copious amounts of decon solution or detergent/water solution. Wash SCBA regulator (if belt-mounted type) with plastic to keep out water. Wash backpack assembly with sponges or cloths. Equipment necessary:

1. Container (30 to 50 gal).
2. Decon solution.
3. Detergent/water solution.
4. Two or three long-handle, soft-bristle scrub brushes.
5. Small buckets.
6. Sponges or cloths.

Station 8: Suit/SCBA/Boot/Glove Rinse

Rinse off decon solution or detergent/water solution using copious amounts of water. Repeat as many times as necessary. Equipment necessary:

1. Container (30 to 50 gal).
2. High-pressure spray unit.
3. Water.
4. Two or three long-handle, soft-bristle scrub brushes.
5. Small buckets.
6. Sponges or cloths.

Station 9: Tank Change

If worker leaves exclusion zone to change air tank, this is the last step in the decontamination procedure. Worker's air tank is exchanged, new outer gloves and boot covers donned, and joints taped. Worker then returns to duty. Equipment necessary:

1. Air tanks.
2. Tape.
3. Boot covers.
4. Gloves.

Station 10: Safety Boot Removal

Remove safety boots and deposit in container with plastic liner. Equipment necessary:

1. Container (30 to 50 gal).
2. Plastic liners.
3. Bench or stool.
4. Boot jack.

Station 11: SCBA Backpack Removal

While still wearing face piece, remove backpack and place on table. Disconnect hose from regulator valve and proceed to next station. Equipment necessary:

1. Table.

Station 12: Slash Suit Removal

With assistance of helper, remove splash suit. Deposit in container with plastic liner. Equipment necessary:

1. Container (30 to 50 gal).
2. Plastic Liners.
3. Bench or stool.

Station 13: Inner Glove Wash

Wash inner gloves with decon solution or detergent/water solution that will not harm skin. Repeat as many times as necessary. Equipment necessary:

1. Basin or bucket.
2. Decon solution.
3. Detergent/water solution.
4. Small table.

Station 14: Inner Glove Rinse

Rinse inner gloves with water. Repeat as many times as necessary. Equipment necessary:

1. Water.
2. Basin or bucket.
3. Small table.

Station 15: Face Piece Removal

Remove face piece. Deposit in container with plastic liner. Avoid touching face with gloves. Equipment necessary:

1. Container (30 to 50 gal).
2. Plastic liners.

Station 16: Inner Glove Removal

Remove inner gloves and deposit in container with plastic liner. Equipment necessary:

1. Container (20 to 30 gal).
2. Plastic liners.

Station 17: Inner Clothing Removal

Remove clothing soaked with perspiration. Place in container with plastic liner. Do not wear inner clothing off the site, since there is a possibility that small amounts of contaminants have been transferred in removing fully encapsulating suit. Equipment necessary:

1. Container (30 to 50 gal).
2. Plastic liners.

Station 18: Field Wash

Shower if highly toxic, skin-corrosive, or skin-absorbable materials are known or suspected to be present. Wash hands and face if shower is not available. Equipment necessary:

1. Water.
2. Soap.
3. Small table.
4. Basin or bucket.
5. Field showers.
6. Towels.

Station 19: Redress

Put on clean clothes. A dressing trailer is needed in inclement weather. Equipment necessary:

1. Tables.
2. Chairs.
3. Lockers.
4. Clothes.

Full Decontamination (Situation 1) and Three Modifications

The preceding description outlines each station that is included in a *complete worst-case decontamination protocol.* It is obvious that different sites will present different hazard levels and, thus, that site-specific modifications of this protocol will be required. The following table illustrates the modification that can be made in response to a variety of conditions.

Situation 1

The individual entering the CRC is observed to be grossly contaminated, or extremely toxic substances are known or suspected to be present.

Situation 2

Same as Situation 1, except that individual needs new air tank and will return to exclusion zone.

Situation 3

Individual entering the CRC is expected to be minimally contaminated. Extremely toxic or skin-corrosive materials are not present. No outer gloves or boot covers are worn. Inner gloves are not contaminated.

Situation 4

Same as Situation 3, except that individual needs new air tank and will return to exclusion zone.

Level C Decontamination (Respiratory Protection)

Equipment Worn

The full decontamination procedure outlined is for workers wearing Level C protection (with taped joints between gloves, boots, and suit). Such protection consists of

1. One-piece, hooded, chemical-resistant splash suit.
2. Canister-equipped full face mask.
3. Hard hat.
4. Chemical-resistant boots with steel toe and shank.
5. Boot covers.
6. Inner and outer gloves.

Procedure for Full Decontamination

Station 1: Segregated Equipment Drop

Deposit equipment used on the site (tools, sampling devices and containers, monitoring instruments, radios, clipboards, etc.) on plastic drop cloths or in different containers with plastic liners. Each will be contaminated to a different degree. Segregation at the drop reduces the probability of cross-contamination. Equipment necessary:

1. Containers of various sizes.
2. Plastic liners.
3. Plastic drop cloths.

Station 2: Boot Cover and Glove Wash

Scrub outer boot covers and gloves with decon solution or detergent/water solution. Equipment necessary:

1. Container (20 to 30 gal).
2. Decon solution.
3. Detergent/water solution.
4. Two or three long-handle, soft-bristle scrub brushes.

Station 3: Boot Cover and Glove Rinse

Rinse off decon solution from Station 2 using copious amounts of water. Repeat as many times as necessary. Equipment necessary:

1. Container (20 to 30 gal).
2. High-pressure spray unit.
3. Water.
4. Two or three long-handled, soft-bristle scrub brushes.

Station 4: Tape Removal

Remove tape around boots and gloves and deposit in container with plastic liner. Equipment necessary:

1. Container (20 to 30 gal).
2. Plastic liners.

Station 5: Boot Cover Removal

Remove boot covers and deposit in container with plastic liner. Equipment necessary:

1. Container (30 to 50 gal).
2. Plastic liner.
3. Bench or stool.

Station 6: Outer Glove Removal

Remove outer gloves and deposit in container with plastic liner. Equipment necessary:

1. Container (20 to 30 gal).
2. Plastic liners.

Station 7: Suit/Safety Boot Wash

Thoroughly wash splash suit and safety boots. Scrub with long-handle, soft-bristle scrub brush and copious amounts of decon solution or detergent/water solution. Repeat as many times as necessary. Equipment necessary:

1. Container (30 to 50 gal).
2. Decon solution.
3. Detergent/water solution.
4. Two or three long-handle, soft-bristle scrub brushes.

Station 8: Suit/Safety Boot Rinse

Rinse off decon solution or detergent/water solution using copious amounts of water. Repeat as many times as necessary. Equipment necessary:

1. Container (30 to 50 gal).
2. High-pressure spray unit.
3. Water.
4. Two or three long-handle, soft-bristle scrub brushes.

Station 9: Canister or Mask Change

If worker leaves exclusion zone to change canister (or mask), this is the last step in the decontamination procedure. Worker's canister is exchanged, new outer gloves and boot covers donned, and joints taped. Worker then returns to duty. Equipment necessary:

1. Canister (or mask).
2. Tape.
3. Boot covers.
4. Gloves.

Station 10: Safety Boot Removal

Remove safety boots and deposit in container with plastic liner. Equipment necessary:

1. Container (30 to 50 gal).
2. Plastic liners.
3. Bench or stool.
4. Boot jack.

Station 11: Splash Suit Removal

With assistance of helper, remove splash suit. Deposit in container with plastic liner. Equipment necessary:

1. Container (30 to 50 gal).
2. Bench or stool.
3. Plastic liner.

Station 12: Inner Glove Wash

Wash inner gloves with decon solution or detergent/water solution that will not harm skin. Repeat as many times as necessary. Equipment necessary:

1. Decon solution.
2. Detergent/water solution.
3. Basin or bucket.

Station 13: Inner Glove Rinse

Rinse inner gloves with water. Repeat as many times as necessary. Equipment necessary:

1. Water.

2. Basin or bucket.
3. Small table.

Station 14: Face Piece Removal

Remove face piece. Deposit in container with plastic liner. Avoid touching face with gloves. Equipment necessary:

1. Container (30 to 50 gal).
2. Plastic liners.

Station 15: Inner Glove Removal

Remove inner gloves and deposit in container with plastic liner. Equipment necessary:

1. Container (20 to 30 gal).
2. Plastic liners.

Station 16: Inner Clothing Removal

Remove clothing soaked with perspiration. Place in container with plastic liner. Do not wear inner clothing off the site, since there is a possibility that small amounts of contaminants have been transferred in removing fully encapsulating suit. Equipment necessary:

1. Container (30 to 50 gal).
2. Plastic liners.

Station 17: Field Wash

Shower if highly toxic, skin-corrosive, or skin-absorbable materials are known or suspected to be present. Wash hands and face if shower is not available. Equipment necessary:

1. Water.
2. Soap.
3. Small table.
4. Basin or bucket.
5. Field showers.
6. Towels.

Station 18: Redress

Put on clean clothes. A dressing trailer is needed in inclement weather. Equipment necessary:

1. Tables.
2. Chairs.
3. Lockers.
4. Clothes.

Full Decontamination (Situation 1) and Three Modifications

The preceding description outlines each station that is included in a complete worst-case decontamination protocol. It is obvious that different sites will present different hazard levels and, thus, that site-specific modifications of this protocol will be required. The following table illustrates the modification that can be made in response to a variety of conditions.

Situation 1:

The individual entering the CRC is observed to be grossly contaminated, or extremely skin-corrosive substances are known or suspected to be present.

Situation 2:

Same as Situation 1, except that individual needs new canister or mask and will return to exclusion zone.

Situation 3:

Individual entering the CRC is expected to be minimally contaminated. Extremely toxic or skin-corrosive materials are not present. No outer gloves or boot covers are worn. Inner gloves are not contaminated.

Situation 4:

Same as Situation 3, except that individual needs new canister or mask and will return to exclusion zone.

Equipment Worn

The full decontamination procedure outlined is for workers wearing Level A protection (with taped joints between gloves, boots, and suit). Such protection consists of:

1. Fully encapsulating suit with integral boots and gloves.
2. SCBA.
3. Hard hat (optional).
4. Chemical-resistant boots with steel toe and shank.
5. Boot covers.
6. Inner and outer gloves.

Procedure for Full Decontamination

Station 1: Segregated Equipment Drop

Deposit equipment used on the site (tools, sampling devices and containers, monitoring instruments, radios, clipboards, etc.) on plastic drop cloths or in different containers with plastic liners. Each will be contaminated to a different degree. Segregation at the drop reduces the probability of cross-contamination. Equipment necessary:

1. Containers of various sizes.
2. Plastic liners.
3. Plastic drop cloths.

Station 2: Outer Garment, Boot, and Glove Wash and Rinse

Scrub outer boot covers and gloves and fully encapsulating suit with decon solution or detergent/water solution. Rinse off using copious amounts of water. Equipment necessary:

1. Container (20 to 30 gal).
2. Decon solution.
3. Detergent/water solution.
4. Two or three long-handle, soft-bristle scrub brushes.
5. Rinse water.

Station 3: Outer Boot and Glove Removal

Remove outer boots and gloves. Deposit in container with plastic liner. Equipment necessary:

1. Container (20 to 30 gal).
2. Plastic liners.
3. Bench or stool.
4. Two or three long-handled, soft-bristle scrub brushes.

Station 4: Tank Changes

If worker leaves exclusive zone to change air tank, this is the last step in the decontamination procedure. Worker's air tank is exchanged, new outer gloves and boot covers are donned, joints taped, and worker returns to duty. Equipment necessary:

1. Air tanks.
2. Tape.
3. Boot covers.
4. Gloves.

Station 5: Boot, Glove, and Outer Garment Removal

Boots, fully encapsulating suit, and inner gloves are removed and deposited in separate containers lined with plastic. Equipment necessary:

1. Container (30 to 50 gal).
2. Plastic liner.
3. Bench or stool.

Station 6: SCBA Removal

SCBA backpack and face piece are removed. Hands and face are thoroughly washed. SCBA is deposited on plastic sheets. Equipment necessary:

1. Plastic sheets.
2. Basin or buckets.
3. Soap and towels.
4. Bench.

Station 7: Field Wash

Thoroughly wash hands and face. Shower as soon as possible. Equipment necessary:

1. Water.
2. Soap.
3. Tables.
4. Wash basin/bucket.

DOWNRANGE CONSIDERATIONS

Introduction

Until one has donned Level B respiratory protective equipment and protective clothing, it is difficult to appreciate the difficulties inherent in working in a hazardous atmosphere. Much preparation is needed in the areas of planning, organization, and equipment needs to compensate for the hardships endured working in toxic environments.

One basic difficulty when working downrange, particularly in Levels A and B, is the ability to identify entry party members. Positive identification is essential for control and safety. Techniques used include the following:

1. Marking outer garments with colored tape.
2. Writing name with felt-tip marker on masking tape.
3. Writing name on hard hat.
4. Writing name on protective clothing.

Whichever method is used, the identification should be located in a spot that is visible when all gear is being worn.

There are simple problems associated with protective clothing, such as the visual acuity problems encountered while wearing a protective mask or dexterity problems encountered while wearing protective gloves. When dressing out, the following points should be noted:

1. As the amount of protective equipment worn is increased
   a. The individual's ability to perform simple tasks is decreased.
   b. The individual's dexterity and mobility is decreased, including his/her ability to carry equipment and to turn knobs and adjust equipment.
   c. Vision problems are encountered, including a loss of visual acuity, a limited

field of vision, and a decrease in the individual's ability to use optical equip ment. These problems increase in cold weather due to the fogging of the mask.

2. Because of the encumbrance of protective clothing, it is necessary to
   a. Keep task simple.
   b. Plan fewer tasks per individual.
3. Wearing of protective equipment will require
   a. More time than is usually required to do a task.
   b. More people than are usually required to do a task.

Site Control

Designation of Work Zones

A primary means of maintaining site control in order to ensure safe, efficient work and to prevent migration of hazardous materials into uncontaminated areas is by designating work areas. The work areas serve to limit site access, contain gross contamination, provide site security, and place a buffer zone between the hazardous site and the community. General considerations for establishing the work zones include:

1. The extent of the contamination, locations of tanks, ponds, leachate streams, discolored soil, stressed vegetation.
2. The probability of airborne contamination — properties of wastes on site, wind speed and direction (predominate).
3. Terrain.
4. Location of local population (schools, residences, business).

There are typically three work zones established: an exclusion zone, a contamination reduction zone, and a support zone, which are defined in the site safety plan (SSP) and controlled by the site safety officer (SSO). Exhibit 11.1 illustrates the general principles of the site work zones.

Exclusion Zone

The exclusion zone is defined as the area which is considered to be contaminated, potentially contaminated, or which could become contaminated in the event of fire, explosion, or spill, and a buffer zone for additional safety. Defining the exclusion or "hot" zone is based on the preliminary site investigation work. Typical considerations in establishing the hot area are

1. Types and quantities of hazardous substances on site.
2. Type of containerization used — drums, ponds, tanks — and its present condition.
3. Properties of the hazardous substances — flammability, reactivity, toxicity, incompatability, radioactivity, corrosivity.
4. Activities that will take place within the zone — heavy equipment operation, drum staging, opening, and sampling.

Once these contaminated or potentially contaminated areas are defined, a buffer zone must be added to ensure the safety of the surrounding populace. Again, the properties of the hazardous substances must be taken into consideration. The buffer zone is to adequately protect the populace from airborne contamination or runoff contamination as the result of an accidental rupture of a container and subsequent release or generation of runoff due to rain. While the delineation of the exclusion area is necessary prior to initiating site work, it can be increased or decreased during the course of project as new data are collected and evaluated. The exclusion area should be adequately secured by fence, warning signs, terrain, or security personnel to prevent any individuals from inadvertently entering the area. Access control to the exclusion area through a rigidly defined entrance/exit point is specified in the SSP and enforced by the SSO. This ensures the proper levels of protection area worn within the area, that adequate personnel are downrange and accounted for, and that site security is maintained. This access point also serves as control of contamination, as it established the "hotline" and contamination reduction zone.

Contamination Reduction Zone

The contamination reduction zone lies upwind of the exclusion zone and is designed to prevent contamination from leaving the site as a result of the work parties' activities. Access control points to both the exclusion zone and support zone are maintained here. This zone provides an area for the decontamination of personnel, equipment, and samples and as an area to assist the work parties (air tank change, equipment staging). It is established in an area assumed to be clean, but is considered potentially contaminated as soon as personnel or equipment are processed from the exclusion area through the PDS.

The access point between the exclusion zone and contamination reduction zone is referred to as the "hotline." It is here that the PDS is set up, insuring the decontamination of all personnel, equipment and samples leaving the exclusion zone. The hotline should be clearly marked by flagging tape, stakes, paint, flags or other prominent means of delineating the control point. This should be the only entrance/exit point to the exclusion zone. However, emergency exit points should be established if there is a significant risk of fire or explosion. Also controlled at the boundary between the contamination reduction zone and exclusion zone is the level of protection. The size of the contamination reduction zone depends on the site activities. It should be large enough to accommodate the PDS and associated activities, but small enough to easily monitor and control these activities. Ideally, personnel in the contamination reduction zone are able to maintain visual contact with both the exclusion and support zone providing line of sight. The other access point between the contamination reduction zone and the support zone is known as the contamination control line. It serves as the entrance/exit point and provides some site security as well as contamination control. Personnel entering the contamination reduction zone must be properly dressed in the level of protection to work within the zone; personnel exiting must be properly decontaminated before entering the support zone.

Support Zone

The support zone is the area furthest away from the hazardous substances. It is where the command post is located and all support activities occur. Besides being located in a known noncontaminated area, several factors are considered in designating this work zone. Wind direction and speed is a prime consideration when locating the command post. Ideally, the support area should be upwind of the exlusion zone with no more than a 20° deviation in direction on either side of a straight line to the exclusion zone. Practically, a deviation of 90° or an arc of 180° can be tolerated. Wind speed variations can create turbulance and/or swirling depending on the site topography. A review of local climatic conditions during the time of year that site activities will take place can provide some insight. If access roads to the site are located on the downwind side, consider obtaining permission to enter from the upwind. Otherwise, the support zone should be located downwind at a distance to allow dispersion of airborne contamination. Continual air monitoring and wearing of respirators by personnel in the support zone may be necessary. Advantage of the terrain should be taken when designating the support zone to provide

1. The best means of access to the exclusion zone and allow sufficient room for the contamination reduction zone.
2. Provide easy access for emergency vehicles.
3. Utilize prominent landmarks for site boundaries and/or access points.
4. Provide line of sight to all activities in the exclusion zone.
5. Control site access.

Also to consider in locating the support zone is the access to a water supply, communications, and power. This is especially important during extended site work, such as remedial action projects. While water can be brought to the support area, availability to phone and power lines is desirable for emergency action and lighting, laboratory work, and personal comfort.

REFERENCES

1. Rosen, M. J. *Surfactants and Interfacial Phenomena* (New York: Wiley-Interscience, 1978), pp. 304.
2. Vo-Dinh, T. "Surface Detection of Contamination: Principles, Applications, and Recent Developments," *J. Environ. Sci.* pp. 40–43 (January/February 1983).
3. Vo-Dinh, T., and R. B. Gammage. "The Use of a Fiberoptics Skin Contamination Monitor in the Workplace," in *Chemcial Hazards in the Workplace* (Washington, DC: American Chemical Society 1981), pp. 269–281.
4. Vo-Dinh, T., and R. B. Gammage. "The Lightpipe Luminoscope for Monitoring Occupational Skin Contamination," *J. Am. Ind. Hyg. Assoc.* 42:112–120 (1981).

5. NIOSH/OSHA/USCG/EPA. "Occupational Safety and Health Guidance Manual for Hazardous Waste Site Activities," U.S. Department of Health and Human Services, U.S. Government Printing Office, Washington, DC, Document No. 85-115 (October 1985).

CHAPTER 14

Community Relations

INTRODUCTION

The subject of community relations involving hazardous materials is much more complex than traditional public relations activities on less controversial topics. A vast and growing number of citizens throughout the country perceive hazardous materials as undesirable elements of their communities. This perception is well documented by the emergence and popularity of the NIMBY (not in my backyard) syndrome. The information presented in this section highlights the process for assessing citizen attitudes toward hazardous materials-related issues and the methods that may be used to address these public concerns. This section includes guidelines for developing a plan of action and provides details on effective implementation of the plan.

Since the late 1970s, citizen awareness and public education regarding hazardous materials has been steadily increasing. The subject is often emotionally charged, with factions of the community disagreeing over the risks and benefits associated with hazardous materials-related operations. A very common example is the conflict generated by perceived health risks vs job security or career opportunities. In some cases, the community will challenge its opponent — either corporation or individual — in court litigation.

The development of an effective community relations program is essential for the successful siting or operation of a hazardous materials facility, for the successful remediation of a hazardous waste site, or for the acceptance of an emergency response plan designed to protect the health and welfare of the community in the event of an accident or malfunction.

Virtually nowhere in the U.S. can a hazardous materials facility be developed or operated without first involving the public. Moreover, the permitting process and other regulatory requirements provide citizens with the legal power, through the de facto veto, to influence current operations and proposed projects in a very real way. Hazardous waste site remediation, expected to accelerate, is increasingly being conducted under the watchful eyes of the public.

It may appear that the NIMBY syndrome is an insurmountable obstacle for projects and operations involving hazardous materials. However, that is not necessarily true. Opposition can be overcome, but only with an organized plan that addresses public concerns in a substantive manner.

ASSESSING THE SITUATION

Just as an organization would not construct a facility without thoroughly evaluating the site from an engineering perspective, the organization should not attempt to implement a community relations program without first evaluating the attitudes and concerns of citizens and interest groups. Accomplishing this task requires sensitivity and, if done correctly, can be the first step on the road to positive interaction with the public. A brief plan of action for the assessment should be approved by the organization's management. The development of a more comprehensive plan for the whole community relations program will be described later in this section.

The assessment process begins in file cabinets. Community relations personnel must search available records for insight on community attitudes toward environmental issues in general and toward the project at hand, specifically. Tangential issues must be noted for a possible translation into interest on this project. For example, although opposition to strip mining in a neighboring county may have no direct bearing on a hazardous material project, the same interest groups may surface for involvement.

Often, a good place to begin the search is in the morgue file of the local newspaper, where all published articles are stored. However, a word of caution is necessary: any movement into the community may carry project visibility with it. Community relations personnel or their equivalent must be discrete and must be careful not to inadvertently attract publicity before the plan has been developed.

In addition, to searching past clippings for information relevant to the project, it is crucial to become familiar with what the local press is currently publishing. A subscription to the area newspapers provides direct access, and the employment of a "clipping service" to supply only those articles of interest offers anonymity and convenience, particularly in instances where many papers are covering the subject. Other information regarding community interest may be obtained from client files as well.

After all possible insight is gained from files, contact can be made with "safe" sources. These sources may include company personnel who live in the area, friendly politicians, or others whose opinions are valid and who represent minimal risk, so that speaking with them does not create ripples in the community.

The purpose of the apparent secretiveness described above is not to attempt to conduct all operations behind the backs of the local residents. On the contrary, there is probably no better way to ensure defeat of the project than to attempt to slip it into the community under the cover of darkness. The purpose of the background research, rather, is to develop a thorough understanding of the situation in the community so that the concerns of the citizens can be addressed openly. Also, the extreme sensitivity of the subject of hazardous materials necessitates caution in the publicity of proposed actions. Groups may exist that would immediately mobilize opposition without waiting to hear the benefits of the proposition. This is further complicated by the phenomenon that people tend to easily form their opinions on the first viable option that is presented and will then tenaciously cling to that opinion even in the face of a better option. This phenomenon makes it very important to have the positive publicity reach the public first.

To complete an effective assessment of the situation from a community relations perspective, it is usually advisable to actually go into the community and meet the key players who have been identified in the background research. However, this must be done with care and planning, moving from the "safest" contacts to those who will most likely be organizing the opposition. An additional angle that should not be overlooked involves contacting the "average" citizens who have not voiced an interest in the project.

In all discussions in the community, the manner in which the interviewers present themselves is of paramount importance. They should begin by stating that the proposing organization is keenly interested in the viewpoints of the citizens and that the purpose of the discussions is to open the lines of communication. A phone number for questions that might later arise should be left with the contacts.

Such visible activity in the community often opens the gates for citizen and media attention. Preliminary publicity must be ready for use to ensure that the project benefits are promoted before any negative aspects are released by the opposition. Also, the interviewers should be prepared to present some of the positive points in their conversations. However, they must be very careful not to be perceived as "salespersons" for the project.

Special sensitivity and caution must be exercised when dealing with elected officials. It is often a good idea to talk with them early in the assessment process because they generally do not like to have people "snooping" around in their communities without their knowledge. A rule to remember is that local politicians do not like surprises. By keeping them informed, you are less likely to encounter unnecessary problems generated by bruised egos. Timing is of critical importance; be aware of hidden agendas in what was apparently a good relationship. It is also wise to know when election sensor begins in local communities. Also, some politicians make it a point to help the media by providing "scoops," so be prepared to handle information leaks.

In reality, the assessment of community attitudes is a never-ending process. However, there is a need to compile the information into a formal plan so that the program can proceed in an orderly and approved fashion. Ideally, the initial assessment is completed quickly to provide for early development of the plan.

PLANNING

The Community Relations Plan (CRP) is a working document that provides an overview of the technical aspects of the project, the history of community involvement to date, and a summary of the major issues that have either already been identified or are anticipated to develop. The CRP should also be a management plan, containing the objectives of the program, the proposed techniques to meet those objectives, and schedules and budgets for the program. Exhibit 14.1 shows a sample outline for a CRP.

Because of the dynamic nature of community relations, the CRP must be revised periodically to reflect new insight and changes in strategy. The plan does not have to be long, but it must be thorough.

Two important points must not be overlooked in the development of the plan. One is that a complete review of regulatory requirements for public participation be included. Second, plans must be made for crisis management. In order to plan effectively for a crisis, worst-case situations must be considered and detailed strategies developed. A list of the names and phone numbers of key personnel can be appended to the plan for quick access during a crisis. It is critical that this list be kept current to reflect personnel changes.

After the plan has been developed, the draft version should be reviewed and approved by those in authority in the organization. You can anticipate a great deal of discussion, because top management is not always amenable to the aggressive techniques that are necessary for an effective program.

IMPLEMENTING THE PLAN

The implementation of the plan — the actual use of the techniques — must take into consideration all other functions that are proceeding concurrently. For example, a regulatory permitting process will have major implications for the community relations program. Also, any site investigations by the technical staff can potentially involve contact with citizens. Internal communications must be established and maintained.

The entire staff that will be involved in the project must be trained on the sensitivities of community relations. Office staff must be included in the training program. Secretaries are sometimes overlooked; yet they are often the first persons to receive or screen telephone calls for reporters or other people desiring information about the project. Special training will be needed for the spokespersons' organization and for any other individuals who will be participating in public hearings, public meetings, or press conferences.

Involvement in community relations is a learning process, and lessons can sometimes carry a very high price tag. It is generally beneficial to obtain professional help in developing and implementing a program. Beware, though, of generalist public relation firms that lack experience with sensitive environmental issues. A program based on inexperience can be more damaging than no organized program at all.

EXHIBIT 14-1: A Sample Work Plan

WORK PLAN
FOR

____

(NAME)

____

(LOCATION)

____

(PROJECT NO.)

PREPARED BY:

NAME	DATE

REVIEWED AND APPROVED BY

	NAME	DATE
PROJECT MANAGER		
HEALTH AND SAFETY MANAGER		

1.0 BACKGROUND

SITE: ______________ EPA REGION: ______________

EPA Contact: ______________ Phone No.: ______________

State Contact: ______________ Phone No.: ______________

Plan Prepared By: ______________ Date: ______________

Proposed Date(s) of Work: ______________

Overall Hazard is: High: ______________ Moderate: ______________

Low: ______________ Unknown: ______________

FACILITY DESCRIPTION: ______________

____

____

____

Principal Disposal Method (type and location): ______________

____

____

____

____

____

____

____

____

____

____

____

____

LEVEL A WORK PLAN

SITE: ______________________ Project No.: ______________

Emergency Information:

Local Resources:		PhoneNumbers:
Office:	______________	__________
Ambulance (Name):	______________	__________
Hospital:	______________	__________
Police (Local or State):	______________	__________
Fire Department: (Name & Volunteer?)	______________	__________
Radio Channel:	______________	__________
Nearest Phone:	______________	__________
Project Manager:	______________	__________
Site Health and Safety Officer:	______________	__________

Emergency Contacts: (Medical and Health):

______________________________ __________

______________________________ __________

______________________________ __________

______________________________ __________

______________________________ __________

Directions to Hospital (Attach Map): ______________________________

____

____

Monitoring used on previous site work: ______________________________

Usual Features (containers, buildings, dikes, power lines, terrain, etc.): ______________

____

____

____

____

____

____

____

____

Status (active, inactive, unknown): ______________________________________

History (worker or non-worker injury; complaints from public; previous agency action): ___

____

____

____

____

____

____

Protection used on previous site work: ______A ______B ______C ______D

Other: ____

____

2.0 SCOPE OF WORK (Task 1, Task 2, etc.)

3.0 TEAM ORGANIZATION

Personnel (in order of Chain of Command):

Name	Discipline	Task Assigned

Characteristics of Waste:

Corrosive ________ Flammable ________ Radioactive ________ Toxic ________

Volatile ________ Reactive ________ Inert ________ Other ________

Physical Hazards of Site

Describe hazards taking into account toxic and pharmacological effects, reactivity, stability, flammability, operational concerns, sampling, decontamination, etc.

____

____

____

____

____

____

____

____

____

Personal Protection

Level of Protection Required: A ________ B________ C________ D________

Modifications: ____

____

____

____

____

LEVEL A protection should be worn when the highest level of respiratory, skin, eye, and mucous membrane protection is needed.

_________ Positive-pressure (pressure demand), self-contained breathing apparatus (MSHA/NIOSH approved) (REQUIRED).

_________ Fully-encapsulating, chemical-resistant suit (REQUIRED).

_________ Chemical resistant inner and outer gloves (REQUIRED).

_________ Chemical resistant boots with steel toe and shank (depending on suit boot construction, worn over or under suit boot) (REQUIRED).

_________ Thermal luminescent detector (TLD) badge for radiation (REQUIRED).

_________ Personal radiation detector (REQUIRED).

_________ Hard hat (under suit).

_________ Coveralls (under suit).

_________ Two-way radio communications (intrinsically safe) (REQUIRED).

LEVEL B protection should be selected when the highest level of respiratory protection is needed, but a lesser level of skin and eye protection. Level Bprotection is the minimum level recommended on initial site entries until hazards have been further identified and defined by monitoring, sampling, and other reliable methods of analysis. Personal equipment corresponding with those findings may be utilized.

_________ Positive-pressure (pressure-demand), self-contained, breathing

_________ apparatus (MSHA/NIOSH approved) (REQUIRED).

_________ Chemical-resistant clothing (overalls and long sleeved jacket,coveralls, hooded, two-piece, chemical splash suit, or disposable, chemical-resistant coveralls) (REQUIRED).

_________ Coveralls (under splash suit).

_________ (TLD) badge for radiation (REQUIRED).

_________ Personal radiation detector (REQUIRED).

_________ Chemical-resistant inner and outer gloves (REQUIRED).

_________ Chemical-resistant boots with steel toe and shank (REQUIRED).

_________ Two-way radio communications (intrinsically safe).

_________ Hard hat.

LEVEL C protection should be selected when the type of hazardous airborne substance is known, concentration measured, criteria for using air-purifying respirators met, and skin and eye exposure is unlikely. Monitoring of the air must be performed to comply with OSHA regulations and ensure respirator effectiveness.

- ________ Full-face, air-purifying respirator (MSHA/NIOSH approved) with cartridge type (REQUIRED).
- ________ Chemical-resistant clothing (overalls and long sleeved jacket, coveralls, hooded, two-piece, chemical splash suit, or disposable, chemical-resistant coveralls) (REQUIRED).
- ________ (TLD) badge for radiation (REQUIRED).
- ________ Personal radiation detector (REQUIRED).
- ________ Chemical resistant inner and outer gloves (REQUIRED).
- ________ Boots, steel toe and shank, chemical-resistant (REQUIRED).
- ________ Two-way radio communications (intrinsically safe).
- ________ Hard hat.
- ________ Escape mask

LEVEL D is primarily a work uniform. It should not be worn on any site where respiratory or skin hazards exist.

- ________ Protective coveralls and protective gloves.
- ________ (TLD) badge for radiation (REQUIRED).
- ________ Personal radiation detector (REQUIRED).
- ________ Boots or shoe with steel toe and shank (REQUIRED).
- ________ Hard hat.
- ________ Safety eye wear.

4.0 PROTECTION AGAINST POTENTIAL HAZARDS ON-SITE

Hazardous/Toxic Material

Known or suspected materials, contaminated media, storage containers, etc. Circle which wastes are present and estimate the amount of waste, with units, by category.

Amount (units)	Amounts (units)	Amount (units)	Amount (units)	Amount (units)	Amount (units)
Paint pigments	Oily wastes	Halogenated solvents	Inorganic acids	Fly ash	Laboratory
			Organic acids		Pharma-ceutical
Metal sludges	Other (Specify)	Non-halogenated solvents	Pickling liquors	Asbestos	Hospital
POTW		Other (Specify)	Caustics	Milling/mine tailings	Radioactive
Aluminum			Pesticides	Ferrous smelting	Municipal
Other (Specify)			Dyes/inks	Non-ferrous smelting	Other (Specify)
			Cyanide	Other (Specify)	
			Phenols		
			Halogens		
			Dioxin (TCDD)		
			Metals		
			Other (Specify)		
			Lagoon Water		

5.0 SITE OPERATIONS

Operations and Monitoring Equipment Checklist

Type of Equipment	Number Needed	Calibrated	Field Ready

Specific Settings of Equipment for Site Work

Equipment	Setting	Chemical Sensitivity Chosen

Start-Up Procedures

____

____

____

____

____

Operating Procedures

Decontamination Procedures (see Section 6.0)

Shut-down Procedures

Work Limitations (time of day, etc.):

Perimeter Establishment:

Site Secured ________ Map of Contamination Zones ________

Perimeter Identified ________ Contamination Zones Identified ________

Medical Surveillance Procedures

Parameters	Action Level	Medium	Method

Explanation:

____

Decontamination Procedures

________ Level A — Segregated equipment drop, boot cover and glove wash, boot cover and glove rinse, tape removal, boot cover removal, outer glove removal, suit/safety hat removal, SCBA backpack removal, inner glove wash, inner glove removal, inner clothing removal, field wash, redress (see Chapter 13).

________ Level B — Segregated equipment drop, boot cover and glove wash, boot cover and glove rinse, tape removal, boot cover removal, outer glove removal, suit/safety boot wash, suit/SCBA/boot/glove rinse (tank change), safety boot removal, splash suit removal, SCBA backpack removal, inner glove wash, inner glove rinse, face piece removal, inner glove removal, inner clothing removal, field wash, redress (see Chapter 13).

________ Level C — Segregated equipment drop, boot cover and glove wash, boot cover and glove rinse, tape removal, boot cover removal, outer glove removal, suit/safety boot wash, suit/safety boot rinse (canister or mask change), safety boot removal, splash suit removal, inner glove wash, inner glove rinse, face piece removal, inner glove removal, inner clothing removal, field wash, redress (see Chapter 13).

________ Level D — Segregated equipment drop, boot and glove wash, boot and glove rinse (see Chapter 13).

________ Level E — No decontamination required.

6.0 DECONTAMINATION PROCEDURES

Map of site showing restricted access zones, protection levels, decontamination areas, equipment layout, and clean zones enclosed?

Yes __________ No__________

If not, sketch below:

Decontamination of Sampling Bottles:

____

____

____

____

____

____

Decontamination of Equipment:

____

____

____

____

____

____

Decontamination Modification (personnel, surfaces, materials, instruments, equipment, etc.):

____

____

____

____

____

____

Disposal Procedures

On-Site

Off-Site

7.0 FIRST AID

BITES

Animal Bites — Thoroughly wash the wound with soap and water. Flush the area with running water and apply a sterile dressing. Immobilize affected part until the victim has been attended by a physician. See that the animal is kept alive and in quarantine. Obtain name and address of the owner of the animal.

Snake Bites — Try to kill snake so hospital will be able to treat victim with correct venom antibodies. Also, proceed with medical emergency suction of venom from victim with a hand snake bite pump. Do not use mouth suction. I

Insect Bites — Remove "stinger" if present. Keep affected part down below the level of the heart. Apply ice bag. For minor bites and stings apply soothing locations, such as calamine. If a wasp or bee sting, wet tobacco with alcohol and place on bite, no swelling and pain will be gone.

BURNS AND SCALDS

Minor Burns — DO NOT APPLY VASELINE OR GREASE OF ANY KIND. Apply cold water applications until pain subsides. Cover with a dry, sterile gauze dressing. Do not break blisters or remove tissue. Seek medical attention.

Severe Burns — Do not remove adhered particles of clothing. Do not apply ice or immerse in cold water. Do not apply ointment, grease, or vaseline. Cover burns with thick sterile dressings. Keep burned feet or legs elevated. Seek medical attention immediately.

Chemical Burns — Wash away the chemical-soaked clothing with large amounts of water. Remove victim's chemical-soaked clothing. If dry lime, brush away before flushing. Apply sterile dressing and seek medical attention.

CRAMPS

Symptoms — Cramps in muscles of abdomen and extremities. Heat exhaustion may also be present.

Treatment — Same as for heat exhaustion.

CUTS — Apply pressure with sterile gauze dressing, and elevate the area until bleeding stops. Apply a bandage and seek medical attention.

EYES

Foreign Objects — Keep the victim from rubbing eye. Flush the eye with water. If flushing fails to remove the object, apply a dry, protective dressing and consult a physician.

Chemicals — Flood the eye thoroughly with water for 15 min. Cover the eye with a dry pad and seek medical attention.

FAINTING — Keep the victim lying down. Loosen tight clothing. If victim vomits, roll him onto his side or turn his head to the side. If necessary, wipe out his mouth.

Maintain an open airway. Bathe his face gently with cool water. Unless recovery is prompt, seek medical attention.

FRACTURES — Deformity of an injured party usually means a fracture. If fracture is suspected, splint the part. DO NOT ATTEMPT TO MOVE INJURED PERSON; seek medical attention immediately.

FROSTBITE

Symptoms — Just before frostbite occurs skin may be flushed, then change to white or grayish-yellow. Pain may be felt early, then subsides. Blisters may appear, affected part feels very cold and numb.

Treatment — Bring victim indoors, cover the frozen area, provide extra clothing and blankets. Rewarm frozen area quickly by immersion in warm water - NOT HOT WATER - DO NOT RUB THE PART. Seek medical attention immediately.

HEAT EXHAUSTION — Caused by exposure to heat — either sun or indoors.

Symptoms — Near-normal body temperature. Skin is pale and clammy. Profuse sweating tiredness, weakness, headache, perhaps cramps, nausea, dizziness, and possible fainting.

Treatment — Keep in lying position and raise victim's feet. Loosen clothing, apply cool wet cloths. If conscious, give sips of salt water (1 teaspoon of salt per glass) over a period of one hour. If vomiting occurs, discontinue salt water. Seek medical attention immediately.

SUNSTROKE

Symptoms — Body temperature is high (106° F or higher). Skin is hot, red, and dry. Pulse is rapid and strong. Victim may be unconscious.

Treatment — Keep victim in lying position with head elevated. Remove clothing and repeatedly sponge the bare skin with cool water or rubbing alcohol. Seek medical attention immediately.

POISONING — Call the poison control center for instruction on immediate care. If victim becomes unconscious, keep the airway open. If breathing stops give artificial respiration by mount-to-mouth breathing. Call an emergency squad as soon as possible.

POISON IVY — Remove contaminated clothing: wash all exposed areas thoroughly with soap and water followed by rubbing alcohol. If rash is mild, apply calamine or other soothing skin lotion. If a sever reaction occurs, seek medical attention.

PUNCTURE WOUNDS — If puncture wound is deeper than skin surface, seek medical attention. Serious infection can arise unless proper treatment is received.

SPRAINS — Elevate injured part and apply ice bag or cold packs. DO NOT SOAK IN HOT WATER. If pain and swelling persist, seek medical attention.

UNCONSCIOUSNESS — Never attempt to give anything by mouth. Keep victim lying flat, maintain open airway. If victim is not breathing, provide artificial respiration by mount-to-mouth breathing and call an emergency squad as soon as possible

8.0 TRAINING

Topic	Hours	Date	Date	Date	Date
Introduction					
First Aid					
CPR					
Decontamination					
OVA					
HNU					
SCBA Review					
Sampling					

Other:

Site Emergency Prodcedures and Responsibilities:

Name	Responsibility	Action

9.0 SITE SAFETY FOLLOW-UP REPORT

(To be Completed for Each Field Change in Plan)

Was the Safety Plan followed as presented? ________ Yes ________ No

Describe, in detail, any changes to the Safety Plan:

____

____

____

____

____

____

Reason for Changes:

____

____

____

____

Approved by: Site Manager ____________________ Date ________________

Site Safety Officer ____________________ Date ________________

Evaluation of Site Safety Plan

Was the Safety Plan adequate? ________ Yes ________ No

What changes would you recommend?

____

____

____

Emergency Contacts

Emergency Routes Hospital:

Hospital: ______

Ionizing Radiation

Normal background is 0.01 to 0.02 mrad/hr.

If less than 3 mrad/hr is found, continue investigation with caution.

If 3 to 10 mrad/hr is found, map 3 mrad/hr contour without entering areas of greater intensity.

If greater than 10 mrad/hr is found, evacuate site than report incident to Health and Safety Manager.

Potential or Actual Fire or Explosion

Call Fire Department; Phone: ____________________

Call Police; Phone: ____________________

Personal Injury

Call Ambulance/Hospital Emergency Room;
Phone: ____________________

Call REMPO Office; Phone: ____________________

Emergency Resources

Poison Control Center ____________________

Airport ____________________

Explosive Ordinance Disposal Unit ____________________

Site Water Supply ____________________

Site Telephone ____________________

Site Radio ____________________

Site Other ____________________

Date Medical Care was Received: ______________________________

Where Medical Care was Received: ______________________________

Address (if off site): ______________________________

Accident Location

Causative agent most directly related to accident (object, substance, material, machinery, equipment, conditions):

____

____

____

Was weather a factor: ______________________________________

Unsafe mechanical/physical/environmental condition at time of accident (be specific):

____

____

____

Unsafe act by injured and/or others contributing to the accident (be specific,must be answered):

____

____

____

Personal factors (improper attitude, lack of knowledge of skill, slow reaction, fatigue):

____

____

____

10.0 ACCIDENT REPORT

Report No. ________________

SITE: ______________________________ PROJECT NO.: ____________________

Location: ____

Date of Report: ____________ Preparer's Name: ________________________

Name and Address of Injured: _________________ SSN: ________________________

____ Age: _________ Sex: _________

Division/Department: _________________ Date of Accident: ________ Time: _________

Accident Category: _____ Motor Vehicle _____ Property Damage _____ Fire

_____ Chemical Exposure _____ Near Miss _____ Other

Severity Of Injury of Illness: _______ Non-disabling _______ Disabling

_______ Medical Treatment _______ Fatality

Amount of Damage: $ _____________ Property Damaged: ______________________

Estimated Number of Days Away from Job: ________________________________

Number of Injury of Illness: ____

Classification of Injury:

_______ Fractures	_______ Heat Burns	_______ Cold Exposure
_______ Dislocations	_______ Chemical Burns	_______ Frostbite
_______ Sprains	_______ Radiation Burns	_______ Heat Stoke
_______ Abrasions	_______ Bruises	_______ Heat Exhaustion
_______ Lacerations	_______ Blisters	_______ Concussion
_______ Punctures	_______ Toxic Respiratory Exposure	_______ Faint/Dizziness
_______ Bites		_______ Toxic Respiratory
_______ Respiratory	_______ Toxic Ingestion	_______ Dermal Allergy

AllergyPart of Body Affected: _______________ Degree of Disability: _______________

Level of personal protection equipment required in Site Safety Plan: ________________

____

____

Modifications: ____

____

____

Was injured using required equipment: ____________________________

____

____

If not, how did actual equipment use differ from plan: ____________________

____

____

What can be done to prevent a recurrence of this type of accident (Modification of machine; mechanical guards; correct environment; training): ________________

____

____

(Use back of sheet, if required)

Witnesses to accident: ________________________________

____

____

Signature of Preparer ________________________

Signature of Site Manager ________________________

Equipment Contributed to This Accident: ______________________________

____

____

____

Emergency Contact Persons Associated With Accident: __________________

____

____

____

____

Health Department Use Only:

Permanent Partial ____

Part of Body ____

Percent Loss or Loss of Use ____________________________________

Time Charge ____

Medical ________________ $ ______________________________________

Total __________________ $ ______________________________________

Name and Address of Physician: __________________________________

____

____

Location of Accident and Address: ________________________________

____

____

Substance	RTEC	Toxicity: Inhalation (mg/m^3)	Toxicity: Oral (mg/kg)	Toxicity: Dermal (mg/kg)	Toxicity: Subcutaneous (mg/kg)	AQIX
Extractable Organics						
Acenaphthene	83329(CAS)					
Acenaphthylene	AB1254000					
Phenanthrene	SF7175000		LD_{50} 700 (MUS)	TD_{10} 71 (MUS)		
Anthracene	CA9350000		TD_{10} 20,000/79 W-1	TD_{10} 3,300		
Fluoranthene	LL4025000		LD_{50} 2,000	LD_{50} 3,180 (rbt)		
Fluorene	86737 (CAS)					
Pyrene	UR2450000			TD_{10} 10 (MUS)		
Benzo(a)anthracene	CV9275000 AP			TD_{10} 18 (MUS)	TD_{10} 2 (MUS)	
Chrysene	GC0700000 AP			TD_{10} 3.6 (MUS)	TD_{10} 200 (MUS)	
Benzo(b)fluoranthene	CU1400000 AP			TD_{10} 72 (MUS)	TD_{10} 72 (MUS)	
Benzo(k)fluoranthene	DF6350000			TD_{10} 2,820 (MUS)	TD_{10} 72 (MUS)	
Benzo(a)pyrene	DJ3675000 AP	TC_{10} 0.00007 (hmn)	TD_{10} 1,000	TD_{10} 17 (rbt)	TD_{10} 0.455	
Indeno (1,2,3-CD)-) pyrene	MK9300000 AP				TD_{10} 72 (MUS)	
Dibenzo(A,H)-) antracene	IM2625000 AP		TD_{10} 360 (MUS)	TD_{10} 1,200 (MUS)	TD 2.4	
Benzo(ghl)perylene	D16200500					
Penlachlorophenol	SH6300000 Indefinite	TLV 0.5 mg/m^3	LD_{50} 50 (rat)	LD_{50} 105 (rat)	LD_{50} 100 (rat)	
Dibenzofuran						
Bis(2-ethylhexyl)-) pthalate	T10350000	TLV 5 mg/m^3	LD_{50} 31 (rat)	LD_{50} (rbt)		
Phenol	SJ3325000	TLV 5 ppm	LD_{50} 414 (rat)	LD_{50} 669 (rat)	LD_{10} 650 (rat)	100–10

Substance	RTEC	Toxicity				
		Inhalation (mg/m^3)	Oral (mg/kg)	Dermal (mg/kg)	Subcutaneous (mg/kg)	AQIX
Napthalene	QJ0525000	TLV 10 ppm	LD_{50} 1,780 (rat)		TD_{10} 3,500 (rat)	10–1
4-Dimethylphenol	ZE5600000		LD_{50} 3,200 (rat)	LD_{50} 1,040 (rat)		
Purgeable Organics						
Benzene	CV1400000 HS	TLV 10 ppm	LD_{50} 3,800 (rat)	TD_{50} 1,200 (rbt)	TD_{50} 600 (MUS)	100–10
Toluene	XS5250000	TLV 100 ppm	LD_{50} 5,000 (rat)	LD_{50} 14,000 (rat)		100–10
Ethyl benzene	DA0700000	TLV 100 ppm	LD_{50} 3'500 (rat)	LD_{50} 5,000 (rbt)		100–10
o-Xylene	ZE2450000	TLV 100 ppm	LD_{10} 5,000 (rat)			100–10
p-Xylene	ZE2625000	TLV 100 ppm	LD_{10} 5,000 (rat)			100–10
1,1,1-Trichloroethene	KJ2475000	TLV 350 ppm	LD_{50} 10,300 (rat)		LD_{50} 500 (rbt)	100–10
	Indefinite					
PCBs						
Endosulfan 1 (alpha)	RB9275000	TLV 0.1 mg/m^3	LD_{50} 18 (rat)	LD_{50} 74 (rat)	TD_{10} 2 (MUS)	Under 1
1242	TQ1356000 HS	TLV 1 mg/m^3	LD_{50} 4,250 (rat)		LD_{10} 345 (gpg)	
1248	TQ1358000 HS	TLV 1 mg/m^3	LD_{50} 11,000 (rat)	LD_{10} 1,269 (rbt)		
1254	TQ1360000 HS	TLV .5 mg/m^3	LD_{50} 1,295 (rat)	LD_{10} 4 (MUS)		
Miscellaneous						
Acetone	AL3150000	TLV 750 ppm	LD_{50} 9,750 (rat)	LD_{50} 20 (rbt)	LD_{10} 5 (dog)	Over 100
2-Methyl naphthaline	QJ9635000		LD_{10} 500 (rat)			
2-Methyl phenol						
4-Methyl phenolStyrene	WL3675000	TLV 50 ppm	LD_{50} 5,000 (rbt)			100–10
Ethyl acetate	AH5425000	TLV 400 ppm	LD_{50} 11 (rat)		LD_{10} 5,000 (rat)	1000–100
Methylnaphthalene	QJ9635000		LD_{50} 4,360 (rat)			

Toxicity Ranking Chart

Acute Ranking	Oral (ID_{50} [rat[)	Inhalation (IC_{10} [rat])	Dermal (LD_{50} [rat])	Probable Oral Lethal Dose
Extremely toxic	<1 mg/kg	<200 ppm/1 hr	<10 mg/kg	A taste
Highly toxic	1–50 mg/kg	200–2000 ppm/1 hr	10–100 mg/kg	Teaspoon
Moderately toxic	50–500 mg/kg	2000–10,000 ppm/1 hr	100–1000 mg/kg	Ounce 1000–5000 ppm/4 hr
Slightly toxic	5–5 g/kg	5000–25,000 ppm/4 hr	1000–5000 mg/kg	Pint
Practically nontoxic	5–15 g/kg	25,000 - 100,000 ppm/4 hr	5000–25,000 mg/kg	Quarttoxic
Relatively harmless	>15 g/kg	>100,000 ppm/4 hrs	>25,000 mg/kg	>1 qt

Chronic Ranking	Listed By
Known human carcinogen	OSHA, IAHC, NCI, RTECS or ACGIH
Suspect human carcinogen	OSHA, IARC, NCI, RTECS or ACGHI
Mutagen	RTECS
Teratogen	RTECS

In Toxicity Column the following format should be used: Example: Extremely Toxic, Inhalation Suspect Human Carcinogen, Oral

Glossary of Terms

Absolute — A chemical substance that is not mixed; pure. An example is absolute alcohol, which is ethyl alcohol containing not more than 1% by weight of water.

ACGIH — American Conference of Governmental Industrial Hygienists; an organization of professional personnel in governmental agencies or educational institutions engaged in occupational safety and health programs. ACGIH develops and publishes recommended occupational exposure limit for hundreds of chemical substances and physical agents. See TLV.

Acid — Any chemical which undergoes dissociation in water with the formation of hydrogen ions. Acids have a sour taste and may cause severe skin burns. Acids turn litmus paper red and have pH values of 0 to 6.

Activity — The number of nuclear disintegrations occurring in a given quantity of material per unit time.

Acute Effect — An adverse effect on a human or animal body, with severe symptoms developing rapidly and coming quickly to a crisis. Examples include dizziness, nausea, skin rashes, inflammation, tearing of the eyes, unconsciousness, and even death.

Acute Toxicity — The adverse (acute) effects resulting from a single dose of or exposure to a substance. Ordinarily used to denote effects in experimental animals.

Aerosol — A fine aerial suspension of particles sufficiently small in size to confer some degree of stability from sedimentation, for example, a smoke or fog.

Air-Line Respirator — A respirator that is connected to a compressed breathing air source by a hose of small inside diameter. The air is developed continuously or intermittently in a sufficient volume to meet the wearer's breathing requirements.

Air-Purifying Respirator — A respirator that uses chemicals to remove specific gases and vapors from the air or that uses a mechanical filter to remove particulate matter. An air-purifying respirator must only be used when there is sufficient oxygen to sustain life and the air contaminant level is below the concentration limits of the device.

Alkali — Any chemical substance which forms soluble soaps with fatty acids. Alkalis are also referred to as bases. They may cause severe burns to the skin. Alkalis turn litmus paper blue and have pH values from 8 to 14.

Allergic Reaction — An abnormal physiological response to a chemical or physical stimuli by a sensitive person.

Anesthetic — A chemical that causes a total or partial loss of sensation. Overexposure to anesthetics can cause impaired judgement, dizziness, drowsiness, headache, unconsciousness, and even death. Examples include alcohol, paint remover, and degreasers.

ANSI — American National Standards Institute; a privately funded, voluntary membership organization that identifies industrial and public needs for national consensus standards and coordinates development of such standards. Many ANSI standards relate to safe design/performance of equipment, such as safety shoes, eyeglasses, smoke detectors, fire pumps, and household appliances; and safe practices or procedures, such as noise measurement, testing of fire extinguishers and flame arrestors, industrial lighting practices, and the use of abrasive wheels.

Antidote — A remedy to relieve, prevent, or counteract the effects of a poison.

API — American Petroleum Institute; voluntary membership organization of the petroleum industry. Among its services, API assists member committees in developing, by the consensus process, and publishing recommended practices for drilling and well servicing, storage tank installation, tank cleaning, piping and fittings, other industry-related design, installation, and operating practices. It also funds and publishes basic reference books and manuals, for example, "Industrial Hygiene Monitoring Manual for Petroleum Refineries and Selected Petrochemical Operations."

Appearance — A description of a substance at normal room temperature and normal atmospheric conditions. Appearance includes the color, size, and consistency of a material.

Aquatic Toxicity — The adverse effects to marine life that result from being exposed to a toxic substance.

Asphyxiant — A vapor or gas which can cause unconsciousness or death by suffocation (lack of oxygen). Most simple asphyxiants are harmful to the body only when they become so concentrated that they reduce oxygen in the air (normally about 21%) to dangerous levels (16% or lower). Asphyxiation is one of the principal potential hazards of working in confined spaces. In addition, some chemicals like carbon monoxide function as chemical asphyxiants by reducing the blood's ability to carry oxygen.

ASTM — American Society for Testing and Materials; voluntary membership organization with members from a broad spectrum of individuals, agencies concerned with materials. As the world's largest source of voluntary consensus standards for materials, products, systems, and services, ASTM is a resource for sampling and testing methods, health and safety aspects of materials, safe performance guidelines, and effects of physical and biological agents and chemicals.

Atmosphere-Supplying Respirator — A respirator that provides breathing air from a source independent of the surrounding atmosphere. There are two types: airline and self-contained breathing apparatus.

Autoignition Temperature — The temperature to which a closed or nearly closed container must be heated in order that the flammable liquid, when introduced into the container, will ignite spontaneously or burn.

Base — See Alkali.

Biodegradable — Capable of being broken down into innocuous products by the action of living things.

Body Burden — The total quantity of a radionuclide present in the body.

Body Burden, Maximum Permissible — That body burden of a radionuclide which, if maintained at a constant level, would produce the maximum permissible dose equivalent in the critical organ.

Boiling Point — The temperature at which a liquid changes to a vapor state at a given pressure; usually expressed in degrees Fahrenheit at sea level pressure (760 mmHg, or one atmosphere). For mixtures, the initial boiling point or the boiling range may be given. Flammable materials with low boiling points generally present special fire hazards. Some approximate boiling points:

Propane	–44°F
Anhydrous ammonia	–28°F
Butane	31°F
Gasoline	100°F
Allyl chloride	113°F
Ethyl glycol	387°F

BOM — Bureau of Mines of the U.S. Department of Interior. BOM began approving air breathing apparatus in 1918, later added all types of respirators. BOM's respirator testing/approval activities have been discontinued. BOM-approved Type 14F gas masks are still acceptable; all other BOM approvals have expired or been replaced by NIOSH approvals.

Bonding — The interconnecting of two objects by means of a clamp and bare wire. Its purpose is to equalize the electrical potential between the objects to prevent a static discharge when transferring a flammable liquid from one container to another. The conductive path is provided by clamps which make contact with the charged object and a low-resistance flexible cable to equalize. See Grounding.

Bhopal — City in India where the people were exposed to toxic gas cloud of isocyanates from a Union Carbide chemical manufacturing plant in December 1985, which engulfed the immediate plant facilities and community.

Bremsstrahlung — The electromagnetic radiation associated with the deceleration of charged particles. The term is also applied to the radiation associated with the acceleration of charged particles.

Bulk Density — Mass of powdered or granulated solid material per unit of volume.

Bumines — See BOM.

C — Ceiling limit; ACGIH terminology. See TLV.

CAA — See Clean Air Act.

Carcinogen — A substance or agent that can cause a growth of abnormal tissue or tumors in humans or animals. A material identified as an animal carcinogen does not necessarily cause cancer in humans. Examples of human carcinogens include coal tar, which can cause skin cancer, and vinyl chloride, which can cause liver cancer.

C.A.S. — Chemical Abstracts Service; a Columbus, Ohio, organization which indexes information published in "Chemical Abstracts" by the American Chemical Society and provides index guides in the "Abstracts" when needed. "C.A.S. Numbers" identify specific chemicals.

Caustic — See Alkali.

cc — Cubic centimeters; a volume measurement in the metric system, equal in capacity to one milliliter (ml) — approximately 20 drops. There are 16.4 cc in one cubic inch.

Ceiling — See C.

Central Nervous System — The brain and spinal cord. These organs supervise and coordinate the activity of the entire nervous system. Sensory impulses are transmitted into the central nervous system. Sensory impulses are transmitted into the central nervous system, and motor impulses are transmitted out.

CERCLA — Comprehensive Environmental Response, Compensation, and Liability Act of 1980. The Act requires that the Coast Guard National Response Center be notified in the event of a hazardous substance release. The Act also provides for a fund (the Superfund) to be used for the cleanup of abandoned hazardous waste disposal sites.

CFR — Code of Federal Regulations. A collection of the regulations that have been promulgated under U.S. law.

Chemical Cartridge Respirator — A respirator that uses various chemical substances to purify inhaled air of certain gases and vapors. This type respirator is effective for concentrations no more than ten times the TLV of the contaminant, if the contaminant has warning properties (odor or irritation) below the TLV.

Chemical Family — A group of single elements or compounds with a common general name. Example: acetone, methyl ethyl ketone (MEK), and methyl isobutyl ketone (MIBK) are of the "ketone" family; acrolein, furfural, and acetaldehyde are of the "aldehyde" family.

Chemical Name — The scientific designation of a chemical in accordance with the nomenclature system as developed by the International Union of Pure and Applied Chemistry (IUPAC) or the Chemical Abstracts Service (CAS) rules of nomenclature.

Chemical Structure — The arrangement within the molecule of atoms and their chemical bonds.

CHEMTREC — Chemical Transportation Emergency Center; a national center established by the Chemical Manufacturers Association (CMA) in Washington, DC, in 1970, to relay pertinent emergency information concerning specific chemicals on request. CHEMTREC has a 24-hr toll-free telephone number (800-424-9300), intended primarily for use by those who respond to chemical transportation emergencies.

Chernobyl — City in Russia where a nuclear power plant experienced a nuclear melt-down which affected adjacent countries and people, emitting radioactive materials in the atmosphere and water.

Chronic Effect — An adverse effect on a human or animal body, with symptoms which develop slowly over a long period of time or which reccur frequently. Examples include cancer and irreversible damage to certain organs.

Chronic Exposure — Long-term contact with a substance.

Chronic Toxicity — Adverse (chronic) effects resulting from repeated doses of or exposures to a substance over a relatively prolonged period of time. Ordinarily used to denote effects in experimental animals.

Clean Air Act — Federal law enacted to regulate/reduce air pollution. Administered by EPA.

Clean Water Act — Federal law enacted to regulate/reduce water pollution. Administered by EPA.

CMA — Chemical Manufacturers Association. See CHEMTREC.

COC — Cleveland Open Cup; a flash point test method.

Code of Federal Regulations — See CFR.

Combustible — Able to catch on fire and burn.

Combustible Liquid — Any liquid having a flashpoint at or above 100°F (37.8°C), but below 200°F (93.3°C), except any mixture having components with flashpoints of 200°F (93.3°C), or higher, the total volume of which make up 99% or more of the total volume of the mixture.

Common Name — Any designation or identification such as code name, code number, trade name, brand name, or generic name used to identify a chemical other than by its chemical name.

Compressed Gas — (1) any gas or mixture of gases having, in a container, an absolute pressure exceeding 40 psi at 70°F (21.1°C); or (2) a gas or mixture of gases having, in a container, an absolute pressure exceeding 104 psi at 130°F (54.4°C), regardless of the pressure at 70°F (21.1°C); or (3) a liquid having a vapor pressure exceeding 40 psi at 100°F (37.8°C) as determined by ASTM D-323-72.

Concentration — The relative amount of a substance when combined or mixed with other substances. Examples: 2 ppm hydrogen sulfide in air, or a 50% caustic solution.

Conditions to Avoid — Conditions encountered during handling or storage which could cause a substance to become unstable.

Confined Space — Any area which has limited openings for energy and exit that would make escape difficult in an emergency, has a lack of ventilation, contains known and potential hazards, and is not intended nor designated for continuous human occupancy.

Conjunctivitis — Inflammation of the conjunctiva, the delicate membrane that lines the eyelids and covers the eyeballs.

Controlled Area — A specified area in which exposure of personnel to radiation or radioactive material is controlled and which is under the supervision of a person who has knowledge of the appropriate radiation protection practices, including pertinent regulations, and who has responsibility for applying them.

Corrosive — As defined by DOT, a corrosive material is a liquid or solid that causes visible destruction or irreversible alterations in human skin tissue at the site of contact or — in the case of leakage from its packaging — a liquid that has a severe corrosion rate on steel. Two common examples are caustic soda and sulfuric acid.

CPSC — Consumer Products Safety Commission; federal agency with responsibility for regulating hazardous materials when they appear in consumer goods. For CPSC purposes, hazards are defined in the Hazardous Substances Act and the Poison Prevention Packaging Act of 1970.

Curie (Ci) — The special unit of activity. One curie equals 3.7×10^{10} disintegrations per second, exactly. By popular usage, the quantity of any radioactive material having an activity of one curie.

Cutaneous Toxicity — See Dermal Toxicity.

CWA — See Clean Water Act.

Dangerous When Wet— See Water Reactive.

Daughter — A nuclide, stable or radioactive, formed by radioactive decay. A synonym for decay product.

Decomposition — Breakdown of a material or substance (by heat, chemical reaction, electrolysis, decay, or other processes) into parts or elements or simpler compounds.

Density — The mass (weight) per unit volume of a substance. For example, lead is much more dense that aluminum.

Depressant — A substance that reduces a bodily functional activity or an instinctive desire, such as appetite.

Dermal — Used on or applied to the skin.

Dermal Toxicity — Adverse effects resulting from a skin exposure to a substance. Ordinarily used to denote effects in experimental animals.

Dermatitis — Inflammation of the skin.

DHHS — U.S. Department of Health and Human Services; created in 1980 to replace the Department of Health, Education, and Welfare (DHEW) as "parent" for NIOSH, Public Health Service, and other agencies related to health and safety.

Dike — A barrier constructed to control or confine hazardous substances and prevent them from entering sewers, ditches, streams, or other flowing waters.

Dilution Ventilation — Air flow designed to dilute contaminants to acceptable levels. Also referred to as general ventilation or exhaust. See General Exhaust.

DOL — U.S. Department of Labor; includes the Occupational Safety and Health Administration (OSHA).

Dose — A general term denoting the quantity of radiation or energy absorbed in a specified mass. For special purposes, its meaning should be appropriately stated, e.g., absorbed dose.

Dose Absorbed — The energy imparted to matter in a volume element by ionizing radiation divided by the mass or irradiated material in that volume element.

Dose Equivalent — The product of absorbed dose, quality factor, and other modifying factors necessary to express on a common scale, for all ionizing radiations, the irradiation incurred by exposed persons.

Dose Equivalent, Maximum Permissible (MPD) — The largest dose equivalent received within a specified period which is permitted by a regulatory agency or other authoritative group on the assumption that receipt of such dose equivalent creates no appreciable somatic or genetic injury. Different levels of MPD may be set for different groups within a population. (By popular usage, dose, maximum permissible, is an accepted synonym.)

DOT — U.S. Department of Transportation; regulates transportation of chemicals and other substances, to aid in the protection of the public as well as fire, law enforcement, and other emergency response personnel, particularly when transportation incidents occur involving hazardous materials.

DOT Hazard Class — DOT requires that hazardous materials offered for shipment be labeled with the proper DOT hazard class. These classes include corrosive, flammable liquid, organic peroxide, ORM-E, poison B, etc. The DOT hazard class may not adequately describe all the hazard properties of the material. A chart of the various DOT hazard classes can be found on the next page.

Dry Chemical — A powdered fire-extinguishing agent usually composed of sodium bicarbonate, potassium bicarbonate, etc.

Edema — An abnormal accumulation of clear watery fluid in the tissues.

Environmental Toxicity — Information obtained as a result of conducting environmental testing designed to study the effects on aquatic and plant life.

EPA — U.S. Environmental Protection Agency; federal agency with environmental protection regulatory and enforcement authority. Administers Clean Air Act, Clean Water Act, FIFRA, RCRA, TSCA, and other federal environmental laws.

Epidemiology — The science which deals with the study of disease in a general population. Determination of the incidence (rate or occurrence) and distribution of a particular disease (as by age, sex, or occupation) may provide information about the causes of the disease.

Exposure — A measure of the ionization produced in air by X- or gamma-radiation. It is the sum of the electrical charges on all of the ions of one sign produced in air when all electrons liberated by photons in a volume element of air are completely stopped in the air, divided by the mass of the air in the volume element.

Evaporation Rate — The rate at which a particular material will vaporize (evaporate) when compared to the rate of vaporization of a known material. The evaporation rate can be useful in evaluating the health and fire hazards of a material. The known material is usually ethyl ether with a vaporization rate designated as 1.0. Vaporization rates of other solvents or materials are then classified as

- **Fast** — Evaporating if greater than 3.0. Examples: methyl ethyl ketone (MEK) = 3.8, acetone = 5.6, hexane = 8.3.
- **Medium** — Evaporating if 0.8 to 3.0. Examples: 190-proof (95%) ethyl alcohol = 1.4, VM&P naphtha = 1.4, MIBK = 1.6.
- **Slow** — Evaporating if less than 0.8. Examples: xylene = 0.6, isobutyl alcohol = 0.6, normal butyl alcohol = 0.4, water = 0.3, mineral spirits = 0.1.

Exothermic Polymerization — A polymerization reaction that involves the production of heat.

Explosive — A chemical that causes a sudden, almost instantaneous release of pressure, gas, and heat when subjected to sudden shock, pressure, or high temperature.

Exposure — Being actually subjected to a hazardous chemical in the course of employment through any route of entry (inhalation, ingestion, skin contact, or absorption, etc.). The federal hazard communication standard also includes both accidental and possible exposures in its definition of exposure.

Extinguishing Media — The fire-fighting substance to be used to control a material in the event of a fire. It is usually named by its generic name, such as fog, foam, water, etc.

Eye Protection — Recommended safety glasses, chemical splash goggles, face shields, etc., to be utilized when handling a hazardous material.

F — Fahrenheit; a scale for measuring temperature. On the Fahrenheit scale water boils at 212°F and freezes at 32°F.

f/cc — Fibers per cubic centimeter of air. The concentration of asbestos in air is reported as fibers of asbestos per cubic centimeter of air.

FDA — The U.S. Food and Drug Administration; under the provisions of the Federal Food, Drug and Cosmetic Act, the FDA establishes requirements for the labeling of foods and drugs to protect consumers from misbranded, unwholesome, ineffective, and hazardous products. FDA also regulates materials for food contact service and the conditions under which such materials are approved.

Federal Register — See FR.

Fibrosis — An abnormal thickening of fibrous connective tissue, usually in the lungs.

FIFRA — Federal Insecticide, Fungicide, and Rodenticide Act; regulations administered by EPA under this Act required that certain useful poisons, such as chemical pesticides, sold to the public contain labels that carry health hazard warnings to protect users.

First Aid — Emergency measures to be taken when a person is suffering from overexposure to a hazardous material, before regular medical help can be obtained.

Flammability Limits — The range of gas or vapor concentration in the air that may ignite or explode if an ignition source is present.

Flammable Aerosol — An aerosol that, when tested by the method described in 16 CFR 1500.45, yields a flame projection exceeding 18 in. at full valve opening, or a flashback (a flame extending back to the valve) at any degree of valve opening.

Flammable Gas — (1) a gas that, at ambient temperature and pressure, forms a flammable mixture with air at a concentration of 13% by volume or less; or (2) a gas that, at ambient temperature and pressure, forms a range of flammable mixtures with air, wider that 12% by volume, regardless of the lower limit.

Flammable Liquid — Any liquid having a flash point below 100°F (37.8°C), except any mixture having components with flashpoints of 100°F (37.8°C) or higher, the total of which make up 99% or more of the total volume of the mixture.

Flammable Solid — A solid, other than a blasting agent or explosive, as defined in CFR 1910.109(a), that is liable to cause fire through friction, absorption of moisture, spontaneous chemical change, or retained heat from manufacturing processing, or which can be ignited readily and, when ignited, burns so vigorously and persistently as to create a serious hazard.

Flashback — Occurs when flame from a torch burns back into the tip, the torch, or the hose. It is often accompanied by a hissing or squealing sound with a smoky or sharp pointed flame.

Flash Ignition — See Flash Point.

Flash Point — The temperature at which a liquid will give off enough flammable vapor to ignite if an ignition source is present. There are several flash point test methods, and flash points may vary for the same material depending on the method used, so the test method is indicated when the flash point is given (150°F PMCC, 200°F TCC, etc.).

Formula — The conventional scientific designation for a material (water is H_2O, sulfuric acid is H_2SO_4, sulfur dioxide is SO_2, etc.).

FR — Federal Register; a daily publication (Monday–Friday) of the federal department regulations that are promulgated under a particular law.

Fume — Airborne dispersion consisting of minute solid particles arising from the heating of a solid body, such as lead. This heating is often accompanied by a chemical reaction, such as oxidation.

g — Gram; a metric unit of mass weight. One ounce U.S. is a about 28 grams and one pound is 454 grams. One teaspoon of sugar weighs about 8 grams.

General Exhaust — A system for exhausting air containing contaminants from a general work area, usually accomplished via dilution.

g/kg — Grams per kilogram; an expression of dose used in oral and dermal toxicology testing to indicate the grams of substance dosed per kilogram of animal body weight. Also see kg.

Generic Name — A nonproprietary name for a material.

Genetically Significant Dose (GSD)— The dose which, if received by every member of the population, would be expected to produce the same total genetic injury to the population as do the actual doses received by the various individuals.

Gram — See g.

Grounding — The procedure used to carry an electrical charge to ground through a conductive path. A typical ground may be connected directly to a conductive water pipe or to a grounding bus and ground rod. See Bonding.

Half-Life, Radioactive — For a single radioactive decay process, the time required for the activity to decrease to half its value by that process.

Half-Value Layer — The thickness of a specific substance which, when introduced into the path of a given beam of radiation, reduces the value of a specific radiation quantity by one-half. It is sometimes expressed in terms of mass per unit area.

Hand Protection — Specific type of gloves or other hand protection required to prevent harmful exposure to hazardous materials.

Hazardous Decomposition Products — Any hazardous materials that may be produced in dangerous amounts if the material reacts with other agents, burns, or is exposed to other processes, such as welding. Examples of hazardous decomposition products formed when certain materials are heated include carbon monoxide and carbon dioxide.

Hazardous Ingredients — The hazardous substances that make up a mixture.

Hazardous Material — Any chemical which is a physical hazard or a health hazard.

Hazardous Polymerization — Polymerization is a chemical reaction in which one or more small molecules combine to form larger molecules. A hazardous polymerization is such a reaction which takes place at a rate which releases large amounts of energy. If hazardous polymerization can occur with a given material, the MSDS usually will list conditions which could start the reaction, and since the material usually contains a polymerization inhibitor, the expected time period before the inhibitor is used up.

Health Hazard — A chemical for which there is statistically significant evidence based on at least one study conducted in accordance with established scientific principles that acute or chronic health effects may occur in exposed employees. The term "health hazard" includes chemicals which are carcinogens, toxic or highly toxic agents, reproductive toxins, irritants, corrosives, sensitizers, hepatotoxins, nephrotoxins, neurotoxins, agents which act on the hematopoietic system, and agents which damage the lungs, skin, eyes, or mucous membranes.

Hematopoietic System — The blood-forming mechanism of the human body.

Hemorrhage — A loss of a large amount of blood in a short period of time, either outside the body or inside.

Hepatotoxin — A substance that causes injury to the liver.

IARC — International Agency for Research on Cancer. IARC publishes monographs, a listing of potential carcinogens.

Ignitable — Capable of being set afire.

Impervious — A material that does not allow another substance to pass through or penetrate it.

Incompatible — Materials which could cause dangerous reactions from direct contact with one another are described as incompatible.

Ingestion — Materials which could cause dangerous reactions from direct contact with one another are described as incompatible.

Inhalation — The breathing in of a substance in the form of a gas, vapor, fume, mist, or dust.

Inhibitor — A chemical which is added to another substance to prevent an unwanted chemical change from occurring.

Insoluble — Incapable of being dissolved in a liquid.

Irritant — A substance which, by contact in sufficient concentration for a sufficient period of time, will cause an inflammatory response or reaction of the eye, skin, or respiratory system. The contact may be a single exposure or multiple exposures. Some primary irritants: chromic acid, nitric acid, sodium hydroxide, calcium chloride, amines, metallic alloys, chlorinated hydrocarbons, ketones, alcohols.

Irritating — An irritating material, as defined by DOT, is a liquid or solid substance which, upon contact with fire or when exposed to air, gives off dangerous or intensely irritating fumes (not including poisonous materials). See Poison, Class A and Poison, Class B.

Isotopes — Nuclides having the same atomic number but different mass numbers. NOTE: this term is often used inaccurately as a synonym for nuclide.

kg — Kilogram; a metric unit of weight, about 2.2 U.S. pounds. See also g/kg, g, and mg.

L — Liter; a metric unit of capacity. A liter is about the same as one U .S. quart.

Lacrimation — Secretion and discharge of tears.

LC — Lethal concentration; a concentration of a substance being tested which will kill a test animal.

LC_{50} — Lethal concentration 50; the concentration of a material in air which, on the basis of laboratory tests, is expected to kill 50% of a group of test animals when administered as a single exposure (usually 1 or 4 hr). The LC_{50} is expressed as parts of material per million parts of air, by volume (ppm) for gases and vapors, or as micrograms of material per liter of air (μg/L) or milligrams of material per cubic meter of air (mg/m^3) for dusts and mists, as well as for gases and vapors.

LD — Lethal dose; a concentration of a substance being tested which will kill a test animal.

LD_{50} — Lethal dose 50; a single dose of a material which, on the basis of laboratory tests, is expected to kill 50% of a group of test animals. The LD_{50} dose is usually expressed as milligrams or grams of material per kilogram of animal body weight (mg/kg or g/kg). If a material has a low LD_{50}, then only a very small amount is needed to cause an adverse effect, and that substance would be considered highly toxic. Generally, substances with low LD_{50}s have high toxicity and vice versa.

LEL — Lower explosive limit or lower flammable limit of a vapor or gas; the lowest concentration (lowest percentage of the substance in air) that will produce a flash of fire when an ignition source (heat, arc, or flame) is present. At concentrations lower than the LEL, the mixture is too "lean" to burn. Also see UEL.

Liter — See L.

Local Exhaust — A system for capturing and exhausting contaminants from the air at the point where the contaminants are produced (welding, grinding, sanding, other processes or operations).

M — Meter; a unit of length in the metric system. One meter is about 39 inches.

m^3 — Cubic meter; a metric measure of volume, about 35.3 cubic feet or 1.3 cubic yards.

Malaise — A feeling of general discomfort, distress, or uneasiness, an out-of-sorts feeling.

mppcf — Million particles per cubic foot; a unit for measuring particles of a substance suspended in air. Exposure limits for mineral dusts (silica, graphite, Portland cement, nuisance dusts, and others), formerly expressed as mppcf, are now more commonly quoted as mg/m^3.

MSHA — The Mining Safety and Health Administration of the U.S. Department of the Interior; federal agency with safety and health regulatory and enforcement authorities for the mining industry.

Mutagen — A material that alters a cell's genetic information and may lead to undesirable inherited conditions.

Narcosis — A state of stupor, unconsciousness, or arrested action produced by the influence of narcotics or other chemicals.

Natural Ventilation — Air movement caused by wind, temperature difference, or other nonmechanical factors.

Nausea — Tendency to vomit, feeling of sickness at the stomach.

NCI — National Cancer Institute; a component of the National Institutes of Health which studies cancer causes and prevention as well as diagnosis, treatment, and rehabilitation of cancer patients.

Nephrotoxin — A substance that causes injury to the kidneys.

Neurotoxin — A material that affects the nerve cells and may produce emotional or behavioral abnormalities.

Neutralize — To eliminate potential hazards by inactivating strong acids, caustics, and oxidizers. For example, acids can be neutralized by adding an appropriate amount of caustic substance to the spill.

NFPA — National Fire Protection Association; an international voluntary membership organization to promote/improve fire protection and prevention and establish safeguards against loss of life and property by fire. Best known on the industrial scene for the National Fire Codes 16 volumes of codes, standards, recommended practices, and manuals developed (and periodically updated) by NFPA technical committees. Among these is NFPA 704M, the code for showing hazards of materials using the familiar diamond-shaped label or placard with appropriate numbers or symbols. The brief explanation on the next page illustrates the NFPA principle of using scales of 0 to 4 (low to high) to classify material hazard.

NIOSH — National Institute for Occupational Safety and Health of the Public Health Service, U.S. Department of Health and Human Services (DHHS); federal agency which recommends occupational exposure limits for various substances and assists OSHA and MSHA in occupational safety and health investigations and research.

Nonflammable — Not easily ignited or, if ignited, not burning rapidly.

Nonsparking Tools — Tools made from beryllium-copper or aluminum-bronze greatly reduce the possibility of igniting dusts, gases, or flammable vapors. Although these tools may emit some sparks when striking metal, the sparks have a low heat content and are not likely to ignite most flammable liquids.

NRC — National Response Center; a notification center in the Coast Guard Building in Washington, DC, with a toll-free telephone number (1-800-424-8802) which must be called when significant oil or chemical spills or other environmentally related accidents occur.

NTP — National Toxicology Program. The NTP publishes an Annual Report on Carcinogens, a listing of potential carcinogens.

Nuclide — A species of atoms characterized by its mass number, atomic number, and energy state of the nucleus, provided that the mean life in that state is long enough to be observable.

Odor — A description of the smell of the substance.

Odor Threshold — The lowest concentration of a substance's vapor, in air, that can be smelled.

Olfactory — Relating to the sense of smell. The olfactory organ in the nasal cavity is the sensing element that detects odors and transmits information to the brain through the olfactory nerves.

Oral — Used in or taken into the body through the mouth.

Oral Toxicity — Adverse effects resulting from taking a substance to the body via the mouth. Ordinarily used to denote effects in experimental animals.

Organic Peroxide — An organic compound that contains the bivalent –O–O– structure and which may be considered to be a structural derivative of hydrogen peroxide where one or both of the hydrogen atoms has been replaced by an organic radical. Some organic peroxides are highly unstable and may decompose with explosive force.

OSHA — Occupational Safety and Health Administration of the U.S. Department of Labor; federal agency with safety and health regulatory and enforcement authorities for most U.S. industries and business.

OSHA 20 Form — Form used during early 1970s by OSHA for information contained on Material Safety Data Sheets (MSDS) before they were invented in 1985. OSHA 20 Forms referenced in Maritime Standards.

OSHA 200 Form — Form used to document the number of illnesses and accidents in a work place, as required by 29 CFR 1919.210. This form must be posted in a general viewing area every year in February.

Overexposure — Exposure to a hazardous material beyond the allowable exposure levels.

Oxidizer — A chemical other than a blasting agent or explosive, as defined in 29 CFR 1910.109(a), that initiates or promotes combustion in other materials, thereby causing fire either of itself or through the release of oxygen or other gases. Chlorate (ClO_3), permanganate (MnO_4), and nitrate (NO_3) compounds are examples of oxiders; note that all contain oxygen (O).

PEL — Permissible exposure limit; the legally enforced exposure limit for a substance established by OSHA regulatory authority. The PEL indicates the permissible concentration of air contaminants to which nearly all workers may be repeatedly exposed 8 hr/day, 40 hr/week, over a working lifetime (30 years) without adverse health effects.

% Volatile — Percent volatile by volume; the percentage of a liquid or solid (by volume) that will evaporate at an ambient temperature of 70°F (unless some other temperature is stated). Examples: butane, gasoline, and paint thinner (mineral spirits) are 100% volatile; their individual evaporation rates vary, but over a period of time each will evaporate completely.

pH — The symbol relating the hydrogen ion (H^-) concentration to that of a given standard solution. A pH of 7 is neutral. Numbers increasing from 7 to 14 indicate greater alkalinity. Numbers decreasing from 7 to 0 indicate greater acidity.

Physical Hazard — A chemical for which there is scientifically valid evidence that it is a combustible liquid, a compressed gas, explosive, flammable, an organic peroxide, an oxidizer, pyrophoric, unstable (reactive), or water reactive.

PMCC — Pensky-Martens Closed Cup; a flash point test method.

Pneumoconiosis — A condition of the lung in which there is permanent deposition of particulate matter and the tissue reaction to its presence. It may range from relatively harmless forms of iron oxide deposition to destructive forms of silicosis.

Poison, Class A — A DOT hazard class for extremely dangerous poisons, that is, poisonous gases or liquids of such nature that a very small amount of the gas or vapor of the liquid, mixed with air, is dangerous to life. Some examples: phosgene, cyanogen, hydrocyanic acid, nitrogen peroxide.

Poison, Class B — A DOT hazard class for liquid, solid, paste, or semisolid substances — other than Class A poisons or irritating materials — which are known (or presumed on the basis of animal tests) to be so toxic to man as to afford a hazard to health during transportation. Some examples: arsenic, beryllium chloride, cyanide, mercuric oxide.

Polymerization — See Hazardous Polymerization.

ppb — Parts per billion; a unit for measuring the concentration of a gas or vapor in air — parts (by volume) of the gas or vapor in a billion parts of air. Usually used to express measurements of extremely low concentrations of unusually toxic gases or vapors. Also used to indicate the concentration of a particular substance in a liquid or solid. For example, one part per billion is analogous to one second every 32 years, or one penny out of $10,000,000.

ppm — Parts per million; a unit for measuring the concentration of a gas or vapor in air — parts (by volume) of the gas of vapor in a million parts of air. Also used to indicate the concentration of a particular substance in a liquid or solid. For example, one part per million is analogous to one inch in 16 miles, one minute every 2 years, or one penny out of $10,000.

psi — Pounds per square inch; for MSDS purposes, a unit for measuring the pressure a material exerts on the walls of a confining vessel or enclosure. For technical accuracy, pressure must be expressed as psig (pounds per square inch gauge) or psia (pounds per square inch absolute; that is, gauge pressure plus sea level atmospheric pressure, of psig plus about 14.7 pounds per square inch). Also see mmHg.

Pulmonary — Relating to, or associated with, the lungs.

Pulmonary Edema — Fluid in the lung tissue.

Pyrophoric — A chemical that will ignite spontaneously in air at a temperature of 130°F (54.4°C) or below.

Quality Factor — A linear energy transfer-dependent factor by which absorbed doses are to be multiplied to obtain the dose equivalent.

Rad — The special unit of absorbed dose. One rad equals 100 ergs per gram.

Radiation Protection Guide (RPG) — The radiation dose which should not be exceeded without careful consideration of the reasons for doing so; every effort should be made to encourage the maintenance of radiation doses as far below this guide as practicable.

Radiation Source — An apparatus or a material emitting or capable of emitting ionizing radiation.

Radioactivity Concentration Guide (RCG) — The concentration of radioactivity in the environment which is determined to result in organ doses equal to the Radiation Protection Guide.

RCRA — Resource Conservation and Recovery Act; federal environmental legislation, administered by EPA, aimed at controlling the generation, treatment, storage, transportation, and disposal of hazardous wastes.

REACTION — A chemical transformation or change; the interaction of two or more substances to form new substances.

REACTIVE — See Unstable.

Reactivity — A description of the tendency of a substance to undergo chemical reaction with the release of energy. Undesirable effects — such as pressure buildup, temperature increase, formation of noxious, toxic, or corrosive by-products — may occur because of the reactivity of substance to heating, burning, direct contact with other materials, or other conditions in use of in storage.

Reducing Agent — In a reduction reaction (which always occurs simultaneously with an oxidation reaction) the reducing agent is the chemical or substance which (1) combines with oxygen or (2) loses electrons to the reaction.

Reproductive Toxin — Substances that affect either male or female reproductive systems and may impair the ability to have children.

Respiratory Protection — Devices that will protect the wearer's respiratory system from overexposure by inhalation to airborne contaminants. Respiratory protection is used when a worker must work in an area where he/she might be exposed to concentrations in excess of the allowable exposure limit.

Respiratory System — The breathing system; includes the lungs and air passages (trachea or "windpipe," larynx, mouth, and nose) to the air outside the body, plus the associated nervous and circulatory supply.

Roentgen (R) — The special unit of exposure. One roentgen equals 2.58×10^{-4} coulomb per kilogram of air.

Routes of Entry — The means by which material may gain access to the body, for example, inhalation, ingestion, and skin contact.

SARA — Superfund Amendment and Reauthorization Act of 1986.

SCBA — See Self-Contained Breathing Apparatus.

Sealed Source — A radioactive source sealed in a container or having a bonded cover where the container or cover has sufficient mechanical strength to prevent contact with the dispersion of the radioactive material under the conditions of use and wear for which it was designed.

Self-Contained Breathing Apparatus — A respiratory protection device that consists of a supply or a means of respirable air, oxygen, or oxygen generating material, carried by the wearer.

Self-Ignition — See Autoignition Temperature.

Sensitizer — A substance which, on first exposure, causes little or no reaction in man or test animals, but which, on repeated exposure, may cause a marked response not necessarily limited to the contact site. Skin sensitization is the most common form of sensitization in the industrial setting, although respiratory sensitization to a few chemicals is also known to occur. Examples of sensitizers include poison ivy, pollen, and some isocyanates and epoxy resin hardeners.

SETA — Setaflash Closed Tester; a flash point test method.

Silicosis — A disease of the lungs caused by the inhalation of silica dust.

"Skin" — A notation, sometimes used with PEL or TLV exposure data; indicates that the stated substance may be absorbed by the skin, mucous membranes, and eyes — either airborne or by direct contact — and that this additional exposure should be considered part of the total exposure.

Skin Absorption — Ability of some hazardous chemicals to pass directly through the skin and enter the bloodstream.

Skin Sensitizer — See Sensitizer.

Skin Toxicity — See Dermal Toxicity.

Solubility in Water — A term expressing the percentage of a material (by weight) that will dissolve in water at ambient temperature. Solubility information can be useful in determining spill cleanup methods and fire-extinguishing agents and methods for a material. Terms used to express solubility are

- Negligible — less than 0.1%
- Slight — 0.1 to 1.0%
- Moderate — 1 to 10%
- Appreciable — more than 10%
- Complete — soluble in all proportions

Solvent — A substance, usually a liquid, in which other substances are dissolved. The most common solvent is water.

Sorbents — Nonreactive materials used to clean up chemical spills. Examples: clay and vermiculite.

Special Fire-Fighting Procedures — Special procedures and/or personal protective equipment that is necessary when a particular substance is involved in a fire.

Species — A biological type; on MSDSs, species refers to the test animals — usually rats, mice, or rabbits — which were used to obtain the toxicity test data reported.

Specific Gravity — The weight of a material compared to the weight of an equal volume of water; an expression of the density (or heaviness) of the material. Example: if a volume of a material weighs 8 lb, and an equal volume of water weighs 10 lb, the material is said to have a specific gravity of 0.8.

$$\frac{8 \text{ lb}}{10 \text{ lb}} = 0.8$$

In soluble materials, specific gravity of less than 1.0 will float in (or on) water. Insoluble materials with specific gravity greater than 1.0 will sink (or go to the bottom) in water. Most (but not all) flammable liquids have specific gravity less than 1.0 and, if not soluble, will float on water — an important consideration for fire suppression and spill cleanup.

Spill or Leak Procedures — The methods, equipment, and precautions that should be used to control or clean up a leak or spill.

Splash-Proof Goggles — Eye protection made of a noncorrosive material that fits snugly against the face, and has indirect ventilation ports.

Spontaneously Combustible — A material that ignites as a result of retained heat from processing, or which will oxidize to generate heat and ignite, or which absorbs moisture to generate heat and ignite.

Stability — An expression of the ability of a material to remain unchanged. For MSDS purposes, a material is stable if it remains in the same form under expected and reasonable conditions of storage or use. Conditions which may cause instability (dangerous change) are stated. Examples: temperatures above 150°F, shock from dropping.

STEL — Short-term exposure limit; ACGIH terminology. See TLV.

Superfund — See CERCLA.

Supplied-Air Respirators — Air-line respirators or self-contained breathing apparatus.

Synonym — Another name or names by which a material is known. Methyl alcohol, for example, is also known as methanol, or wood alcohol.

Systemic Poison — A poison which spreads throughout the body, affecting all body systems and organs. Its adverse effect is not localized in one spot or area.

Systemic Toxicity — Adverse effects caused by a substance which affects the body in a general rather than local manner.

Target Organ Toxic — A toxic substance that attacks a specific organ of the body. For example, overexposure to carbon tetrachloride can cause liver damage.

TCC — Tag (Tagliabue) Closed Cup; a flash point test method.

Teratogen — A substance or agent to which exposure of a pregnant female can result in malformations in the fetus. An example is thalidomide.

Tissue Necrosis — Local tissue death that occurs in groups of cells because of disease or injury.

TLV — Threshold limit value; a term used by ACGIH to express the airborne concentration of a material to which nearly all persons can be exposed day after day, without adverse effects. ACGIH expresses TLVs in three ways:

- **TLV-TWA:** the allowable Time Weighted Average concentration for a normal 8-hr work day or 40-hr work week.
- **TLV-STEL:** the Short-Term Exposure Limit, or maximum concentration for a continuous 15-min exposure period (maximum of four such periods per day, with at least 60 min between exposure periods, provided that the daily TLV-TWA is not exceeded).
- **TLV-C:** the Ceiling limit — the concentration that should not be exceeded even instantaneously.

See also "Skin."

TOC — TAG Open Cup; a flash point test method.

Toxicity — The sum of adverse effects resulting from exposure to a material, generally by the mouth, skin, or respiratory tract.

Toxic Substance — Any substance which can cause acute or chronic injury to the human body, or which is suspected of being able to cause diseases or injury under some conditions.

Trade Name — The trademark name or commercial trade name for a material.

Trade Secret — Any confidential formula patten, process, device, information, or compilation of information (including chemical name or other unique chemical identifier) that is used in an employer's business and that gives the employer an opportunity to obtain an advantage over competitors who do not know or use it.

TSCA — Toxic Substance Control Act; federal environmental legislation, administered by EPA, for regulating the manufacturer, handling, and use of materials classified as "toxic substances."

TWA — Time-weighted average exposure; ACGIH terminology. See TLV.

UEL — Upper explosive limit or upper flammable limit of a vapor or gas; the highest concentration (highest percentage of the substance in air) that will produce a flash of fire when an ignition source (heat, arc, or flame) is present. At higher concentrations, the mixture is too "rich" to burn. See also LEL.

UFL — See UEL.

UN Number — A registry number assigned to dangerous commonly carried goods by the United Nations Committee of Experts on the Transport of Dangerous Goods. This UN number is required in shipping documentation and on packaging as part of the DOT regulations for shipping hazardous materials.

Unstable — A chemical which, in the pure state, or as produced or transported, will vigorously polymerize, decompose, condense, or will become self-reactive under conditions of shock, pressure, or temperature. These chemicals are also referred to as reactive.

Unusual Fire and Explosion Hazards — Hazards that might occur as a result of overheating or burning a specific material. These include chemical reactions or changes in chemical composition or any special hazards involved in extinguishing the burning material.

USDA — U.S. Department of Agriculture; prior to 1971, USDA performed tests and issued approvals on respirators for use with pesticides. In 1971, the Bureau of Mines took over the pesticide respirator testing/approval functions — procedures later delegated to the Testing and Certification Branch (TCB) of NIOSH.

Vapor — The gaseous form of a solid or liquid substance as it evaporates.

Vapor Density — The weight of a vapor or gas compared to the weight of an equal volume of air; an expression of the density of the vapor or gas. Materials lighter than air have vapor densities less than 1.0 (examples: acetylene, methane, hydrogen). Materials heavier than air (examples: propane, hydrogen sulfide, ethane, butane, chlorine, sulfur dioxide) have vapor densities greater than 1.0. All vapors and gases will mix with air, but the lighter materials will tend to rise and dissipate (unless confined). Heavier vapors and gases are likely to concentrate in low places — along or under floors, in slumps, sewers and manholes, in trenches and ditches — where they may create fire or health hazards.

Vapor Pressure — The pressure exerted by a saturated vapor above its own liquid in a closed container. When quality control tests are performed on products, the test temperature is usually 100°F and the vapor pressure is expressed as pounds per square inch (psig or psia), but vapor pressures reported on MSDSs are in millimeters of mercury (mmHg) at 68°F (20°C), unless stated otherwise. Three facts are important to remember:

1. Vapor pressure of a substance at 100°F will always be higher than the vapor pressure of the substance at 68°F (20°C).
2. Vapor pressure reported on MSDSs in mmHg are usually very low pressures; 760 mmHg is equivalent to 14.7 lb/in^2.
3. The lower the boiling point of a substance, the higher its vapor pressure.

Ventilation — Circulating fresh air to replace contaminated air.

Vermiculite — An expanded mica (hydrated magnesium-aluminum-iron silicate) used as sorbent for spill control and cleanup.

Viscosity — The tendency of a fluid to resist internal flow without regard to its density.

Volatility — A measure of how quickly a substance forms a vapor at ordinary temperatures.

Waste Disposal Methods — Proper disposal methods for contaminated material, recovered liquids or solids, and their containers.

Water Reactive — A chemical that reacts with water to release a gas that is either flammable or presents a health hazard.

Woburn — City in Massachusetts where there was an increase in incidences of leukemia. Leukemia was caused by chemicals being emitted from the W.A. Trace Corporation fertilizing plant into adjacent water sources and getting into public drinking water. The primary target group were children 13 and under. Major toxic tort case in U.S. history. Settlement of $6 million to parents (children died) out of court.

Appendix

RADIATION HAZARDS

Prepared by Dr. Loren L. Hatch, Medical Officer
(Occupational Medicine), NIOSH, 11/81
Now retired in St. Augustine, Florida, 1989

Study of radioactivity in the Florida phosphate surface mining industry has been going on since 1974. That radiation is a concern and arouses skepticism in the minds of the public and cynicism among miners.

The phosphate industry does not manufacturer any radionuclides; however, the industry does redistribute in the environment large quantities of naturally occurring radionuclides. Nor is the phosphate industry the only group that redistributes naturally occurring radionuclides. Other industries include titanium, coal, and some building materials.

The primary focus of the EPA studies are in phosphate mining and milling, uranium mining and milling, and construction materials, and to some extent, western coal mining.

As discussed earlier, huge quantities of gypsum as a by-product of phosphte mining and phosphoric acid plants are stored at the plant site. These gypsum piles frequently cover 300 acres and are up to 150 ft high. These piles are a potential source of radioactive radiation.

The mining, milling, and processing of phosphate can cause varying amounts of technologically enhanced natural radioactivity (TENR). The amount of radioactivity is dependent on the inherent uranium and thorium content of the phosphate. About three million person-rem is attributed to TENR.

The EPA office of radiation programs estimates that the largest source of ionizing radiation dose to the population is ten million person-rem from ambient ionizing radiation.

Background radiation levels in the geographic area of the phosphate mines are caused by low-level, naturally occurring radiation that has been brought to the surface as a result of the phosphate mining and processing industry. Phosphate miners work among widely varying environmental conditions, including temperature, pressure, magnetic fields, gravitational fields, and radiation.

In the case of radiation, we know that high levels are harmful. Ideally, radioactive contamination should be removed to reduce the radiation values at least to the naturally occurring background level or lower.

One way of reducing radiation concerns the recovery of uranium-bearing materials from the waste portion of the phosphate matrix. The phosphatic clay waste and slime is primarily the suspension of clay particles and water. This slurry contains approximately 25% of unrecovered phosphatic ore where a significant quantity of uranium and radium is found.

Recovery of uranium-bearing minerals from the slime would not only recover phosphatic materials, but would result in a decrease in the radiation problem and increase the production of recoverable uranium.

Another area in the phosphate mine with low-level radioactivity occurs in the wet process phosphoric acid production used to produce fertilizers. This process results in the generation of large quantities of gypsum byproduct, which is at present stockpiled at the mine sites, as indicated earlier.

This gypsum contains radium-226, which is a parent isotope of radan-222. Radan and daughter isotopes are of concern to miners.

Removal of radium-226 economically by recovery or removal is ideal.

Radiation

Ionizing Radiation

Ionizing radiation is energy in the form of alpha, beta, or neutron particles or X- or gamma rays. This energy is emitted from radioactive materials. Ionization radiation means that the energy separates atoms into ions. Miners at the time of exposure cannot sense the absorption of energy and thus cannot easily recognize the danger. In order to properly assess the potential ionizing radiation hazard, miners must be able to properly identify sources and determine the necessary protection.

The health effects on miners to ionizing radiation depend upon several factors:

1. Type of radiation.
2. Energy of the radiation.
3. Time period over which the radiation is received.
4. Relative biological which the radiation is received.
5. Time/dose and point of exposure.
6. The area of the body exposed.

Generally, alpha particles pose no external hazards, as they generally do not penetrate the skin.

To minimize miner exposure to external radiation, limiting the length of exposure is essential. If the exposure rate is 100 milliroentgens per hour (mR/hr) then after 1 h, an employee would have been exposed to 100 mR of radiation. If the time of the miner's exposure was reduced to half an hour, then the exposure would be halved, or 50 mR.

The second most important method of reducing radiation is increasing the distance from the miner to the source of radiation. As one doubles the distance between the source and the miner, the exposure is reduced by a factor of four.

The third method of protecting miners is by shielding, that is placing materials such as lead or concrete between the miner and the source of radiation.

Ionizing radiation levels are measured by a meter that is calibrated for the type of radiation present. These generally include

- Ionization chamber
- Geiger-Mueller counter
- Proportional counter
- Solid-state detector
- Scintillation counter

Filter paper smears of surfaces coated with radioactive material are counted by the above instruments.

Some miners wear film badges to detect gamma radiation. These badges are worn over a period of time and used to estimate the accumulated dose of radiation to the whole body or to certain parts of the body.

It is very difficult to estimate the internal radiation doses from alpha emitters. Body wastes such as urine can be analyzed, or whole-body counters can be utilized.

When controls such as time, distance, shielding, and ventilation are not adequate, personal protective equipment may be required to protect miners; this includes specialized breathing apparatus, gloves, use of tools or tongs, and protective clothing.

Records should be maintained showing individual miners' accumulated dose. If the accumulated doses are high for miners, they should be moved to areas of lower potential exposure. Pregnant women should not be exposed to ionizing radiation (Hatch).

Regular medical examination should be performed on miners exposed to radiation including urine analysis, blood tests, and sometimes eye and skin examinations.

Nonionizing Radiation

Nonionizing radiation includes radiation that cannot transmit enough energy to break and atom into ions. This source of radiation generally comes from sunlight, certain types of lamps, some hot objects, ultraviolet light, infrared light, microwaves, radar, radio, TV, and energy emissions.

Radio waves, microwaves, and electric power waves are described in terms of frequency. The higher the frequency of the radiation, the shorter the wavelength and the higher the energy transmitted. Miners should know each type of radiation and the parts of the body potentially effected:

1. Ultraviolet — skin and eyes. Sunlight, arc welding, germicidal lamps, black light lamps, incandescent lamps and lasers.
2. Infrared — eyes and body temperature. Industrial operation involves temperatures of 1000 K, such as hot metals and welding.
3. Microwaves — body tissue and eyes. Radio broadcasting, television broadcasting, radar, microwave ovens, alarm systems, concrete shearing.
4. Laser — eye and skin. Cutting, welding, drilling, and surveying.

Miners should have on-going educational programs to remind them of the potential hazards of ionizing and non-ionizing radiation.

Biological Aspects of Irradiation*

There are many exposures to radiation and radioactive materials in mining.

According to the former U.S. Atomic Energy Commission, the average annual occupational exposure for all industry is estimated at about 500 mrem per person.

Miners continually receive a dose of natural radiation that comes both from sources external to the body and from naturally occurring radionuclides deposited within the body. External sources are primarily cosmic radiation and gamma radiation from materials naturally present in the ground and in building materials. Several radionuclides are deposited in the body from foods, drinking water, and air, including uranium and its decay product thorium and radiopotassium and radiocarbon.

Some radiation exposure comes from man-made sources such as fallout from nuclear weapons testing, effluence from nuclear and other facilities processing or using radionuclides, luminous dial clocks, watches, and signs, and electronic devices, such as television sets, using high voltages.

Certain diagnostic and therapeutic procedures involve external radiation with beta-, gamma-, or X-radiation and internal radiation from ingestion or injected radionuclides.

Practically all occupational radiation involves chronic exposures occurring over many months and years. These exposures are usually less than 100 mrem but occasionally an accident or an acute exposure may give a miner a dose as high as 25 rem for a very short period of time.

As with other occupational exposures, there may be a very long lag time between exposure and the human effects.

Irradiation of miners produces two types of effects: genetic and somatic. Independent of any somatic effect, irradiation of the testicles may cause genetic effects, since mutations which are caused by heritable changes in the germ plasm may occur. Permanent sterility occurs only with absorbed doses to the testicles of

* Some of this section was taken from a paper by Edgar C. Barnes in "The Industrial Environment — Its Evaluation and Control," U.S. Department of Health, Education, and Welfare, Public Health Service, Center for Disease Control, National Institute for Occupational Safety and Health (1973).

500 to 600 rads of X- or gamma-radiation. A single dose of 50 rads may induce brief sterility in miners.

Lag effects occurring many years after exposure may be caused by acute exposure resulting in leukemia.

In the case of chronic occupational exposures, if they are within the recommended dose limits the possibility of determining any late effects is minimal, because the normal lifetime of humans might preclude discovering any latent effects. Effects including those of particular concern, cancers, cataracts, and shortened life may also be delayed for long periods of time.

In the study of biological effects, the variation due to different kinds of radiation is referred to as relative biological effectiveness (RBE). This is the ratio of absorbed doses that produce equal effects with cobalt 60 gamma-rays or 200- to 250-kV X-rays used as a reference.

Various tissues and organs of the body are not affected equally by equal irradiation, but vary considerably.

For uniform, whole-body irradiation, the blood-forming organs (such as the red bone narrow), the lens of the eye, and the testicles are more susceptible to significant effects, and these are designated as critical organs.

For individuals past reproductive age, the gonads are not a critical organ.

For those internal radiation sources that do not irradiate the body uniformly, the distribution and metabolic pattern for each radionuclide will determine which organs and tissues receive the larger dose. Other critical organs and tissues include lung, gastro-intestinal tract, bone, muscle, fatty tissue, thyroid, kidney, spleen, pancreas, and prostate.

The total curies of a radionuclide is designated as the "body burden."

Irradiation Exposure, Control, and Evaluation

The basic concept in radiation protection is the establishment of a controlled area where access is limited, and when a person must enter these areas, adequate supervision is available.

The second principle is control of the escape of radioactive materials from these controlled areas.

Many miners have seen the warning notice symbol which identifies these controlled areas. It is a purple trefoil on a yellow background. The third basic method which has already been discussed is increasing the distance between the workers and the radiation source. The fourth method, shielding, was already discussed.

Many gauges used in the mining process have beta-radiation sources built in and they must be adequately shielded and identified.

Evaluation to establish the dose equivalent from external radiation sources is usually accomplished by conducting radiation surveys. Proper instruments must be selected for the survey, and they must be conducted by competent personnel.

A discussion of radiation surveys and equipment is beyond the scope of this learning module.

Methods of examining miners for exposure is more difficult and a routine physical examination may lead to a false sense of security. A relatively new cytogenetic technique involves the determination of chromosome irregularities found in somatic human blood cells. Examination of urine and feces may be an effective technique if performed by a knowledgeable nuclear physician. Many radiologist have this extra qualification as do some occupational physicians.

Radiation is governed by a variety of government and private agencies. Many state agencies have their own codes for radiation which are equal to or exceed federal regulations.

RANGE VS. ENERGY FOR ALPHA PARTICLES IN AIR (STP)

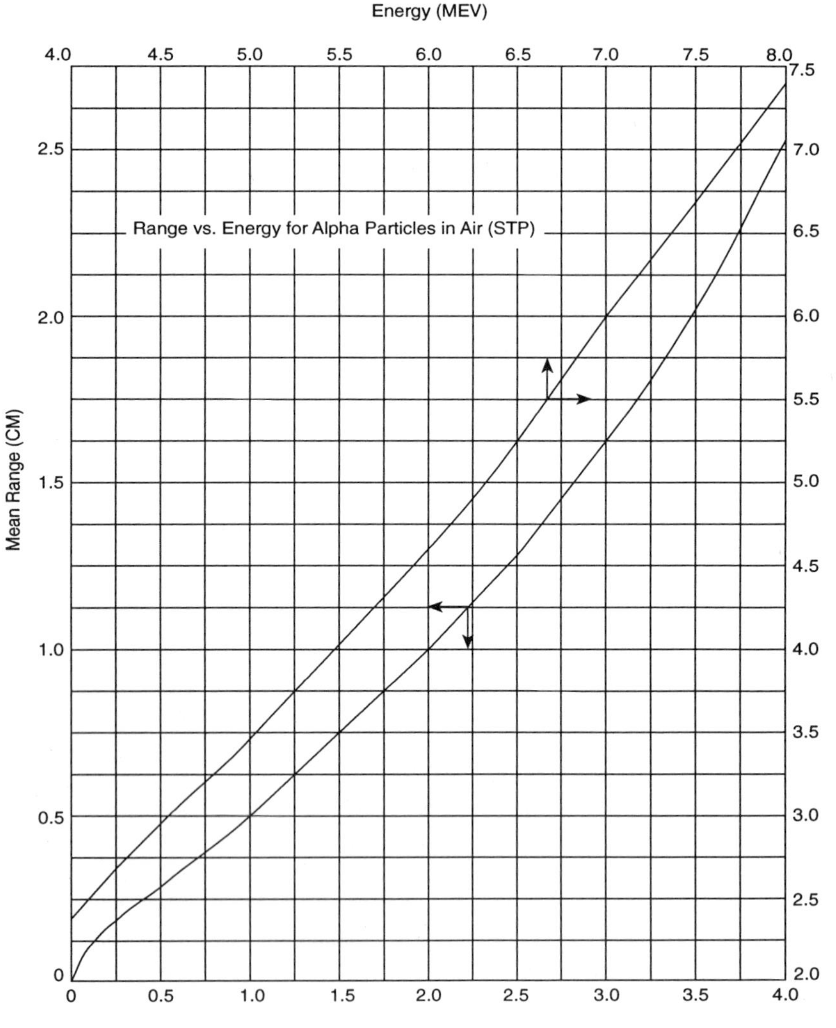

Mean quality factors, QF[a], and values of neutron flus density which in a period of 40 hours results in a maximum dose equivalent of 100 mrem.

Neutron Energy McV	$\overline{QF}$	Neutron Flux Density cm^{-2} s^{-2}
2.5×10^{-6} (thermal)	2	680
1×10^{-7}	2	680
1×10^{-6}	2	560
1×10^{-5}	2	560
1×10^{-4}	2	580
1×10^{-3}	2	680
1×10^{-2}	2.5	700
1×10^{-1}	7.5	115
5×10^{-1}	11	27
1	11	19
2.5	9	20
5	8	16
7	7	17
10	6.5	17
14	7.5	12
20	8	11
40	7	10
60	5.5	11
1×10^{2}	4	14
2×10^{2}	3.5	13
3×10^{2}	3.5	11
4×10^{2}	3.5	10

[a] Maximum value of $\overline{QF}$ in a 30-cm phantom.

CARCINOGENIC CHEMICAL LIST

Developed by Janet McCloud, M.P.H.
Private Consultant in Environmental Public Health
198 Tollgate Trail, Longwood, Florida 32750

CHEMICAL NAME — Acrylonitrile
TLV (if known) — 2 ppm, 4.5 mg/m^3
STATUS — probable
ATTRIBUTED CANCER(S) — colon and lung
EXPOSURE ROUTE — inhalation
LATENCY — 10–20 years

CHEMICAL NAME — Acetylaminofluorene-2
TLV (if known)
STATUS — + in rats
ATTRIBUTED CANCER(S) — liver and bladder
EXPOSURE ROUTE — oral, skin, inhalation
LATENCY

CHEMICAL NAME — Allyl chloride
TLV (if known) -2 ppm, 5 mg/m^3
STATUS — questionable
ATTRIBUTED CANCER(S)
EXPOSURE ROUTE — oral, skin
LATENCY

CHEMICAL NAME — Aluminum oxide

TLV (if known) — 10 mg/m^3
STATUS — questionable
ATTRIBUTED CANCER(S) — in rats
EXPOSURE ROUTE — inhalation
LATENCY

CHEMICAL NAME — Aluminum silicate
TLV (if known) — 2 mg/m^3
STATUS — questionable
ATTRIBUTED CANCER(S) — in rats
EXPOSURE ROUTE — inhalation
LATENCY

CHEMICAL NAME — Aminobiphenyl-4
TLV (if known)
STATUS — Conclusive, mice, rats, rabbits, and dogs
ATTRIBUTED CANCER(S) — bladder
EXPOSURE ROUTE — inhalation and skin
LATENCY — 15–35 years

CHEMICAL NAME — Aniline
TLV (if known) — 5 ppm, 19 mg/m^3
STATUS — not definite
ATTRIBUTED CANCER(S) — in rats
EXPOSURE ROUTE — inhalation and skin
LATENCY

CHEMICAL NAME — Arsenic
TLV (if known) — 0.05 mg/m^3
STATUS — indefinite
ATTRIBUTED CANCER(S) — skin, lung, liver, mouth, esophogus, bladder paranasal sinus
EXPOSURE ROUTE — inhalation and ingestion
LATENCY

CHEMICAL NAME — Arsenic trisulfide
TLV (if known) — 0.05 ppm
STATUS — indefinite
ATTRIBUTED CANCER(S) — in rats
EXPOSURE ROUTE — inhalation and ingestion
LATENCY

CHEMICAL NAME — Asbestos
TLV (if known) — 0.2 fibers/cc or air
STATUS — + human

ATTRIBUTED CANCER(S) — GI, lung, intestinal esophagus, colon, stomach
EXPOSURE ROUTE — inhalation and ingestion
LATENCY — 15–50 years

CHEMICAL NAME — Asphalt
TLV (if known)
STATUS — possible
ATTRIBUTED CANCER(S) — skin, scrotum, possible lung and larynx
EXPOSURE ROUTE — inhalation and skin
LATENCY

CHEMICAL NAME — Barium sulfate
TLV (if known) — 0.5 mg/m^3
STATUS — questionable
ATTRIBUTED CANCER(S) — in rats
EXPOSURE ROUTE — inhalation, skin, ingestion
LATENCY

CHEMICAL NAME — Benzene
TLV (if known) — 1 ppm
STATUS — conclusive
ATTRIBUTED CANCER(S) — hematotoxic-leukemia poss. lympho-sarcomas
EXPOSURE ROUTE — inhalation and skin
LATENCY — 18 mg/m^3 for 13 years old. EPA SNARLS indicate one-time exposure needed high or low dose.

CHEMICAL NAME — Benzidine
TLV (if known) — 0.1%
STATUS — + human
ATTRIBUTED CANCER(S) — bladder, liver, ear duct, intestines
EXPOSURE ROUTE — inhalation, skin, ingestion
LATENCY — 16 years

CHEMICAL NAME — Benzoyl peroxide
TLV (if known) — 5 mg/m^3
STATUS — suggestive
ATTRIBUTED CANCER(S) — skin in mice
EXPOSURE ROUTE — skin
LATENCY

CHEMICAL NAME — Beryllium
TLV (if known) — 0.002 mg/m^3
STATUS — + human

ATTRIBUTED CANCER(S) — lungs
EXPOSURE ROUTE — inhalation
LATENCY — 10 years

CHEMICAL NAME — Bipyridine, 2, -2
TLV (if known)
STATUS — suggestive
ATTRIBUTED CANCER(S) — s.c. in mice
EXPOSURE ROUTE — skin
LATENCY

CHEMICAL NAME — Cadmium sulfide
TLV (if known) — 0.05 mg/m^3
STATUS — animal +
ATTRIBUTED CANCER(S) — connective tissue, prostate, lung, liver in humans
EXPOSURE ROUTE — ingestion
LATENCY

CHEMICAL NAME — Carbaryl
TLV (if known) — 5 mg/m^3
STATUS — animal +
ATTRIBUTED CANCER(S) — in rats
EXPOSURE ROUTE — ingestion
LATENCY

CHEMICAL NAME — Carbon tetrachloride
TLV (if known) — 10 ppm, 65 mg/m^3
STATUS — susp. in humans
ATTRIBUTED CANCER(S) — liver
EXPOSURE ROUTE — oral and inhalation
LATENCY

CHEMICAL NAME — Ceramic dust
TLV (if known) — 2 mg/m^3
STATUS — questionable
ATTRIBUTED CANCER(S) — in rats
EXPOSURE ROUTE — inhalation and oral
LATENCY

CHEMICAL NAME — Chloroform
TLV (if known) — 50 ppm, 240 mg/m^3
STATUS — high susp. in humans
ATTRIBUTED CANCER(S) — liver (mice), kidneys (rats)
EXPOSURE ROUTE — oral, ingestion
LATENCY

CHEMICAL NAME — Chloromethyl ether-bis
TLV (if known) — not known
STATUS — susp. in humans
ATTRIBUTED CANCER(S) — lung, small cell carcinoma, skin, sarcoma
EXPOSURE ROUTE — inhalation, ingestion, skin
LATENCY — 10–15 years

CHEMICAL NAME — Chromic oxide
TLV (if known) — 0.1 mg/m^3
STATUS — animal (susp.) susp. in humans
ATTRIBUTED CANCER(S) — lung, nasal cavity, paranasal sinus, poss. stomach and larynx
EXPOSURE ROUTE — inhalation
LATENCY — 10–30 years

CHEMICAL NAME — Chromium (VI)
TLV (if known) — 0.001 mg/m^3
STATUS — susp. humans
ATTRIBUTED CANCER(S) — lung
EXPOSURE ROUTE — inhalation, ingestion, skin
LATENCY — 10–30 years

CHEMICAL NAME — Chromium oxide
TLV (if known) — TWA 0.5 mg/m^3
STATUS — susp. humans
ATTRIBUTED CANCER(S) — lung, nasal
EXPOSURE ROUTE — inhalation, ingestion, skin
LATENCY — 10–30 years

CHEMICAL NAME — Chromium trioxide
TLV (if known) — TWA 0.1 mg/m^3
STATUS — susp. humans
ATTRIBUTED CANCER(S) — lung, nasal
EXPOSURE ROUTE — inhalation, ingestion, skin
LATENCY — 10–30 years

CHEMICAL NAME — Chromium (elemental)
TLV (if known) — 1 mg/m^3
STATUS — susp. nasal
ATTRIBUTED CANCER(S) — lung, nasal
EXPOSURE ROUTE — inhalation, ingestion, skin
LATENCY — 10–30 years

CHEMICAL NAME — Coal tar products
TLV (if known) — 2 mg/m^3
STATUS — + human

ATTRIBUTED CANCER(S) — skin, scrotum, lip, larynx, lung
EXPOSURE ROUTE — inhalation, skin, ingestion
LATENCY — 9–25 years

CHEMICAL NAME — Cobalt, cobalt chloride
TLV (if known) — 0.1 mg/m^3
STATUS — susp. humans
ATTRIBUTED CANCER(S) — lung, connective tissue
EXPOSURE ROUTE — inhalation, skin, ingestion
LATENCY

CHEMICAL NAME — Coke oven emission
TLV (if known) — 0.2 $\mu g/m^3$
STATUS — + human
ATTRIBUTED CANCER(S) — skin, scrotum, lip, larynx, lung
EXPOSURE ROUTE — inhalation, skin, ingestion
LATENCY — 9–25 years

CHEMICAL NAME — Copper sulfate
TLV (if known) — 0.1 mg/m^3
STATUS — questionable
ATTRIBUTED CANCER(S) — lung
EXPOSURE ROUTE — inhalation, skin, ingestion
LATENCY

CHEMICAL NAME — Crystalline silica
TLV (if known) — 0.1 mg/l^2, respirable dust
STATUS — conclusive animal
ATTRIBUTED CANCER(S) — lung, respiratory system
EXPOSURE ROUTE — inhalation, ingestion, skin
LATENCY

CHEMICAL NAME — Cupferron
TLV (if known)
STATUS — conclusive, animal
ATTRIBUTED CANCER(S) — in rats and mice
EXPOSURE ROUTE — oral
LATENCY

CHEMICAL NAME — DBCP
TLV (if known) — 10 ppm 1 mg/m^3
STATUS — + human
ATTRIBUTED CANCER(S) — squamous cell carcinomas
EXPOSURE ROUTE — inhalation
LATENCY

CHEMICAL NAME — Dichlorobenzene
TLV (if known) — 50 ppm 300 mg/m^3
STATUS — indefinite
ATTRIBUTED CANCER(S) — multiple sites including bladder in rats
EXPOSURE ROUTE — inhalation, skin
LATENCY

CHEMICAL NAME — Dichlorobenzidine
TLV (if known) — all contact avoided
STATUS — + animal
ATTRIBUTED CANCER(S) — bladder and other
EXPOSURE ROUTE — inhalation, skin
LATENCY

CHEMICAL NAME — Dichloromethane
TLV (if known) — TWA 100 ppm 210 mg/m^3
STATUS — potential
ATTRIBUTED CANCER(S) — in rats
EXPOSURE ROUTE — inhalation, skin
LATENCY

CHEMICAL NAME — Diglycidyl ether or biphenol
TLV (if known)
STATUS — conclusive in animals
ATTRIBUTED CANCER(S) — in rats
EXPOSURE ROUTE — skin
LATENCY

CHEMICAL NAME — diisobutylphenoxy-ethoxyethyldimethyl-benzylammonium chloride monohydrate
TLV (if known)
STATUS — suggestive
ATTRIBUTED CANCER(S) — in rats
EXPOSURE ROUTE — skin
LATENCY

CHEMICAL NAME — Dimethyl aniline
TLV (if known)
STATUS — suggestive
ATTRIBUTED CANCER(S) — in rats
EXPOSURE ROUTE — oral
LATENCY

CHEMICAL NAME — Dioxane
TLV (if known) — 25 ppm 90 mg/m^3

STATUS — conclusive
ATTRIBUTED CANCER(S) — multiple sites nasal cavities, liver
EXPOSURE ROUTE — skin, ingestion, inhalation
LATENCY

CHEMICAL NAME — Epichlorohydrin
TLV (if known) — 5 ppm 19 mg/m^3
STATUS — + animal
ATTRIBUTED CANCER(S) — two-stage skin carcinoma, local sarcomas
EXPOSURE ROUTE — inhalation, skin, ingestion
LATENCY

CHEMICAL NAME — Ethyl alcohol
TLV (if known) — 1000 ppm 1900 mg/m^3
STATUS — + animal
ATTRIBUTED CANCER(S) — mouth, larynx, esophagus, tongue, liver
EXPOSURE ROUTE — oral
LATENCY

CHEMICAL NAME — Ethylene dibromide
TLV (if known) — 0.13 ppm 1 mg/m^3
STATUS — + animal
ATTRIBUTED CANCER(S) — in mice and rats
EXPOSURE ROUTE — oral
LATENCY

CHEMICAL NAME — Ethylene dichloride
TLV (if known) — TWA 5 ppm
STATUS — positive
ATTRIBUTED CANCER(S) — in mice and rats
EXPOSURE ROUTE — oral
LATENCY

CHEMICAL NAME — Ethylene oxide
TLV (if known) — 50 ppm 90 mg/m^3
STATUS — suggestive
ATTRIBUTED CANCER(S) — in mice
EXPOSURE ROUTE — oral, inhalation
LATENCY

CHEMICAL NAME — Ethyleneimine
TLV (if known)
STATUS — + animal
ATTRIBUTED CANCER(S) — sarcoma, kidney, liver, lung
EXPOSURE ROUTE — inhalation, skin
LATENCY

CHEMICAL NAME — Ferrous sulfate
TLV (if known)
STATUS — suggestive
ATTRIBUTED CANCER(S) — in mice
EXPOSURE ROUTE — oral
LATENCY

CHEMICAL NAME — Fiberglass
TLV (if known) — 3 fibers/cc air
STATUS — conclusive
ATTRIBUTED CANCER(S) — in rat, hamster, questionable in mice
EXPOSURE ROUTE — inhalation
LATENCY

CHEMICAL NAME — Fluoride
TLV (if known) — 2.5 mg/m^3
STATUS — possible
ATTRIBUTED CANCER(S) — respiratory
EXPOSURE ROUTE — inhalation
LATENCY

CHEMICAL NAME — Fluorocarbon polymers
TLV (if known)
STATUS — indefinite
ATTRIBUTED CANCER(S) — in mice
EXPOSURE ROUTE — oral
LATENCY

CHEMICAL NAME — Formaldehyde
TLV (if known) — 1 ppm 3 mg/m^3
STATUS — susp. in humans
ATTRIBUTED CANCER(S) — lung, nasal/nose
EXPOSURE ROUTE — inhalation
LATENCY

CHEMICAL NAME — Gold
TLV (if known)
STATUS — conclusive
ATTRIBUTED CANCER(S) — in rats and mice
EXPOSURE ROUTE — implant
LATENCY

CHEMICAL NAME — Gum
TLV (if known)
STATUS — suggestive
ATTRIBUTED CANCER(S) — in rats

EXPOSURE ROUTE
LATENCY

CHEMICAL NAME — high-voltage X-ray
TLV (if known)
STATUS — conclusive
ATTRIBUTED CANCER(S) — leukemia, epithelial tumors
EXPOSURE ROUTE — irradiation
LATENCY — 2–25, 15–25 years

CHEMICAL NAME — hydrazine
TLV (if known) — 1 ppm 1.3 mg/m^3
STATUS — + animal
ATTRIBUTED CANCER(S) — lung and liver
EXPOSURE ROUTE — oral, inhalation, skin
LATENCY

CHEMICAL NAME — hydroquione
TLV (if known) — 2 mg/m^3
STATUS — indefinite
ATTRIBUTED CANCER(S) — in mice
EXPOSURE ROUTE — i.p.
LATENCY

CHEMICAL NAME — Iron, iron oxide
TLV (if known) — 10 mg/m^3
STATUS — susp. humans
ATTRIBUTED CANCER(S) — respiratory tract, liver, connective tissues
EXPOSURE ROUTE — inhalation
LATENCY — 10+ years

CHEMICAL NAME — Isobutyl alcohol
TLV (if known) — 100 ppm
STATUS — conclusive
ATTRIBUTED CANCER(S) — in rats
EXPOSURE ROUTE — s.c.
LATENCY

CHEMICAL NAME — Kepone
TLV (if known) — TWA 1 $\mu g/m^3$
STATUS — experimental
ATTRIBUTED CANCER(S) — transplacental cancer, liver, thyroid
EXPOSURE ROUTE — inhalation, ingestion skin
LATENCY

CHEMICAL NAME — Lead acetate
TLV (if known)
STATUS — + animal
ATTRIBUTED CANCER(S) — in rats
EXPOSURE ROUTE — oral
LATENCY

CHEMICAL NAME — Lead naphthenate
TLV (if known)
STATUS — questionable
ATTRIBUTED CANCER(S) — in animals
EXPOSURE ROUTE
LATENCY

CHEMICAL NAME — malathion
TLV (if known) — TWA 10 mg/m^3
STATUS — questionable
ATTRIBUTED CANCER(S) — in animals
EXPOSURE ROUTE — oral
LATENCY

CHEMICAL NAME — maleic anhydride
TLV (if known) — 0.25 ppm 1 mg/m^3
STATUS — questionable
ATTRIBUTED CANCER(S) — in rats
EXPOSURE ROUTE — s.c.
LATENCY

CHEMICAL NAME — Manganese
TLV (if known) — 5 mg/m^3
STATUS — questionable
ATTRIBUTED CANCER(S) — in rats
EXPOSURE ROUTE — implant
LATENCY

CHEMICAL NAME — Mercury
TLV (if known) — 0.1 mg/m^3
STATUS — suggestive
ATTRIBUTED CANCER(S) — in rats
EXPOSURE ROUTE — i.p.
LATENCY

CHEMICAL NAME — Methyl chloromethyl ether
TLV (if known) — no contact
STATUS — susp. humans

ATTRIBUTED CANCER(S) — carcinoma
EXPOSURE ROUTE — inhalation, skin
LATENCY — 10–15 years

CHEMICAL NAME — Methyl orange
TLV (if known)
STATUS — indefinite
ATTRIBUTED CANCER(S) — sarcoma
EXPOSURE ROUTE — oral
LATENCY

CHEMICAL NAME — Methyl red
TLV (if known)
STATUS — questionable
ATTRIBUTED CANCER(S) — in rats
EXPOSURE ROUTE
LATENCY

CHEMICAL NAME — Methylene chloride
TLV (if known) — 500 ppm
STATUS — potential
ATTRIBUTED CANCER(S) — in animals
EXPOSURE ROUTE — oral
LATENCY

CHEMICAL NAME — Napthylamine-alpha
TLV (if known)
STATUS — + human
ATTRIBUTED CANCER(S) — bladder
EXPOSURE ROUTE — inhalation, precutaneous
LATENCY — 22 years

CHEMICAL NAME — Naphthylamine beta
TLV (if known) — no contact
STATUS — + human
ATTRIBUTED CANCER(S) — bladder
EXPOSURE ROUTE — inhalation, percutaneous
LATENCY — 16 years

CHEMICAL NAME — Nickel oxide
TLV (if known) — 1 mg/m^3
STATUS — + animal
ATTRIBUTED CANCER(S) — nasal cavities, lung, skin, sarcoma
EXPOSURE ROUTE — inhalation, ingestion, skin
LATENCY — 5–40 years

CHEMICAL NAME — Nickel sulfate
TLV (if known) — 1 mg/m^3
STATUS — suggestive
ATTRIBUTED CANCER(S) — rats
EXPOSURE ROUTE
LATENCY

CHEMICAL NAME — Nickel titanate
TLV (if known) — 1 mg/m^3
STATUS — conclusive
ATTRIBUTED CANCER(S) — rats
EXPOSURE ROUTE — i.m.
LATENCY

CHEMICAL NAME — Nickel (elemental)
TLV (if known) — 1 mg/m^3
STATUS — animal poss.
ATTRIBUTED CANCER(S) — 1 mg/m^3
EXPOSURE ROUTE — i.m.
LATENCY

CHEMICAL NAME — Oleic acid
TLV (if known)
STATUS — questionable
ATTRIBUTED CANCER(S) — animals
EXPOSURE ROUTE
LATENCY

CHEMICAL NAME — Ozone 0.1
TLV (if known) — 0.1 ppm 0.2 mg/m^3
STATUS — suggestive
ATTRIBUTED CANCER(S) — mice
EXPOSURE ROUTE — inhalation
LATENCY

CHEMICAL NAME — Parathion
TLV (if known) — 0.11 mg/m^3
STATUS — questionable
ATTRIBUTED CANCER(S) — animals
EXPOSURE ROUTE — skin
LATENCY

CHEMICAL NAME — Phenol
TLV (if known) — 5 ppm
STATUS — conclusive

ATTRIBUTED CANCER(S) — mice
EXPOSURE ROUTE — skin
LATENCY

CHEMICAL NAME — Platinum
TLV (if known) — 0.002 mg/m^3
STATUS — conclusive
ATTRIBUTED CANCER(S) — rats and mice
EXPOSURE ROUTE — s.c.
LATENCY

CHEMICAL NAME — Polychlorinated biphenols
TLV (if known) — 1 mg/m^3 (42% chlorine) 0.5 mg/m^3 (54% chlorine)
STATUS — + animal
ATTRIBUTED CANCER(S) — triplacentally, liver
EXPOSURE ROUTE — skin, inhalation
LATENCY

CHEMICAL NAME — Polydimethyl siloxane
TLV (if known)
STATUS — suggestive
ATTRIBUTED CANCER(S) — rat
EXPOSURE ROUTE — i.m.
LATENCY

CHEMICAL NAME — Polyethylene glycol
TLV (if known) — 5 ppm
STATUS — questionable
ATTRIBUTED CANCER(S)
EXPOSURE ROUTE — i.v.
LATENCY

CHEMICAL NAME — Polyurethane dust
TLV (if known)
STATUS — + animal
ATTRIBUTED CANCER(S) — fibrosarcomas
EXPOSURE ROUTE — inhalation, ingestion
LATENCY

CHEMICAL NAME — Potassium chromate
TLV (if known)
STATUS — animal suggested
ATTRIBUTED CANCER(S) — mice
EXPOSURE ROUTE — oral
LATENCY

CHEMICAL NAME — Propiolactone-beta
TLV (if known) — no contact
STATUS — + animal
ATTRIBUTED CANCER(S) — skin, lungs, connective tissue
EXPOSURE ROUTE — inhalation percutaneous
LATENCY

CHEMICAL NAME — Propyl alcohol
TLV (if known) — 200 ppm (skin)
STATUS — conclusive, animal
ATTRIBUTED CANCER(S) — rat
EXPOSURE ROUTE — oral
LATENCY

CHEMICAL NAME — Resorcinol
TLV (if known) — TWA 10 ppm
STATUS — suggestive
ATTRIBUTED CANCER(S) — mouse
EXPOSURE ROUTE — skin
LATENCY

CHEMICAL NAME — Resorcinol diglycidyl ether
TLV (if known) — 45 mg/m^3
STATUS — indefinite
ATTRIBUTED CANCER(S) — rats
EXPOSURE ROUTE — s.c. injextion
LATENCY

CHEMICAL NAME — Rhodamine beta, rhodamine
TLV (if known)
STATUS — + animal
ATTRIBUTED CANCER(S) — local sarcomas
EXPOSURE ROUTE — s.c.
LATENCY

CHEMICAL NAME — Silica dust
TLV (if known) — see crystaline silica (0.1 mg/m^2, respirable dust)
STATUS
ATTRIBUTED CANCER(S)
EXPOSURE ROUTE
LATENCY

CHEMICAL NAME — Silver
TLV (if known) — .01 mg/m^3
STATUS — + animal

ATTRIBUTED CANCER(S) — mice and rats
EXPOSURE ROUTE — implant
LATENCY

CHEMICAL NAME — Silver nitrate
TLV (if known) — 0.01 mg/m^3
STATUS — suggestive
ATTRIBUTED CANCER(S) — mice
EXPOSURE ROUTE — skin
LATENCY

CHEMICAL NAME — Sodium arsenate
TLV (if known)
STATUS — indefinite
ATTRIBUTED CANCER(S) — mice
EXPOSURE ROUTE — skin
LATENCY

CHEMICAL NAME — Sodium arsenite
TLV (if known)
STATUS — human susp.
ATTRIBUTED CANCER(S)
EXPOSURE ROUTE
LATENCY

CHEMICAL NAME — Sodium chromate
TLV (if known) — 30 mg/m^3
STATUS — man, conclusive
ATTRIBUTED CANCER(S) — lung
EXPOSURE ROUTE — inhalation
LATENCY

CHEMICAL NAME — Sodium dichromate
TLV (if known)
STATUS — potential
ATTRIBUTED CANCER(S) — lung
EXPOSURE ROUTE — inhalation
LATENCY

CHEMICAL NAME — Stearic acid
TLV (if known)
STATUS — questionable
ATTRIBUTED CANCER(S) — mice
EXPOSURE ROUTE — implant
LATENCY

CHEMICAL NAME — Stronium chromate
TLV (if known) — 30 mg/m^3
STATUS — + animal
ATTRIBUTED CANCER(S) — lung, nasal cavity, paranasal sinus, exp. stomach and larynx
EXPOSURE ROUTE — inhalation
LATENCY

CHEMICAL NAME — Styrene
TLV (if known) — 50 ppm
STATUS — + mouse
ATTRIBUTED CANCER(S)
EXPOSURE ROUTE — oral
LATENCY

CHEMICAL NAME — Sulfur dioxide
TLV (if known) — 5 ppm 13 mg/m^3
STATUS — questionable
ATTRIBUTED CANCER(S) — mice
EXPOSURE ROUTE — inhalation
LATENCY

CHEMICAL NAME — Tantalum pentoxide
TLV (if known) — 5 mg/m^3
STATUS — suggestive
ATTRIBUTED CANCER(S) — rat
EXPOSURE ROUTE — implant
LATENCY

CHEMICAL NAME — Tetrachlorothane
TLV (if known) — 5 ppm 35 mg/m^3
STATUS — positive
ATTRIBUTED CANCER(S) — lung
EXPOSURE ROUTE — inhalation, ingestion
LATENCY

CHEMICAL NAME — Tetrachloroethylene
TLV (if known) — 50 ppm
STATUS — positive
ATTRIBUTED CANCER(S) — liver
EXPOSURE ROUTE — inhalation skin
LATENCY

CHEMICAL NAME — Tetrafluoroethylene
TLV (if known) — 5 ppm
STATUS — + animal

ATTRIBUTED CANCER(S) — sarcomas
EXPOSURE ROUTE — implant
LATENCY

CHEMICAL NAME — Thiourea
TLV (if known)
STATUS — susp.
ATTRIBUTED CANCER(S) — liver thyroid
EXPOSURE ROUTE
LATENCY

CHEMICAL NAME — Titanium dioxide
TLV (if known) — 15 mg/m^3
STATUS — conclusive
ATTRIBUTED CANCER(S) — in rats
EXPOSURE ROUTE — implant
LATENCY

CHEMICAL NAME — Toluene diisocyanate
TLV (if known) — 0.02 ppm 0.14 mg/m^3
STATUS — conclusive, rats and female mice
ATTRIBUTED CANCER(S) — lung
EXPOSURE ROUTE — respiratory system
LATENCY

CHEMICAL NAME — Toluidine toner
TLV (if known) — 5 ppm
STATUS — human susp.
ATTRIBUTED CANCER(S)
EXPOSURE ROUTE
LATENCY

CHEMICAL NAME — Trichloroethane-1,1,1
TLV (if known) — 350 ppm
STATUS — negative final results
ATTRIBUTED CANCER(S)
EXPOSURE ROUTE
LATENCY

CHEMICAL NAME — Trichloroethylene
TLV (if known) — 100 ppm
STATUS — weak, experimental
ATTRIBUTED CANCER(S) — liver
EXPOSURE ROUTE

LATENCY

CHEMICAL NAME — Trichlorofiuromethane
TLV (if known) — C 1000
STATUS — indefinite
ATTRIBUTED CANCER(S)
EXPOSURE ROUTE
LATENCY

CHEMICAL NAME — Triethanolamine
TLV (if known)
STATUS — conclusive
ATTRIBUTED CANCER(S) — in mice
EXPOSURE ROUTE — oral
LATENCY

CHEMICAL NAME — Vanadium
TLV (if known) — 0.05 mg/m^3
STATUS — questionable
ATTRIBUTED CANCER(S) — in rats
EXPOSURE ROUTE — i.m.
LATENCY

CHEMICAL NAME — Vinyl chloride
TLV (if known) — 5 ppm
STATUS — + human
ATTRIBUTED CANCER(S) — liver, lung, skin, blood-forming tissue, brain, mammary, carcin. transplacental, carcinoma of skin, lung, esteochondrama
EXPOSURE ROUTE — inhalation
LATENCY — 12–29 years

CHEMICAL NAME — Vinylcylohexane
TLV (if known)
STATUS — + animal
ATTRIBUTED CANCER(S)
EXPOSURE ROUTE
LATENCY

CHEMICAL NAME — Wood dust
TLV (if known) — 1 mg/m^3
STATUS — potential human
ATTRIBUTED CANCER(S) — local sarcomas, nasal sinus and cavity, rectum, pancreas, lung, small intest. testes
EXPOSURE ROUTE — inhalation, ingestion
LATENCY — 24–49 years

CHEMICAL NAME — Zinc chloride
TLV (if known) — 1 mg/m^3
STATUS — suggestive
ATTRIBUTED CANCER(S) — hamster
EXPOSURE ROUTE
LATENCY

CHEMICAL NAME — Zinc chromate
TLV (if known) — 0.05 mg/m^3
STATUS — susp. human
ATTRIBUTED CANCER(S) — lung
EXPOSURE ROUTE — inhalation, ingestion
LATENCY

CHEMICAL NAME — Zinc sulfate
TLV (if known)
STATUS — suggestive
ATTRIBUTED CANCER(S) — in rabbit
EXPOSURE ROUTE — s.c..
LATENCY

This table was compiled using the following resources: "Chemical/Physical Agent Codes"- supplied by Martin Marietta, *Handbook of Toxic and Hazardous Chemicals*, by Martin Sittig, Noyes Publications (1981), *Cancer Causing Chemicals*, by Irving Sax,Van Nostrand ReinHold Co. (1981), *Dangerous Properties of Industrial Materials*, by N. Irving Sax, Van Nostrand Reinhold Co. (1979), *Environmental and Industrial Health Hazards*, by R. A. Trevethick, Wm. Heinimann Medical Books Ltd., London (1976), *Occupational Diseases — A Guide to Their Recognition*, Marcus M. Key, Editor, U.S. Department of Health, Education and Welfare, Public Health Service, CDC, NIOSH, 9/1978, *Index of Signs and Symptoms of Industrial Diseases*, B.A. Fay and C. E. Billings, 5/1980, U.S. Department of Health and Human Services, "Health Letter" Vol. 1 (March 1985) Spec. Suppl. "Dangerous Properties of Industrial Materials Report," Vol. 3 (6), Vol. 3(4), Vol. 3 (5) (July-September).

Revised 1987 by Dr. H. Sacarello, Ref.: 1987–88 Threshold Limit Values and Biological Exposure Indices. (Cincinnati, OH, ACGIH, 1987).

Fourth Annual Report on Carcinogens, Summary, 1987, U.S. Department of Health and Human Services, Public Health Service, NTP85-002.

Field Data Sheet for Collecting Airborne Vapors

Date ______

Sample ______

______

Location ______

______

Collection Device ______

Pump Flow Rate ______ liters/min.

Sampling Time ______ min.

Pressure ______ mm Hg

Temperature (T) ______ °C

Compound Sampled ______

OSHA-TWA Limit (L) ______ µl/L or mg/m³•

Density (P) ______ g/ml

Molecular Weight (M) ______ g/mole

Volume Desorbing Solvent (D) ______ ml

Volume Air Sampled (V) ______ liters

Corrected Volume (V_s) ______ liters
at 25°C, 760mm Hg

$$V_s = V \frac{\text{Pressure}}{760\text{mm}} \frac{298°\text{K}}{T + 273}$$

Standard Preparation

Using Equation 1, determine the volume of the airborne contaminant necessary to prepare standards equivalent to the OSHA-TWA limit (or fraction F thereof).

Equation 1:

$$\text{volume } (\mu l) = (F) \frac{(L) \quad (V_s) \quad (M)}{(24450) \quad (P) \quad (D)}$$

$$= (F) \frac{\frac{\mu l}{\text{liter}} \quad \_\_\text{liters} \quad \frac{g}{\text{mole}}}{\frac{24450\text{ml}}{\text{molo}} \quad \frac{g}{\text{ml}} \quad \_\_\text{ ml desorbing solvent}}$$

Sample Analysis

From the calibration curve, determine the concentration $\frac{\mu l}{\text{ml desorbing solvent}}$ of the airborne vapor sampled.

Using Equation 3, convert the concentration to ppm of compound in the air sampled. Correct for less than 100% desorption (see page 7).

Equation 3:

$$\text{ppm} = \frac{\mu l}{\text{liter}} = \frac{(\mu l \text{ of compound}) \ (24450) \ (P) \ (D)}{(V_s) \ (M)}$$

$$= \frac{\frac{\mu l \text{ compound}}{\_\_\text{ml desorbing solvent}} \quad \frac{24450\text{ml}}{\text{mole}} \quad \frac{g}{\text{ml}} \quad \_\_\text{ ml desorbing solvent}}{\_\_\text{ liters air} \quad \frac{g}{\text{mole}}} = \_\_\_\text{ppm}$$

•Volume/volume and weight/volume concentrations can be easily interconverted, using the formula:

$$\text{ppm to mg/m}^3 = (x)\text{mg/m}^3 = \frac{[(y)\text{ ppm}] \ (\text{molecular wt. of contaminant})}{24.5}$$

SUPELCO

Worksheet may be reproduced without written permission from Supelco, Inc.

Supelco Park • Bellefonte, PA 16823-0048 USA

Field Data Sheet for Collecting Airborne Particles

Date ____________________

Sample ____________________

____

Location ____________________

____

Collection Device ____________________

Pump Flow Rate ____________ liters/min.

Sampling Time ____________ min.

Pressure ____________ mm Hg

Temperature (T) ____________ °C

Compound Sampled ____________________

OSHA-TWA Limit (L) ____________ µl/L or mg/m³*

Density (P) ____________ g/ml

Molecular Weight (M) ____________ g/mole

Volume Desorbing Solvent (D) ____________ ml

Volume Air Sampled (V) ____________ liters

Corrected Volume (V_s) ____________ liters
at 25°C, 760mm Hg

$$V_s = V \frac{\text{Pressure}}{760\text{mm}} \frac{298°K}{T + 273}$$

Standard Preparation

Using Equation 2, determine the volume of the airborne contaminant necessary to prepare standards equivalent to the OSHA-TWA limit (or fraction F thereof).

Equation 2:

$$\text{weight (mg)} = (F) \frac{(L)\ (V_s)}{(D)} \times \frac{m^3}{1000L}$$

$$= (F) \frac{\frac{mg}{m^3} \quad \underline{\quad}\ \text{liters} \quad \frac{m^3}{1000L}}{\underline{\quad}\ \text{ml desorbing solvent}}$$

Prepare a calibration curve of concentration $\frac{mg}{\text{ml desorbing solvent}}$ versus area response from the chromatograph for the compound sampled.

Sample Analysis

From the calibration curve, determine the concentration (mg) of the airborne solid sampled.
Using Equation 4, convert the concentration to mg/m³ for the collected sample. Correct for less than 100% desorption (see page 7).

Equation 4:

$$\frac{mg}{m^3} = \frac{(\text{mg contaminant/ml desorbing solvent})\ (D)}{(V_s)\ (m^3/1000L)}$$

$$= \frac{(\underline{\quad}\ \text{mg/ml})\ (\underline{\quad}\ \text{ml desorbing solvent})}{\underline{\quad}\ \text{liters} \quad \frac{m^3}{1000L}} = \underline{\quad}\ \text{mg/m}^3$$

*Volume/volume and weight/volume concentrations can be easily interconverted, using the formula:

$$\text{mg/m}^3 \text{ to ppm} = (y)\text{ppm} = \frac{[(x)\ \text{mg/m}^3]\ (24.5)}{\text{molecular weight of contaminant}}$$

SUPELCO

Supelco Park • Bellefonte, PA 16823-0048 USA

CLINICAL APPLICATIONS OF AMINO ACID ANALYSIS

IMPLICATIONS FOR NUTRITIONAL MEDICINE

By
J. Alexander Bralley, Ph.D.
MetaMetrix Medical Research Laboratory
Norcross, Georgia

The body conducts a marvelously complicated series of chemical reactions in precisely controlled ways to give us what we term health. With over 5000 reactions occurring each second in a cell, it is a wonder that we do not get sick more often. The strength of nutritional/molecular medicine is that it deals with this fundamental aspect of health, the human biochemistry. By using the actual natural molecules in optimal quantities to re-establish a normal balance in a particular set of critical reactions, we can help correct the cause of some disease processes in a non-toxic way.

This, of course, is the desired goal of nutritionally-oriented health professionals. The problem is in the precise analysis of what might be wrong with the body's chemistry that can be corrected nutritionally. Early nutritional medicine advocates tended to use shotgun approaches, loading the patient with anything and everything they felt would benefit them. However, with the advent of more sophisticated testing procedures, it became possible to assess nutritional status more precisely and tailor make treatment programs. Tests include serum and intracellular enzyme, vitamin, and mineral analyses of various bodily components, essential fatty acid profiles, and metabolite levels--the list can go on. The obvious problem is that it becomes very expensive to do exhaustive nutrient status testing routinely on each patient. Consequently, these tests have been reserved for the more difficult cases. Ideally, and in the future, the nutritional/molecular medicine practitioner will have available low-cost assessment techniques to analyze a wide variety of nutrient-dependent biochemical processes which influence the risk for degenerative disease in the patient.

> *"The broad applications for this test make it particularly attractive to nutritionally oriented health practitioners because of the wide spectrum of vital nutritional information present in the report."*

VITAMIN & MINERAL ASSESSMENT: Amino Acid Analysis of plasma and/or urine goes a long way toward achieving this goal. Not only does it measure levels of 41 amino acids in the body that affect many important processes, it also gives insights into the patient's functional needs for a wide variety of vitamins and minerals. Many of the enzymes which catalyze the interconversion of amino acids require vitamin and mineral co-factors to function optimally. In many cases, where incomplete conversion of one product to another is due to sluggish enzymes, this actually indicates a functional need for increased intake of a vitamin or mineral co-factor.

For example, the enzyme converting the essential amino acid phenylalanine to tyrosine requires iron as an activating co-factor. Low levels of tyrosine and high phenylalanine would, therefore, be indicative of a functional iron deficiency. A block in the metabolism of phenylalanine due to an iron deficiency could adversely affect thyroid hormone and melanin synthesis, and catecholamine neurotransmitter production. Low endogenous epinephrine and norepinephrine affect behavior, learning and memory, autonomic function, and the ability to respond properly to stress. This situation could then be obviated by supplementation of iron and tyrosine. Low iron levels have recently been associated with decreased cognitive performance and electrophysiological asymmetries. (1,19) These results are thought to be due to the influence of iron on neurotransmitter synthesis.

This is just one instance of how functional mineral deficiency is reflected in the Amino Acid Analysis. Other minerals whose functional levels might also be assessed with this test include magnesium, manganese, copper, zinc, and (indirectly) selenium. Vitamins play a vital role as enzyme co-factors, such as pyridoxal-5-phosphate (P-5-P) derived from B6 and requiring zinc for adequate phosphorylation. P-5-P is widely used in amino acid metabolism, along with alpha-ketoglutaric acid (a-Kg) in the transfer of amine groups. Functional need for B6 or P-5-P and a-Kg is accurately assessed with Amino Acid Analysis. Valuable information can be obtained on the functional need for other vitamins and nutrients such as: B1, B2, B3, B5, B12, Folic Acid, C, betaine, biotin, choline, and antioxidant systems in general.

It is important to note that optimal dosages or delivery forms of these nutrients can be monitored using Amino Acid Analysis. Whereas serum levels of these nutrients may be normal, functional concentrations of the nutrient may not be present at the enzymatic site. A recent case reported in the New England Journal of Medicine cites such a situation; serum B12 was normal yet Amino Acid Analysis revealed a functional B12 deficiency that was corrected by intramuscular injection of B12 with subsequent reversal of the mental retardation. (2) Functional assessment of nutritional status is the key to nutritional/molecular medicine and Amino Acid Analysis is proving a useful tool in this regard.

CLINICAL STATES: In addition to helping define nutrient status, Amino Acid Analysis can be useful in the detection and targeted treatment of a wide variety of disease states. This is because amino acids are involved in so many basic chemical reactions in tissues. Here are a few examples:

Chronic Fatigue: This is a common complaint in patients who may have a variety of other symptoms, too. Although a growing number of people come to their doctor because of severe, debilitating chronic fatigue, general fatigue is such a consistent finding in patients seeking help from the nutritional/molecular medicine practitioner that it deserves special comment. Certain amino acids can be used as an energy source. Concentrations of these as well as their metabolites can give indications of the body's ability to generate cellular energy. Consistent findings of inadequate cellular energy generation in Amino Acid Analyses are commonplace. This is the basis for the complaint of chronic fatigue. These people are unable to generate energy at the most basic cellular level. They often need diet and lifestyle counseling, and supplementation of alpha-ketoglutaric acid, citric and aspartic acid, pantethine, coenzyme Q10, B6 and the Branched Chain amino acids, and possibly alanine and threonine. (3,4) Recent research at our lab has shown substantial improvement in chronic fatigue syndrome patients using a customized formulation of amino acids based upon their fasting plasma amino acid analysis. We are presently persuing this study to carefully document and define the essential parameters in the treatment.

Allergies: This field is receiving more attention presently because of the increased awareness of the broad effects of food and chemical allergy and/or hypersensitivity on health. Useful clinical treatments based on Amino Acid Analysis have been pointed out by Pangborn. (5) The mechanisms include disordered methionine metabolism, taurine depletion, and free radical pathology. Many patients with food allergies report improved tolerance of foods with amino acid supplementation, particularly when plasma taurine and histidine levels are low. Taurine is a potent anti-oxidant and helps to control chemical sensitivity reactions reactions.

Seizures: The amino acid taurine has been used clinically to control seizures. Because of its ability to regulate intracellular potassium concentration, it is also useful in congestive heart failure. (6,7)

Cardiovascular Disease: Amino Acid Analysis can monitor homocystine levels. Moderate elevations of this amino acid are now being associated with significant increased risk to atherosclerosis.(8) This can be corrected in most cases with adequate supplementation of B6, betaine, folic acid and B12, and cysteine, combined with a low meat protein diet. (9) Taurine has recently been found to have powerful anti-platelet aggregation properties. (20) This study suggests taurine supplementation to be as effective in preventing cardiovascular disease in regards to platelet aggregation as is aspirin taken on a regular basis.

Rheumatoid Arthritis: This inflammatory process has been related to free radical pathology secondary to allergic hyperimmune reactivity.(5) It also has been correlated with low histidine levels. (10) Amino Acid Analysis has proven useful in evaluating both of these situations in the patient.

Candida Albicans and Bowel Toxemia: Bowel overgrowth of Candida and foreign bacteria produce the amino acids Beta-alanine and Beta-aminoisobutyric acid. These beta amino acids increase metabolic requirements of B6 and impair renal conservation of other amino acids (especially taurine). This situation often results in hyperaminoacidurias with protein intolerances and excessive free radical activity characteristic of many Candida patients. Monitoring the normalization of urinary amino acid levels parallels the improvement in the patient and can indicate specific supplements useful at different times in the recovery. (11)

Digestive Disorders: Improper digestion and assimilation are often reflected as low levels of essential amino acids in plasma or urine. Food proteins are hydrolyzed in the stomach and small intestine. Low hydrochloric acid in the stomach inhibits the activity of pepsin that cleaves protein at the aromatic amino acids phenylalanine, tyrosine, and tryptophan. Low levels of these would be indicative of inadequate stomach acid production. The intestinal mucosa peptidases are activated by zinc and manganese. Functional deficiencies of these minerals can lead to low levels of the branched chain amino acids leucine, isoleucine, and valine, which are cleaved by these peptidases. High small intestine acidity will also inhibit pancreatic proteolytic enzyme activity. This will often be seen as low levels of essential amino acids.

Cancer Prevention: Recent research has noted that tumor cell DNA consistently lacks normal methylation on its base pairs. It is this methylation which is thought to inhibit gene expression. Therefore, inadequate methyl group transfer may be responsible for the transformation of normal cells to tumor lines. Methylation is a primary function of methionine metabolism and can be readily evaluated with Amino Acid Analysis, and in many cases improved nutritionally. This may be an important application of the analysis in lowering potential risk for this major disease (12,13).

Athletic Performance: A new and very intriguing application of Amino Acid Analysis we have been pursuing at our laboratory has been towards improving muscle development and endurance in athletes. Based on results of an amino acid analysis, individual supplement and diet programs were designed to enhance cellular energy generation and nitrogen retention. The results have been excellent and may very well point to a potent and healthful substitute for anabolic steroids commonly used by today's athletes. (14)

Behavior Disorders: The effects of amino acids on neurotransmitters in the brain have been known for some time. (15) Tryptophan, tyrosine, and phenylalanine along with B6 have been shown to be useful in treating depression (16, 17, 18). Amino Acid Analysis can help evaluate the potential need for and effectiveness of this type of treatment. Learning and behavioral disorders in children are often ameliorated by correcting cellular energy generation deficits revealed in their analyses.

Elderly: A recent study examined fasting plasma amino acid patterns in healthy young and elderly men. The elderly exhibited consistent low levels of essential amino acids. This finding has significant implications for the elderly who often exhibit psycho/neurological problems. Low levels of phenylalanine, tyrosine, tryptophan, branched chain amino acids and histidine were frequently seen in the elderly in this study. These amino acids are intimately involved in nervous system function and may significantly influence disorders in this area (21). Routine assessment of plasma amino acids accompanied by appropriate supplementation may be extremely useful in optimizing the health of individuals as they age and prevent the debilitating and dehumanizing effects of these psycho/neurological problems.

Other equally useful areas of application of Amino Acid Analysis include blood sugar disorders, immune dysfunction, trauma, post-surgical recovery, sclerosis, eating disorders, connective tissue disorders and subacute ammonia toxicities. The broad applications for this test make it particularly attractive to nutritionally oriented health practitioners because of the wide spectrum of vital nutritional information present in the report.

This, however, has been one of the problems in the general use of the test.... the amount and complexity of the information can appear overwhelming. This situation will be changing as health professionals become more familiar with the chemistry involved. The newer computer interpretations of these tests will help greatly here. As more published material on applications of this advanced technology becomes available, Amino acid Analysis may become a routine test performed on all patients coming in for a nutritional/molecular checkup.

REFERENCES

1. Tucker, D. M. et al. Iron status and brain function: Am. J. Clin. Nutr. 39:105-113(1984).
2. Shinnar, D.L. Cobalamin C mutation in adolescence. N. Eng. J. Med. 311:451(1984).
3. Harper et al. Review of Physiological Chemistry, 17th Ed. Lange Medical Publications (1979).
4. Cherrington, A.D. et al. Amino acids in gluconeogenesis. In Amino Acids: Metabolism and Medical Applications, pp. 63-76. (Ed.) Blackburn et al., Wright PSG, Boston, 1983.
5. Pangborn, J.B. Nutritional and anti-inflammatory aspects of amino acids. In Yearbook of Nutritional Medicine, 1984-85, pp. 153-178. (ed.) Bland, J.S., Keats, New Canaan, 1985.
6. Huxtable, R. J., and Pasanter-Moralies, H. Taurine in Nutrition and Neurology, Plenum, New York, 1982.
7. Azuma, J. et al. Taurine for treatment of congestive heart failure. Int. J. Cardiol. 2:303(1982).
8. McCully, K.S. and Wilson, R. B. Homocysteine theory of atherosclerosis. Atherosclerosis, 22:2(1975).
9. Wikken, D. E. L. Homocysteinuria: The effects of betaine in treatment of patients not responsive to pyridoxine. N. Engl. J. Med., 309:448(1983).
10. Gerber, D. A. Low free serum histidine concentration in rheumatoid arthritis. A measure of disease activity. J. Clin. Invest. 55:1164-73(1975).
11. Scriver, C. D. et al. Disorders of B-alanine, carnosine and homocarnosine metabolism. In The Metabolic Basis of Inherited Disease, 5th Ed. (Ed.)Stanbury et al, Chap. 28, McGraw-Hill, NY, 1983.
12. Finkelstein, D. A. Regulation of methionine metabolism in mammals. Arch. Biochem. Biophys. 159:160(1973).
13. Hoffman, R. M. Altered methionine metabolism, DNA, methylation and oncogene expression in carcinogenesis, Biocheml. Biophys. Acta 738:49-87(1984).
14. Bralley, J.A., Murray, M. Research suggests amino acid supplements can be as potent as anabolic steroids. Sports Fitness, Sept. 1985, 54-60.
15. Wurtman and Wurtman. Nutrition and the Brain, Vol 1-4, Raven Press, NY(1977-83).
16. Adams, P.W. et al. Effect of pyridoxine upon depression associated with oral contraceptives. Lancet, 1:897(1973).
17. Gelenberg, A.J. et al. Tyrosine treatment of depression. Am. J. Psychiat. 137:622(1980).
18. Thomson, J. et al. The treatment of depression in general practice: A comparison of L-tryptophan, amitriptyline and combination with placebo. Psychol. Med. 12:741(1982).
19. Youdin MBH et al. Putative biological mechanisms of the effect of iron deficiency on brain chemistry and behavior. Am. J. Clin. Nutr. 50:607-17(1989).
20. Hayes, KC et al. Taurine modulates platelet aggregation in cats and humans. Am. J. Clin. Nutr. 49:1211-16(1989).
21. Rudman, D. et al. Fasting plasma amino acids in elderly men. Am. J. Clin. Nutr. 49:559-66(1989).

INDEX

A

I

N

O

Q

R

S

T

U

V

W

X

Z